Studies in Big Data

Volume 24

Series editor

Janusz Kacprzyk, Polish Academy of Sciences, Warsaw, Poland
e-mail: kacprzyk@ibspan.waw.pl

About this Series

The series "Studies in Big Data" (SBD) publishes new developments and advances in the various areas of Big Data- quickly and with a high quality. The intent is to cover the theory, research, development, and applications of Big Data, as embedded in the fields of engineering, computer science, physics, economics and life sciences. The books of the series refer to the analysis and understanding of large, complex, and/or distributed data sets generated from recent digital sources coming from sensors or other physical instruments as well as simulations, crowd sourcing, social networks or other internet transactions, such as emails or video click streams and other. The series contains monographs, lecture notes and edited volumes in Big Data spanning the areas of computational intelligence incl. neural networks, evolutionary computation, soft computing, fuzzy systems, as well as artificial intelligence, data mining, modern statistics and Operations research, as well as self-organizing systems. Of particular value to both the contributors and the readership are the short publication timeframe and the world-wide distribution, which enable both wide and rapid dissemination of research output.

More information about this series at http://www.springer.com/series/11970

Witold Pedrycz · Shyi-Ming Chen
Editors

Data Science and Big Data: An Environment of Computational Intelligence

Springer

Editors
Witold Pedrycz
Department of Electrical and Computer Engineering
University of Alberta
Edmonton, AB
Canada

Shyi-Ming Chen
Department of Computer Science and Information Engineering
National Taiwan University of Science and Technology
Taipei
Taiwan

ISSN 2197-6503 ISSN 2197-6511 (electronic)
Studies in Big Data
ISBN 978-3-319-85162-4 ISBN 978-3-319-53474-9 (eBook)
DOI 10.1007/978-3-319-53474-9

Softcover reprint of the hardcover 1st edition 2017

Printed on acid-free paper

This Springer imprint is published by Springer Nature
The registered company is Springer International Publishing AG
The registered company address is: Gewerbestrasse 11, 6330 Cham, Switzerland

Preface

The disciplines of Data Science and Big Data, coming hand in hand, form one of the rapidly growing areas of research, have already attracted attention of industry and business. The prominent characterization of the area highlighting the essence of the problems encountered there comes as a 3V (volume, variety, variability) or 4V characteristics (with veracity being added to the original list). The area itself has initialized new directions of fundamental and applied research as well as led to interesting applications, especially those being drawn by the immediate needs to deal with large repositories of data and building some tangible, user-centric models of relationships in data.

A general scheme of Data Science involves various facets: descriptive (concerning reporting—identifying what happened and answering a question why it has happened), predictive (embracing all the investigations of describing what will happen), and prescriptive (focusing on acting—make it happen) contributing to the development of its schemes and implying consecutive ways of the usage of the developed technologies. The investigated models of Data Science are visibly oriented to the end-user, and along with the regular requirements of accuracy (which are present in any modeling) come the requirements of abilities to process huge and varying data sets and the needs for robustness, interpretability, and simplicity.

Computational intelligence (CI) with its armamentarium of methodologies and tools is located in a unique position to address the inherently present needs of Data Analytics in several ways by coping with a sheer volume of data, setting a suitable level of abstraction, dealing with distributed nature of data along with associated requirements of privacy and security, and building interpretable findings at a suitable level of abstraction.

This volume consists of twelve chapters and is structured into two main parts: The first part elaborates on the fundamentals of Data Analytics and covers a number of essential topics such as large scale clustering, search and learning in highly dimensional spaces, over-sampling for imbalanced data, online anomaly detection, CI-based classifiers for Big Data, Machine Learning for processing Big Data and event detection. The second part of this book focuses on applications demonstrating the use of the paradigms of Data Analytics and CI to safety assessment, management of smart grids, real-time data, and power systems.

Given the timely theme of this project and its scope, this book is aimed at a broad audience of researchers and practitioners. Owing to the nature of the material being covered and a way it has been organized, one can envision with high confidence that it will appeal to the well-established communities including those active in various disciplines in which Data Analytics plays a pivotal role.

Considering a way in which the edited volume is structured, this book could serve as a useful reference material for graduate students and senior undergraduate students in courses such as those on Big Data, Data Analytics, intelligent systems, data mining, computational intelligence, management, and operations research.

We would like to take this opportunity to express our sincere thanks to the authors for presenting advanced results of their innovative research and delivering their insights into the area. The reviewers deserve our thanks for their constructive and timely input. We greatly appreciate a continuous support and encouragement coming from the Editor-in-Chief, Prof. Janusz Kacprzyk, whose leadership and vision makes this book series a unique vehicle to disseminate the most recent, highly relevant, and far-reaching publications in the domain of Computational Intelligence and its various applications.

We hope that the readers will find this volume of genuine interest, and the research reported here will help foster further progress in research, education, and numerous practical endeavors.

Edmonton, Canada
Taipei, Taiwan

Witold Pedrycz
Shyi-Ming Chen

Contents

Part II Applications

Part I
Fundamentals

Large-Scale Clustering Algorithms

Rocco Langone, Vilen Jumutc and Johan A. K. Suykens

Abstract Computational tools in modern data analysis must be scalable to satisfy business and research time constraints. In this regard, two alternatives are possible: (i) adapt available algorithms or design new approaches such that they can run on a distributed computing environment (ii) develop model-based learning techniques that can be trained efficiently on a small subset of the data and make reliable predictions. In this chapter two recent algorithms following these different directions are reviewed. In particular, in the first part a scalable in-memory spectral clustering algorithm is described. This technique relies on a kernel-based formulation of the spectral clustering problem also known as kernel spectral clustering. More precisely, a finite dimensional approximation of the feature map via the Nyström method is used to solve the primal optimization problem, which decreases the computational time from cubic to linear. In the second part, a distributed clustering approach with fixed computational budget is illustrated. This method extends the k-means algorithm by applying regularization at the level of prototype vectors. An optimal stochastic gradient descent scheme for learning with l_1 and l_2 norms is utilized, which makes the approach less sensitive to the influence of outliers while computing the prototype vectors.

Keywords Data clustering · Big data · Kernel methods · Nyström approximation · Stochastic optimization · K-means · Map-Reduce · Regularization · In-memory algorithms · scalability

R. Langone (✉) · V. Jumutc · J.A.K. Suykens
KU Leuven ESAT-STADIUS, Kasteelpark Arenberg 10, B-3001 Leuven, Belgium
e-mail: rocco.langone@esat.kuleuven.be

V. Jumutc
e-mail: vilen.jumutc@esat.kuleuven.be

J.A.K. Suykens
e-mail: johan.suykens@esat.kuleuven.be

W. Pedrycz and S.-M. Chen (eds.), *Data Science and Big Data: An Environment of Computational Intelligence*, Studies in Big Data 24,
DOI 10.1007/978-3-319-53474-9_1

1 Introduction

Data clustering allows to partition a set of points into groups called clusters which are as similar as possible. It plays a key role in computational intelligence because of its diverse applications in various domains. Examples include collaborative filtering and market segmentation, where clustering is used to provide personalized recommendations to users, trend detection which allows to discover key trends events in streaming data, community detection in social networks, and many others [1].

With the advent of the big data era, a key challenge for data clustering lies in its scalability, that is, how to speed-up a clustering algorithm without affecting its performance. To this purpose, two main directions have been explored [1]: (i) sampling-based algorithms or techniques using random projections (ii) parallel and distributed methods. The first type of algorithms allows to tackle the computational complexity due either to the large amount of data instances or their high dimensionality. More precisely, sampling-based algorithms perform clustering on a sample of the datasets and then generalize it to whole dataset. As a consequence, execution time and memory space decrease. Examples of such algorithms are CLARANS [2], which tries to find the best medoids representing the clusters, BIRCH [3], where a new data structure called clustering feature is introduced in order to reduce the I/O cost in the in-memory computational time, CURE [4], which uses a set of well-scattered data points to represent a cluster in order to detect general shapes. Randomized techniques reduce the dimension of the input data matrix by transforming it into a lower dimensional space and then perform clustering on this reduced space. In this framework, [5] uses random projections to speed-up the k-means algorithm. In [6], a method called Colibri allows to cluster large static and dynamic graphs. In contrast to the typical single machine clustering, parallel algorithms use multiple machines or multiple cores in a single machine to speed up the computation and increase the scalability. Furthermore, they can be either memory-based if the data fit in the memory and each machine/core can load it, or disk-based algorithm which use Map-Reduce [7] to process huge amounts of disk-resident data in a massively parallel way. An example of memory-based algorithm is ParMETIS [8], which is a parallel graph-partitioning approach. Disk-based methods include parallel k-means [9], a k-means algorithm implemented on Map-Reduce and a distributed co-clustering algorithm named DisCO [10]. Finally, the interested reader may refer to [11, 12] for some recent surveys on clustering algorithms for big data.

In this chapter two algorithms for large-scale data clustering are reviewed. The first one, named fixed-size kernel spectral clustering (FSKSC), is a sampling-based spectral clustering method. Spectral clustering (SC) [13–16] has been shown to be among the most effective clustering algorithms. This is mainly due to its ability of detecting complex nonlinear structures thanks to the mapping of the original data into the space spanned by the eigenvectors of the Laplacian matrix. By formulating the spectral clustering problem within a least squares support vector machine setting [17], kernel spectral clustering [18, 19] (KSC) allows to tackle its main drawbacks represented by the lack of a rigorous model selection procedure and a systematic out-of-sample property. However, when the number of training data is large the complexity of constructing the Laplacian matrix and computing its eigendecomposition can become intractable. In this respect, the FSKSC algorithm represents a solution to this issue which exploits the Nyström method [20] to avoid the construction of the kernel matrix and therefore reduces the time and space costs. The second algorithm that will be described is a distributed k-means approach which extends the k-means algorithm by applying l_1 and l_2 regularization to enforce the norm of the prototype vectors to be small. This allows to decrease the sensitivity of the algorithm to both the initialization and the presence of outliers. Furthermore, either stochastic gradient descent [21] or dual averaging [22] are used to learn the prototype vectors, which are computed in parallel on a multi-core machine.[1]

The remainder of the chapter is organized as follows. Section 3 summarizes the standard spectral clustering and k-means approaches. In Sect. 4 the fixed-size KSC method will be presented. Section 5 is devoted to summarize the regularized stochastic k-means algorithm. Afterwards, some experimental results will be illustrated in Sect. 6. Finally some conclusions are given.

2 Notation

$\mathbf{x}^T$	Transpose of the vector $\mathbf{x}$
$\mathbf{A}^T$	Transpose of the matrix $\mathbf{A}$
$\mathbf{I}_N$	$N \times N$ Identity matrix
$\mathbf{1}_N$	$N \times 1$ Vector of ones
$\mathscr{D}_{\text{tr}} = \{\mathbf{x}_i\}_{i=1}^{N_{\text{tr}}}$	Training sample of N_{tr} data points
$\varphi(\cdot)$	Feature map
$\mathscr{F}$	Feature space of dimension d_h
$\{\mathscr{C}_p\}_{p=1}^{k}$	Partitioning composed of k clusters
$\lvert \cdot \rvert$	Cardinality of a set
$\lvert\lvert \cdot \rvert\rvert_p$	p-norm of a vector
∇f	Gradient of function f

3 Standard Clustering Approaches

3.1 Spectral Clustering

Spectral clustering represents a solution to the graph partitioning problem. More precisely, it allows to divide a graph into weakly connected sub-graphs by making use of the spectral properties of the graph Laplacian matrix [13–15].

[1] The same schemes can be extended with little effort to a multiple machine framework.

A graph (or network) $\mathscr{G} = (\mathscr{V}, \mathscr{E})$ is a mathematical structure used to model pairwise relations between certain objects. It refers to a set of N vertices or nodes $\mathscr{V} = \{v_i\}_{i=1}^N$ and a collection of edges $\mathscr{E}$ that connect pairs of vertices. If the edges are provided with weights the corresponding graph is weighted, otherwise it is referred as an unweighted graph. The topology of a graph is described by the similarity or affinity matrix, which is an $N \times N$ matrix $\boldsymbol{S}$, where S_{ij} indicates the link between the vertices i and j. Associated to the similarity matrix there is the degree matrix $\boldsymbol{D} = \text{diag}(\boldsymbol{d}) \in \mathbb{R}^{N \times N}$, with $\boldsymbol{d} = [d_1, \dots, d_N]^T = \boldsymbol{S}\boldsymbol{1}_N \in \mathbb{R}^{N \times 1}$ and $\boldsymbol{1}_N$ indicating the $N \times 1$ vector of ones. Basically the degree d_i of node i is the sum of all the edges (or weights) connecting node i with the other vertices: $d_i = \sum_{j=1}^N S_{ij}$.

The most basic formulation of the graph partitioning problem seeks to split an unweighted graph into k non-overlapping sets $\mathscr{C}_1, \dots, \mathscr{C}_k$ with similar cardinality in order to minimize the cut size, which is the number of edges running between the groups. The related optimization problem is referred as the normalized cut (NC) objective defined as:

$$\begin{aligned} \min_{\boldsymbol{G}} \quad & k - \text{tr}(\boldsymbol{G}^T \boldsymbol{L}_n \boldsymbol{G}) \\ \text{subject to} \quad & \boldsymbol{G}^T \boldsymbol{G} = \boldsymbol{I} \end{aligned} \tag{1}$$

where:

- $\boldsymbol{L}_n = \boldsymbol{I} - \boldsymbol{D}^{-\frac{1}{2}} \boldsymbol{S} \boldsymbol{D}^{-\frac{1}{2}}$ is called the normalized Laplacian
- $\boldsymbol{G} = [\boldsymbol{g}_1, \dots, \boldsymbol{g}_k]$ is the matrix containing the normalized cluster indicator vectors $\boldsymbol{g}_l = \frac{\boldsymbol{D}^{\frac{1}{2}} \boldsymbol{f}_l}{||\boldsymbol{D}^{\frac{1}{2}} \boldsymbol{f}_l||_2}$
- $\boldsymbol{f}_l$, with $l = 1, \dots, k$, is the cluster indicator vector for the l-th cluster. It has a 1 in the entries corresponding to the nodes in the l-th cluster and 0 otherwise. Moreover, the cluster indicator matrix can be defined as $\boldsymbol{F} = [\boldsymbol{f}_1, \dots, \boldsymbol{f}_k] \in \{0, 1\}^{N \times k}$
- $\boldsymbol{I}$ denotes the identity matrix.

Unfortunately this is a NP-hard problem. However, approximate solutions in polynomial time can be obtained by relaxing the entries of $\boldsymbol{G}$ to take continuous values:

$$\begin{aligned} \min_{\hat{\boldsymbol{G}}} \quad & k - \text{tr}(\hat{\boldsymbol{G}}^T \boldsymbol{L}_n \hat{\boldsymbol{G}}) \\ \text{subject to} \quad & \hat{\boldsymbol{G}}^T \hat{\boldsymbol{G}} = \boldsymbol{I}. \end{aligned} \tag{2}$$

with $\hat{\boldsymbol{G}} \in \mathbb{R}^{N \times k}$ Solving problem (2) is equivalent to finding the solution to the following eigenvalue problem:

$$\boldsymbol{L}_n \boldsymbol{g} = \lambda \boldsymbol{g}. \tag{3}$$

Basically, the relaxed clustering information is contained in the eigenvectors corresponding to the k smallest eigenvalues of the normalized Laplacian $\boldsymbol{L}_n$. In addition to the normalized Laplacian, other Laplacians can be defined, like the unnormalized Laplacian $\boldsymbol{L} = \boldsymbol{D} - \boldsymbol{S}$ and the random walk Laplacian $\boldsymbol{L}_{rw} = \boldsymbol{D}^{-1}\boldsymbol{S}$. The latter owes its name to the fact that it represents the transition matrix of a random walk associated to the graph, whose stationary distribution describes the situation in which the random walker remains most of the time in the same cluster with rare jumps to the other clusters [23].

Spectral clustering suffers from a scalability problem in both memory usage and computational time when the number of data instances N is large. In particular, time complexity is $O(N^3)$, which is needed to solve eigenvalue problem (3), and space complexity is $O(N^2)$, which is required to store the Laplacian matrix. In Sect. 4 the fixed-size KSC method will be thoroughly discussed, and some related works representing different solutions to this scalability issue will be briefly reviewed in Sect. 4.1.

3.2 K-Means

Given a set of observations $\mathscr{D} = \{\mathbf{x}_i\}_{i=1}^N$, with $\mathbf{x}_i \in \mathbb{R}^d$, k-means clustering [24] aims to partition the data sets into k subsets $\mathscr{S}_1, \dots, \mathscr{S}_k$, so as to minimize the distortion function, that is the sum of distances of each point in every cluster to the corresponding center. This optimization problem can be expressed as follows:

$$\min_{\boldsymbol{\mu}^{(1)},\dots,\boldsymbol{\mu}^{(k)}} \sum_{l=1}^{k} \left[\frac{1}{2N_l} \sum_{\mathbf{x}\in\mathscr{S}_l} \|\boldsymbol{\mu}^{(l)} - \mathbf{x}\|_2^2 \right], \quad (4)$$

where $\boldsymbol{\mu}^{(l)}$ is the mean of the points in $\mathscr{S}_l$. Since this problem is NP-hard, an alternate optimization procedure similar to the expectation-maximization algorithm is employed, which converges quickly to a local optimum. In practice, after randomly initializing the cluster centers, an assignment and an update step are repeated until the cluster memberships no longer change. In the assignment step each point is assigned to the closest center, i.e. the cluster whose mean yields the least within-cluster sum of squares. In the update step, the new cluster centroids are calculated.

The outcomes produced by the standard k-means algorithm are highly sensitive to the initialization of the cluster centers and the presence of outliers. In Sect. 5 we further discuss the regularized stochastic k-means approach which, similarly to other methods briefly reviewed in Sect. 5.1, allows to tackle these issues through stochastic optimization approaches.

4 Fixed-Size Kernel Spectral Clustering (FSKSC)

In this section we review an alternative approach to scale-up spectral clustering named fixed-size kernel spectral clustering, which was recently proposed in [25]. Compared to the existing techniques, the major advantages of this method are the possibility to extend the clustering model to new out-of-sample points and a precise model selection scheme.

4.1 Related Work

Several algorithms have been devised to speed-up spectral clustering. Examples include power iteration clustering [26], spectral grouping using the Nyström method [27], incremental algorithms where some initial clusters computed on an initial subset of the data are modified in different ways [28–30], parallel spectral clustering [31], methods based on the incomplete Cholesky decomposition [32–34], landmark-based spectral clustering [35], consensus spectral clustering [36], vector quantization based approximate spectral clustering [37], approximate pairwise clustering [38].

4.2 KSC Overview

The multiway kernel spectral clustering (KSC) formulation is stated as a combination of $k-1$ binary problems, where k denotes the number of clusters [19]. More precisely, given a set of training data $\mathscr{D}_{\text{tr}} = \{\mathbf{x}_i\}_{i=1}^{N_{\text{tr}}}$, the primal problem is expressed by the following objective:

$$\begin{aligned} \min_{\mathbf{w}^{(l)},\mathbf{e}^{(l)},b_l} \quad & \frac{1}{2}\sum_{l=1}^{k-1} \mathbf{w}^{(l)^T}\mathbf{w}^{(l)} - \frac{1}{2}\sum_{l=1}^{k-1} \gamma_l \mathbf{e}^{(l)^T} V \mathbf{e}^{(l)} \\ \text{subject to} \quad & \mathbf{e}^{(l)} = \boldsymbol{\Phi}\mathbf{w}^{(l)} + b_l \mathbf{1}_{N_{\text{tr}}}, l = 1,\dots,k-1. \end{aligned} \tag{5}$$

The $\mathbf{e}^{(l)} = [e_1^{(l)}, \dots, e_i^{(l)}, \dots, e_{N_{\text{tr}}}^{(l)}]^T$ denotes the projections of the training data mapped in the feature space along the direction $\mathbf{w}^{(l)}$. For a given point $\mathbf{x}_i$, the corresponding clustering score is given by:

$$e_i^{(l)} = \mathbf{w}^{(l)^T}\varphi(\mathbf{x}_i) + b_l. \tag{6}$$

In fact, as in a classification setting, the binary clustering model is expressed by an hyperplane passing through the origin, that is $e_i^{(l)} - \mathbf{w}^{(l)^T}\varphi(\mathbf{x}_i) - b_l = 0$. Problem (5) is nothing but a weighted kernel PCA in the feature space $\varphi : \mathbb{R}^d \to \mathbb{R}^{d_h}$, where the aim is to maximize the weighted variances of the scores, i.e. $\mathbf{e}^{(l)^T} V \mathbf{e}^{(l)}$ while keeping the squared norm of the vector $\mathbf{w}^{(l)}$ small. The constants $\gamma_l \in \mathbb{R}^+$ are regularization parameters, $\mathbf{V} \in \mathbb{R}^{N_{\text{tr}} \times N_{\text{tr}}}$ is the weighting matrix and $\boldsymbol{\Phi}$ is the $N_{\text{tr}} \times d_h$ feature matrix $\boldsymbol{\Phi} = [\varphi(\mathbf{x}_1)^T; \dots; \varphi(\mathbf{x}_{N_{\text{tr}}})^T]$, b_l are bias terms.

The dual problem associated to (5) is given by:

$$\mathbf{V}\mathbf{M}_V\boldsymbol{\Omega}\boldsymbol{\alpha}^{(l)} = \lambda_l\boldsymbol{\alpha}^{(l)} \tag{7}$$

where $\boldsymbol{\Omega}$ denotes the kernel matrix with ij-th entry $\boldsymbol{\Omega}_{ij} = K(\mathbf{x}_i, \mathbf{x}_j) = \varphi(\mathbf{x}_i)^T\varphi(\mathbf{x}_j)$. $K : \mathbb{R}^d \times \mathbb{R}^d \rightarrow \mathbb{R}$ means the kernel function. $\mathbf{M}_V$ is a centering matrix defined as $\mathbf{M}_V = \mathbf{I}_{N_{\text{tr}}} - \frac{1}{\mathbf{1}_{N_{\text{tr}}}^T \mathbf{V}\mathbf{1}_{N_{\text{tr}}}}\mathbf{1}_{N_{\text{tr}}}\mathbf{1}_{N_{\text{tr}}}^T\mathbf{V}$, the $\boldsymbol{\alpha}^{(l)}$ are vectors of dual variables, $\boldsymbol{\lambda}_l = \frac{N_{\text{tr}}}{\gamma_l}$. By setting[2] $\mathbf{V} = \mathbf{D}^{-1}$, being $\mathbf{D}$ the graph degree matrix which is diagonal with positive elements $D_{ii} = \sum_j \Omega_{ij}$, problem (7) is closely related to spectral clustering with random walk Laplacian [23, 42, 43], and objective (5) is referred as the kernel spectral clustering problem.

The dual clustering model for the i-th training point can be expressed as follows:

$$e_i^{(l)} = \sum_{j=1}^{N_{\text{tr}}} \alpha_j^{(l)} K(\mathbf{x}_j, \mathbf{x}_i) + b_l, j = 1, \ldots, N_{\text{tr}}, l = 1, \ldots, k-1. \tag{8}$$

By binarizing the projections $e_i^{(l)}$ as $\text{sign}(e_i^{(l)})$ and selecting the most frequent binary indicators, a code-book $\mathcal{CB} = \{c_p\}_{p=1}^k$ with the k cluster prototypes can be formed. Then, for any given point (either training or test), its cluster membership can be computed by taking the sign of the corresponding projection and assigning to the cluster represented by the closest prototype in terms of hamming distance. The KSC method is summarized in algorithm 1, and the related Matlab package is freely available on the Web.[3] Finally, the interested reader can refer to the recent review [18] for more details on the KSC approach and its applications.

Algorithm 1: KSC algorithm [19]

Data: Training set $\mathcal{D}_{\text{tr}} = \{\mathbf{x}_i\}_{i=1}^{N_{\text{tr}}}$, test set $\mathcal{D}_{\text{test}} = \{\mathbf{x}_r^{\text{test}}\}_{r=1}^{N_{\text{test}}}$ kernel function $K : \mathbb{R}^d \times \mathbb{R}^d \rightarrow \mathbb{R}$, kernel parameters (if any), number of clusters k.

Result: Clusters $\{\mathcal{C}_1, \ldots, \mathcal{C}_k\}$, codebook $\mathcal{CB} = \{c_p\}_{p=1}^k$ with $\{c_p\} \in \{-1, 1\}^{k-1}$.

1 compute the training eigenvectors $\boldsymbol{\alpha}^{(l)}$, $l = 1, \ldots, k-1$, corresponding to the $k-1$ largest eigenvalues of problem (7)

2 let $\mathbf{A} \in \mathbb{R}^{N_{\text{tr}} \times (k-1)}$ be the matrix containing the vectors $\boldsymbol{\alpha}^{(1)}, \ldots, \boldsymbol{\alpha}^{(k-1)}$ as columns

3 binarize $\mathbf{A}$ and let the code-book $\mathcal{CB} = \{c_p\}_{p=1}^k$ be composed by the k encodings of $\mathbf{Q} = \text{sign}(A)$ with the most occurrences

4 $\forall i, i = 1, \ldots, N_{\text{tr}}$, assign $\mathbf{x}_i$ to A_{p^*} where $p^* = \text{argmin}_p d_H(\text{sign}(\boldsymbol{\alpha}_i), c_p)$ and $d_H(\cdot, \cdot)$ is the Hamming distance

5 binarize the test data projections $\text{sign}(\mathbf{e}_r^{(l)})$, $r = 1, \ldots, N_{\text{test}}$, and let $\text{sign}(\mathbf{e}_r) \in \{-1, 1\}^{k-1}$ be the encoding vector of $\mathbf{x}_r^{\text{test}}$

6 $\forall r$, assign $\mathbf{x}_r^{\text{test}}$ to A_{p^*}, where $p^* = \text{argmin}_p d_H(\text{sign}(\mathbf{e}_r), c_p)$.

[2]By choosing $\mathbf{V} = \mathbf{I}$, problem (7) represents a kernel PCA objective [39–41].

[3]http://www.esat.kuleuven.be/stadius/ADB/alzate/softwareKSClab.php.

4.3 Fixed-Size KSC Approach

When the number of training datapoints N_{tr} is large, problem (7) can become intractable both in terms of memory bottleneck and execution time. A solution to this issue is offered by the fixed-size kernel spectral clustering (FSKSC) method where the primal problem instead of the dual is solved, as proposed in [17] in case of classification and regression. In particular, as discussed in [25], the FSKSC approach is based on the following unconstrained re-formulation of the KSC primal objective (5), where $\mathbf{V} = \mathbf{D}^{-1}$:

$$\min_{\hat{\mathbf{w}}^{(l)},\hat{b}_l} \quad \frac{1}{2}\sum_{l=1}^{k-1} \hat{\mathbf{w}}^{(l)^T}\hat{\mathbf{w}}^{(l)} - \frac{1}{2}\sum_{l=1}^{k-1} \gamma_l(\hat{\boldsymbol{\Phi}}\hat{\mathbf{w}}^{(l)} + \hat{b}_l\mathbf{1}_{N_{\text{tr}}})^T\hat{\boldsymbol{D}}^{-1}(\hat{\boldsymbol{\Phi}}\hat{\mathbf{w}}^{(l)} + \hat{b}_l\mathbf{1}_{N_{\text{tr}}}) \tag{9}$$

where $\hat{\boldsymbol{\Phi}} = [\hat{\varphi}(\mathbf{x}_1)^T; \ldots; \hat{\varphi}(\mathbf{x}_{N_{\text{tr}}})^T] \in \mathbb{R}^{N_{\text{tr}}\times m}$ is the approximated feature matrix, $\hat{\boldsymbol{D}} \in \mathbb{R}^{N_{\text{tr}}\times N_{\text{tr}}}$ is the corresponding degree matrix, and $\hat{\varphi} : \mathbb{R}^d \to \mathbb{R}^m$ indicates a finite dimensional approximation of the feature[4] map $\varphi(\cdot)$ which can be obtained through the Nyström method [44]. The minimizer of (9) can be found by computing $\nabla J(\mathbf{w}^l, b_l) = 0$, that is:

$$\frac{\partial \mathcal{J}}{\partial \hat{\mathbf{w}}^{(l)}} = 0 \quad \rightarrow \quad \hat{\mathbf{w}}^{(l)} = \gamma_l(\hat{\boldsymbol{\Phi}}^T\hat{\boldsymbol{D}}^{-1}\hat{\boldsymbol{\Phi}}\hat{\mathbf{w}}^{(l)} + \hat{\boldsymbol{\Phi}}^T\hat{\boldsymbol{D}}^{-1}\mathbf{1}_{N_{\text{tr}}}\hat{b}_l)$$

$$\frac{\partial \mathcal{J}}{\partial \hat{b}_l} = 0 \quad \rightarrow \quad \mathbf{1}_{N_{\text{tr}}}^T\hat{\boldsymbol{D}}^{-1}\hat{\boldsymbol{\Phi}}\hat{\mathbf{w}}^{(l)} = -\mathbf{1}_{N_{\text{tr}}}^T\hat{\boldsymbol{D}}^{-1}\mathbf{1}_{N_{\text{tr}}}\hat{b}_l.$$

These optimality conditions lead to the following eigenvalue problem to solve in order to find the model parameters:

$$\mathbf{R}\hat{\mathbf{w}}^{(l)} = \hat{\lambda}_l\hat{\mathbf{w}}^{(l)} \tag{10}$$

with $\hat{\lambda}_l = \frac{1}{\gamma_l}$, $\mathbf{R} = \hat{\boldsymbol{\Phi}}^T\hat{\mathbf{D}}^{-1}\hat{\boldsymbol{\Phi}} - \frac{(\mathbf{1}_{N_{\text{tr}}}^T\hat{\mathbf{D}}^{-1}\hat{\boldsymbol{\Phi}})^T(\mathbf{1}_{N_{\text{tr}}}^T\hat{\mathbf{D}}^{-1}\hat{\boldsymbol{\Phi}})}{\mathbf{1}_{N_{\text{tr}}}^T\hat{\mathbf{D}}^{-1}\mathbf{1}_{N_{\text{tr}}}}$ and $\hat{b}_l = -\frac{\mathbf{1}_{N_{\text{tr}}}^T\hat{\mathbf{D}}^{-1}\hat{\boldsymbol{\Phi}}}{\mathbf{1}_{N_{\text{tr}}}^T\hat{\mathbf{D}}^{-1}\mathbf{1}_{N_{\text{tr}}}}\hat{\mathbf{w}}^{(l)}$. Notice that we now have to solve an eigenvalue problem of size $m \times m$, which can be done very efficiently by choosing m such that $m \ll N_{\text{tr}}$. Furthermore, the diagonal of matrix $\hat{\mathbf{D}}$ can be calculated as $\hat{\mathbf{d}} = \hat{\boldsymbol{\Phi}}(\hat{\boldsymbol{\Phi}}^T\mathbf{1}_m)$, i.e. without constructing the full matrix $\hat{\boldsymbol{\Phi}}\hat{\boldsymbol{\Phi}}^T$.

Once $\hat{\mathbf{w}}^{(l)}, \hat{b}_l$ have been computed, the cluster memberships can be obtained by applying the k-means algorithm on the projections $\hat{e}_i^{(l)} = \hat{\mathbf{w}}^{(l)^T}\hat{\varphi}(\mathbf{x}_i) + \hat{b}_l$ for training data and $\hat{e}_r^{(l),\text{test}} = \hat{\mathbf{w}}^{(l)^T}\hat{\varphi}(\mathbf{x}_i^{\text{test}}) + \hat{b}_l$ in case of test points, as for the classical spectral clustering technique. The entire algorithm is depicted in Fig. 2, and a Matlab implementation is freely available for download.[5] Finally, Fig. 1 illustrates examples of clustering obtained in case of the *Iris*, *Dermatology* and *S1* datasets available at the UCI machine learning repository.

---

[4]The m points needed to estimate the components of $\hat{\varphi}$ are selected at random.

[5]http://www.esat.kuleuven.be/stadius/ADB/langone/softwareKSCFSlab.php.

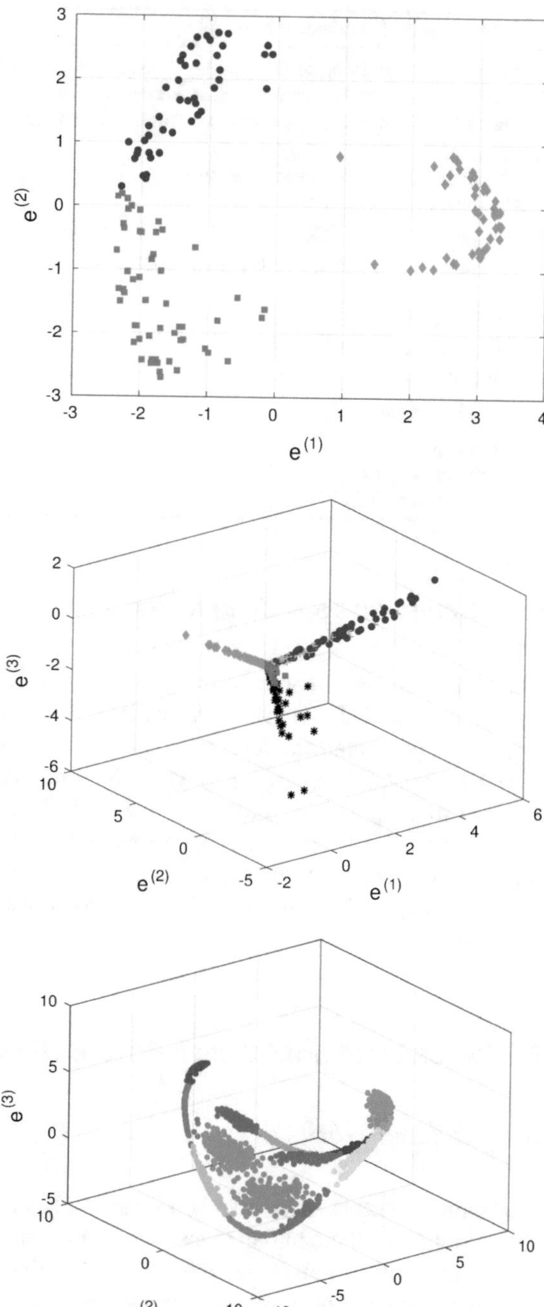

Fig. 1 FSKSC embedding illustrative example. Data points represented in the space of the projections in case of the Iris, Dermatology and S1 datasets. The different colors relate to the various clusters detected by the FSKSC algorithm

Algorithm 2: Fixed-size KSC [25]

Input : training set $\mathcal{D} = \{\mathbf{x}_i\}_{i=1}^{N_{tr}}$, Test set $\mathcal{D}_{test} = \{\mathbf{x}_i\}_{r=1}^{N_{test}}$.
Settings : size Nyström subset m, kernel parameter σ, number of clusters k
Output : $\mathbf{q}$ and $\mathbf{q}_{test}$ vectors of predicted cluster memberships.

`/* Approximate feature map: */`
Compute $\boldsymbol{\Omega}_{m \times m}$
Compute $[\mathbf{U},\boldsymbol{\Lambda}] = \text{SVD}(\boldsymbol{\Omega}_{m \times m})$
Compute $\hat{\boldsymbol{\Phi}}$ by means of the Nyström method

`/* Training: */`
Solve $\mathbf{R}\hat{\mathbf{w}}^{(l)} = \hat{\lambda}_l \hat{\mathbf{w}}^{(l)}$
Compute $\mathbf{E} = [\mathbf{e}^{(1)}, \ldots, \mathbf{e}^{k-1}]$
$[\mathbf{q},\mathbf{C}_{tr}] = \text{kmeans}(\mathbf{E},\text{k})$

`/* Test: */`
Compute $\mathbf{E}_{test} = [\mathbf{e}_{test}^{(1)}, \ldots, \mathbf{e}_{test}^{k-1}]$
$\mathbf{q}_{test} = \text{kmeans}(\mathbf{E}_{test},\text{k,'start'},\mathbf{C}_{tr})$

4.4 Computational Complexity

The computational complexity of the fixed-size KSC algorithm depends mainly on the size m of the Nyström subset used to construct the approximate feature map $\hat{\boldsymbol{\Phi}}$. In particular, the total time complexity (training + test) is approximately $O(m^3) + O(mN_{tr}) + O(mN_{test})$, which is the time needed to solve (10) and to compute the training and test clustering scores. Furthermore, the space complexity is $O(m^2) + O(mN_{tr}) + O(mN_{test})$, which is needed to construct matrix $\mathbf{R}$ and to build the training and test feature matrices $\hat{\boldsymbol{\Phi}}$ and $\hat{\boldsymbol{\Phi}}_{test}$. Since we can choose $m \ll N_{tr} < N_{test}$ [25], the complexity of the algorithm is approximately linear, as can be evinced also from Fig. 6.

5 Regularized Stochastic K-Means (RSKM)

5.1 Related Work

The main drawbacks of the standard k-means algorithm are the instability caused by the randomness in the initialization and the presence of outliers, which can bias the computation of the cluster centroids and hence the final memberships. To stabilize the performance of the k-means algorithm [45] applies the stochastic learning paradigm relying on the probabilistic draw of some specific random variable dependent upon the distribution of per-sample distances to the centroids. In [21] one seeks to find a new cluster centroid by observing one or a small mini-batch sample at iterate t and calculating the corresponding gradient descent step. Recent developments [46, 47] indicate that the regularization with different norms might be useful when one deals with high-dimensional datasets and seeks for a sparse solution. In particular, [46] proposes to use an adaptive group Lasso penalty [48] and obtain a solution per prototype vector in a closed-form. In [49] the authors are studying the problem of overlapping clusters where there are possible outliers in data. They propose an objective function which can be viewed as a reformulation of the traditional k-means objective which captures also the degrees of overlap and non-exhaustiveness.

5.2 *Generalities*

Given a dataset $\mathcal{D} = \{\mathbf{x}_i\}_{i=1}^N$ with N independent observations, the regularized k-means objective can be expressed as follows:

$$\min_{\boldsymbol{\mu}^{(1)},\ldots,\boldsymbol{\mu}^{(k)}} \sum_{l=1}^{k} \left[\frac{1}{2N_l} \sum_{\mathbf{x}\in\mathcal{S}_l} \|\boldsymbol{\mu}^{(l)} - \mathbf{x}\|_2^2 + C\psi(\boldsymbol{\mu}^{(l)}) \right], \tag{11}$$

where $\psi(\boldsymbol{\mu}^{(l)})$ represents the regularizer, C is the trade-off parameter, $N_l = |\mathcal{S}_l|$ is the cardinality of the corresponding set $\mathcal{S}_l$ corresponding to the l-th individual cluster. In a stochastic optimization paradigm objective (11) can be optimized through gradient descent, meaning that one takes at any step t some gradient $g_t \in \partial f(\boldsymbol{\mu}_t^{(l)})$ w.r.t. only one sample $\mathbf{x}_t$ from $\mathcal{S}_l$ and the current iterate $\boldsymbol{\mu}_t^{(l)}$ at hand. This online learning problem is usually terminated until some ϵ-tolerance criterion is met or the total number of iterations is exceeded. In the above setting one deals with a simple clustering model $c(\mathbf{x}) = \arg\min_l \|\boldsymbol{\mu}^{(l)} - \mathbf{x}\|_2$ and updates cluster memberships of the entire dataset $\hat{\mathcal{S}}$ after individual solutions $\boldsymbol{\mu}^{(l)}$, i.e. the centroids, are computed. From a practical point of view, we denote this update as an outer iteration or synchronization step and use it to fix $\mathcal{S}_l$ for learning each individual prototype vector $\boldsymbol{\mu}^{(l)}$ in parallel through a Map-Reduce scheme. This algorithmic procedure is depicted in Fig. 2. As we can notice the Map-Reduce framework is needed to parallelize learning of individual prototype vectors using either the SGD-based approach or the adaptive dual averaging scheme. In each outer p-th iteration we `Reduce()` all learned centroids to the matrix $\mathbf{W}_p$ and re-partition the data again with `Map()`. After we reach T_{out} iterations we stop and re-partition the data according to the final solution and proximity to the prototype vectors.

5.3 *l_2-Regularization*

In this section the Stochastic Gradient Descent (SGD) scheme for learning objective (11) with $\psi(\boldsymbol{\mu}^{(l)}) = \frac{1}{2}\|\boldsymbol{\mu}^{(l)}\|_2^2$ is presented. If we use the l_2 regularization, the optimization problem becomes:

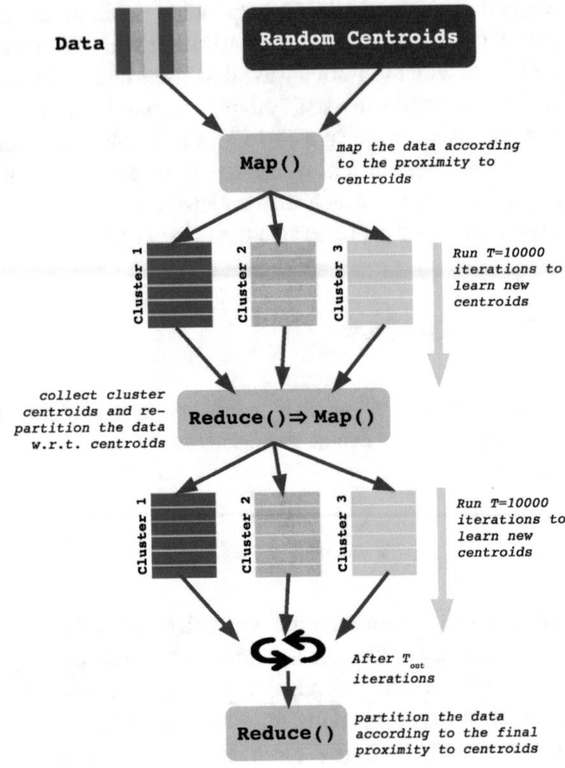

Fig. 2 Schematic visualization of the Map-Reduce scheme

$$\min_{\boldsymbol{\mu}^{(l)}} f(\boldsymbol{\mu}^{(l)}) \triangleq \frac{1}{2N} \sum_{j=1}^{N} \|\boldsymbol{\mu}^{(l)} - \mathbf{x}_j\|_2^2 + \frac{C}{2} \|\boldsymbol{\mu}^{(l)}\|_2^2, \tag{12}$$

where function $f(\boldsymbol{\mu}^{(l)})$ is λ-strongly convex with Lipschitz continuous gradient and Lipschitz constant equal to L. It can be easily verified that $\lambda = L = C + 1$ by observing basic inequalities which $f(\boldsymbol{\mu}^{(l)})$ should satisfy in this case [50, 51]:

$$\|\nabla f(\boldsymbol{\mu}^{(l)}) - \nabla f(\boldsymbol{\mu}^{(l)})\|_2 \geq \lambda \|\boldsymbol{\mu}_1^{(l)} - \boldsymbol{\mu}_2^{(l)}\|_2 \Longrightarrow$$
$$\|(C+1)\boldsymbol{\mu}_1^{(l)} - (C+1)\boldsymbol{\mu}_2^{(l)}\|_2 \geq \lambda \|\boldsymbol{\mu}_1^{(l)} - \boldsymbol{\mu}_2^{(l)}\|_2$$

and

$$\|\nabla f(\boldsymbol{\mu}_1^{(l)}) - \nabla f(\boldsymbol{\mu}_2^{(l)})\|_2 \leq L \|\boldsymbol{\mu}_1^{(l)} - \boldsymbol{\mu}_2^{(l)}\|_2 \Longrightarrow$$
$$\|(C+1)\boldsymbol{\mu}_1^{(l)} - (C+1)\boldsymbol{\mu}_2^{(l)}\|_2 \leq L \|\boldsymbol{\mu}_1^{(l)} - \boldsymbol{\mu}_2^{(l)}\|_2$$

which can be satisfied if and only if $\lambda = L = C + 1$. In this case a proper sequence of SGD step-sizes η_t should be applied in order to achieve optimal convergence rate [52]. As a consequence, we set $\eta_t = \frac{1}{Ct}$ such that the convergence rate to the ε-optimal solution would be $\mathcal{O}(\frac{1}{T})$, being T the total number of iterations, i.e. $1 \leq t \leq T$. This leads to a cheap, robust and stable to perturbation learning procedure with a fixed computational budget imposed on the total number of iterations and gradient re-computations needed to find a feasible solution.

The complete algorithm is illustrated in Algorithm 3. The first step is the initialization of a random matrix $\mathbf{M}_0$ of size $d \times k$, where d is the input dimension and k is the number of clusters. After initialization T_{out} outer synchronization iterations are performed in which, based on previously learned individual prototype vectors $\boldsymbol{\mu}^{(l)}$, the cluster memberships and re-partition $\hat{\mathcal{S}}$ are calculated (line 4). Afterwards we run in parallel a basic SGD scheme for the l_2-regularized optimization objective (12) and concatenate the result with $\mathbf{M}_p$ by the `Append` function. When the total number of outer iterations T_{out} is exceeded we exit with the final partitioning of $\hat{\mathcal{S}}$ by $c(x) = \arg\min_i \|\mathbf{M}^{(l)}_{T_{out}} - \mathbf{x}\|_2$ where l denotes the l-th column of $\mathbf{M}_{T_{out}}$.

Algorithm 3: l_2-Regularized stochastic k-means

Data: $\hat{\mathcal{S}}, C > 0, T \geq 1, T_{out} \geq 1, k \geq 2, \varepsilon > 0$
1 Initialize $\mathbf{M}_0$ randomly for all clusters $(1 \leq l \leq k)$
2 **for** $p \leftarrow 1$ **to** T_{out} **do**
3   Initialize empty matrix $\mathbf{M}_p$
4   Partition $\hat{\mathcal{S}}$ by $c(x) = \arg\min_l \|\mathbf{M}^{(l)}_{p-1} - \mathbf{x}\|_2$
5   **for** $\mathcal{S}_l \subset \hat{\mathcal{S}}$ ***in parallel*** **do**
6   Initialize $\boldsymbol{\mu}^{(l)}_0$ randomly
7   **for** $t \leftarrow 1$ **to** T **do**
8   Draw a sample $\mathbf{x}_t \in \mathcal{S}_l$
9   Set $\eta_t = 1/(Ct)$
10             $\boldsymbol{\mu}^{(l)}_t = \boldsymbol{\mu}^{(l)}_{t-1} - \eta_t(C\boldsymbol{\mu}^{(l)}_{t-1} + \boldsymbol{\mu}^{(l)}_{t-1} - \mathbf{x}_t)$
11   **if** $\|\boldsymbol{\mu}^{(l)}_t - \boldsymbol{\mu}^{(l)}_{t-1}\|_2 \leq \varepsilon$ **then**
12   `Append` $(\boldsymbol{\mu}^{(l)}_t, \mathbf{M}_p)$
13   **return**
14   **end**
15   **end**
16   `Append` $(\boldsymbol{\mu}^{(l)}_T, \mathbf{M}_p)$
17   **end**
18 **end**
19 **return** $\hat{\mathcal{S}}$ *is partitioned by* $c(x) = \arg\min_l \|\mathbf{M}^{(l)}_{T_{out}} - \mathbf{x}\|_2$

5.4 l_1-Regularization

In this section we present a different learning scheme induced by l_1-norm regularization and corresponding regularized dual averaging methods [53] with adaptive primal-dual iterate updates [54]. The main optimization objective is given by [55]:

$$\min_{\boldsymbol{\mu}^{(l)}} f(\boldsymbol{\mu}^{(l)}) \triangleq \frac{1}{2N} \sum_{j=1}^{N} \|\boldsymbol{\mu}^{(l)} - \mathbf{x}_j\|_2^2 + C\|\boldsymbol{\mu}^{(l)}\|_1. \tag{13}$$

By using a *simple dual averaging* scheme [22] and adaptive strategy from [54] problem (13) can be solved effectively by the following sequence of iterates $\boldsymbol{\mu}_{t+1}^{(l)}$:

$$\boldsymbol{\mu}_{t+1}^{(l)} = \arg\min_{\boldsymbol{\mu}^{(l)}} \left\{ \frac{\eta}{t} \sum_{\tau=1}^{t} \langle g_\tau, \boldsymbol{\mu}^{(l)} \rangle + \eta C \|\boldsymbol{\mu}^{(l)}\|_1 + \frac{1}{t} h(\boldsymbol{\mu}^{(l)}) \right\}, \tag{14}$$

where $h_t(\boldsymbol{\mu}^{(l)})$ is an adaptive strongly convex proximal term, g_t represents a gradient of the $\|\boldsymbol{\mu}^{(l)} - \mathbf{x}_t\|^2$ term w.r.t. only one randomly drawn sample $\mathbf{x}_t \in \mathscr{S}_l$ and current iterate $\boldsymbol{\mu}_t^{(l)}$, while η is a fixed step-size. In the regularized Adaptive Dual Averaging (ADA) scheme [54] one is interested in finding a corresponding step-size for each coordinate which is inversely proportional to the time-based norm of the coordinate in the sequence $\{g_t\}_{t\geq 1}$ of gradients. In case of our algorithm, the coordinate-wise update of the $\boldsymbol{\mu}_t^{(l)}$ iterate in the adaptive dual averaging scheme can be summarized as follows:

$$\boldsymbol{\mu}_{t+1,q}^{(l)} = \text{sign}(-\hat{g}_{t,q}) \frac{\eta t}{H_{t,qq}} [|\hat{g}_{t,q}| - \lambda]_+, \tag{15}$$

where $\hat{g}_{t,q} = \frac{1}{t}\sum_{\tau=1}^{t} g_{\tau,q}$ is the coordinate-wise mean across $\{g_t\}_{t\geq 1}$ sequence, $H_{t,qq} = \rho + \|g_{1:t,q}\|_2$ is the time-based norm of the q-th coordinate across the same sequence and $[x]_+ = \max(0, x)$. In Eq. (15) two important parameters are present: C which controls the importance of the l_1-norm regularization and η which is necessary for the proper convergence of the entire sequence of $\boldsymbol{\mu}_t^{(l)}$ iterates.

An outline of our distributed stochastic l_1-regularized k-means algorithm is depicted in Algorithm 4. Compared to the l_2 regularization, the iterate $\boldsymbol{\mu}_t^{(l)}$ now has a closed form solution and depends on the dual average (and the sequence of gradients $\{g_t\}_{t\geq 1}$). Another important difference is the presence of some additional parameters: the fixed step-size η and the additive constant ρ for making $H_{t,qq}$ term non-zero. These additional degrees of freedom might be beneficial from the generalization perspective. However, an increased computational cost has to be expected due to the cross-validation needed for their selection. Both versions of the regularized stochastic k-means method presented in Sects. 5.3 and 5.4 are available for download.[6]

[6] http://www.esat.kuleuven.be/stadius/ADB/jumutc/softwareSALSA.php.

Algorithm 4: l_1-Regularized stochastic k-means [55]

Data: $\hat{\mathscr{S}}, C > 0, \eta > 0, \rho > 0, T \geq 1, T_{out} \geq 1, k \geq 2, \varepsilon > 0$
Initialize $\mathbf{M}_0$ randomly for all clusters $(1 \leq l \leq k)$
for $p \leftarrow 1$ **to** T_{out} **do**
  Initialize empty matrix $\mathbf{M}_p$
  Partition $\hat{\mathscr{S}}$ by $c(x) = \arg\min_l \|\mathbf{M}_{p-1}^{(l)} - \mathbf{x}\|_2$
  for $\mathscr{S}_l \subset \hat{\mathscr{S}}$ ***in parallel*** **do**
    Initialize $\boldsymbol{\mu}_1^{(l)}$ randomly, $\hat{g}_0 = 0$
    for $t \leftarrow 1$ **to** T **do**
      Draw a sample $\mathbf{x}_t \in \mathscr{S}_l$
      Calculate gradient $g_t = \boldsymbol{\mu}_t^{(l)} - \mathbf{x}_t$
      Find the average $\hat{g}_t = \frac{t-1}{t}\hat{g}_{t-1} + \frac{1}{t}g_t$
      Calculate $H_{t,qq} = \rho + \|g_{1:t,q}\|_2$
      $\boldsymbol{\mu}_{t+1,q}^{(l)} = \text{sign}(-\hat{g}_{t,q})\frac{\eta t}{H_{t,qq}}[|\hat{g}_{t,q}| - C]_+$
      if $\|\boldsymbol{\mu}_t^{(l)} - \boldsymbol{\mu}_{t+1}^{(l)}\|_2 \leq \varepsilon$ **then**
        `Append` $(\boldsymbol{\mu}_{t+1}^{(l)}, \mathbf{M}_p)$
        return
      end
    end
    `Append` $(\boldsymbol{\mu}_{T+1}^{(l)}, \mathbf{M}_p)$
  end
end
return $\hat{\mathscr{S}}$ *is partitioned by* $c(x) = \arg\min_l \|\mathbf{M}_{T_{out}}^{(l)} - \mathbf{x}\|_2$

5.5 *Influence of Outliers*

Thanks to the regularization terms that have been added to the k-means objective in Eqs. (13) and (12), the regularized stochastic k-means becomes less sensitive to the influence of the outliers. Furthermore, the stochastic optimization schemes allow to reduce also the sensitivity to the initialization. In order to illustrate this aspects, a synthetic dataset consisting of three Gaussian clouds corrupted by outliers is used as benchmark. As shown in Fig. 3, while k-means can fail to recover the true cluster centroids and, as a consequence, produces a wrong partitioning, the regularized schemes are always able to correctly identify the three clouds of points.

5.6 *Theoretical Guarantees*

In this section a theoretical analysis of the algorithms described previously is discussed. In case of the l_2-norm, two results in expectation obtained by [52] for smooth and strongly convex functions are properly reformulated. Regarding the l_1-norm, our

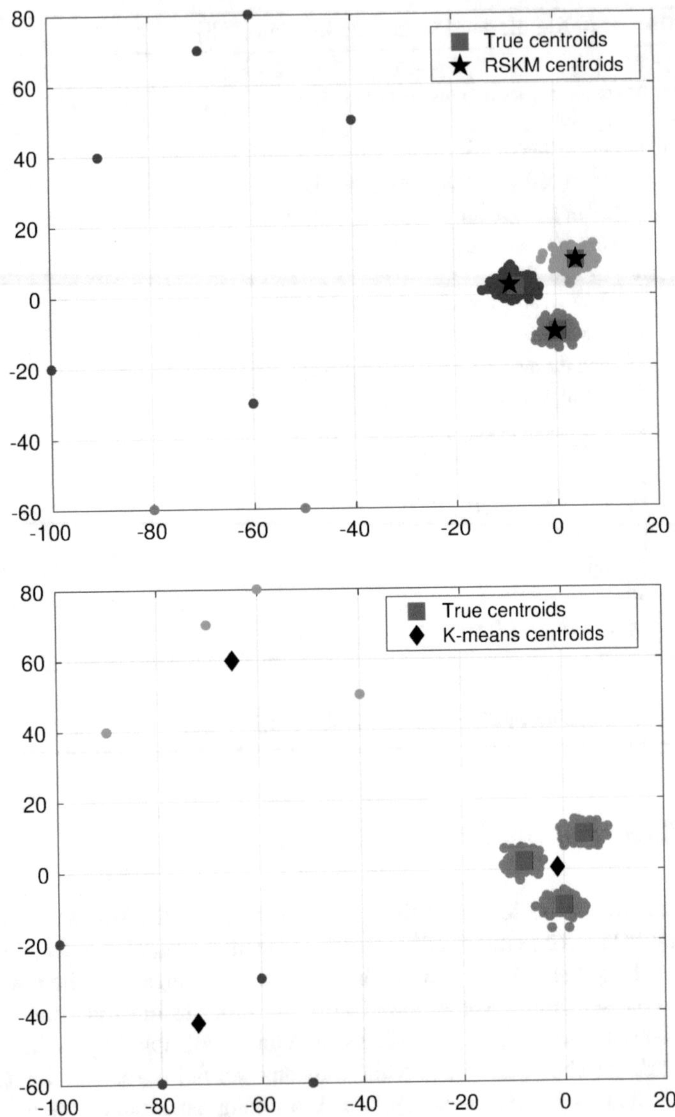

Fig. 3 Influence of outliers. *(Top)* K-means clustering of a synthetic dataset with three clusters corrupted by outliers. *(Bottom)* In this case RSKM is insensitive to the outliers and allows to perfectly detect the three Gaussians, while K-means only yields a reasonable result 4 times out of 10 runs

theoretical results are stemmed directly from various lemmas and corollaries related to the adaptive subgradient method presented in [54].

5.6.1 l_2-norm

As it was shown in Sect. 5.3 the l_2-regularized k-means objective (12) is a smooth strongly convex function with Lipschitz continuous gradient. Based on this, an upper bound on $f(\boldsymbol{\mu}_T^{(l)}) - f(\boldsymbol{\mu}_*^{(l)})$ in expectation can be derived, where $\boldsymbol{\mu}_*^{(l)}$ denotes the optimal center for the l-th cluster, where $l = 1, \ldots, k$.

Theorem 1 *Consider strongly convex function $f(\boldsymbol{\mu}^{(l)})$ in Eq. (12) which is ν-smooth with respect to $\boldsymbol{\mu}_*^{(l)}$ over the convex set $\mathscr{W}$. Suppose that $\mathbb{E}\|\hat{g}_t\|^2 \leq G^2$. Then if we take any $C > 0$ and pick the step-size $\eta = \frac{1}{C}t$, it holds for any T that:*

$$\mathbb{E}[f(\boldsymbol{\mu}_T^{(l)}) - f(\boldsymbol{\mu}_*^{(l)})] \leq \frac{2G^2}{(C+1)T}. \tag{16}$$

Proof This result follows directly from Theorem 1 in [52] where the ν-smoothness is defined as $f(\boldsymbol{\mu}^{(l)}) - f(\boldsymbol{\mu}_*^{(l)}) \leq \frac{\nu}{2}\|\boldsymbol{\mu}^{(l)} - \boldsymbol{\mu}_*^{(l)}\|$. From the theory of convex optimization we know that this inequality is a particular case of a more general inequality for functions with Lipschitz continuous gradients. From Sect. 5.3 we know that our Lipschitz constant is $L = C + 1$. Plugging the already known constants into the aforementioned Theorem 1 completes our proof.

Furthermore, an upper bound on $\|\boldsymbol{\mu}_T - \boldsymbol{\mu}_*\|^2$ in expectation can be obtained:

Theorem 2 *Consider strongly convex function $f(\boldsymbol{\mu})$ in Eq.(12) over the convex set $\mathscr{W}$. Suppose that $\mathbb{E}\|\hat{g}_t\|^2 \leq G^2$. Then if we take any $C > 0$ and pick the step-size $\eta = \frac{1}{C}t$, it holds for any T that:*

$$\mathbb{E}[\|\boldsymbol{\mu}_T - \boldsymbol{\mu}_*\|^2)] \leq \frac{4G^2}{(C+1)^2T}. \tag{17}$$

Proof This result directly follows from Lemma 1 in [52] if we take into account that $f(\boldsymbol{\mu})$ is $(C+1)$-strongly convex.

5.6.2 l_1-norm

First consider the following implication of Lemma 4 in [54] over the running subgradient $g_t = \boldsymbol{\mu}_t^{(l)} - \boldsymbol{x}_t$ of the first term in the optimization objective defined in Eq. (13):

$$\sum_{t=1}^{T} \|f_t'(\boldsymbol{\mu}_t^{(l)})\|^2 \leq 2\sum_{l=1}^{d} \|g_{1:T,q}\|_2. \tag{18}$$

Table 1 Datasets

Size	Dataset	N	d
Small	Iris	150	4
	Ecoli	336	8
	Libras	360	91
	Dermatology	366	33
	Vowel	528	10
	Spambase	4601	57
	S1	5000	2
	S2	5000	2
	S3	5000	2
	S4	5000	2
Medium	Opt digits	5620	64
	Pen digits	10992	16
	Magic	19020	11
	RCV1	20242	1960
	Shuttle	58000	9
Large	Skin	245057	3
	Covertype	581012	54
	GalaxyZoo	667944	9
	Poker	1025010	10
	Susy	5000000	18
	Higgs	11000000	28

where $\|g_{1:T,q}\|_2$ is the time-based norm of the q-th coordinate. Here we can see a direct link to some of our previously presented results in Theorem 2 where we operate over the bounds of iterate specific subgradients.

Theorem 3 *By defining the following infinity norm* $D_\infty = \sup_{\boldsymbol{\mu}^{(l)} \in \boldsymbol{M}} \|\boldsymbol{\mu}^{(l)} - \boldsymbol{\mu}_*^{(l)}\|_\infty$ *w.r.t. the optimal solution* $\boldsymbol{\mu}_*^{(l)}$*, setting the learning rate* $\eta = D_\infty/\sqrt{2}$ *and applying update steps to* $\boldsymbol{\mu}_t^{(l)}$ *in Algorithm 4 we get:*

$$\mathbb{E}_{t\in\{1\ldots T\}}[f(\boldsymbol{\mu}_t^{(l)}) - f(\boldsymbol{\mu}_*^{(l)})] \leq \frac{\sqrt{2}D_\infty}{T} \sum_{l=1}^{d} \|g_{1:T,l}\|_2. \tag{19}$$

Proof Our result directly follows from Corollary 6 in [54] and averaging the regret term $R_\phi(T)$ (defining an expectation over the running index t) *w.r.t.* the optimal solution $f(\boldsymbol{\mu}_*^{(l)})$.

Our bounds imply faster convergence rates than non-adaptive algorithms on sparse data, though this depends on the geometry of the underlying optimization space of $\boldsymbol{M}$.

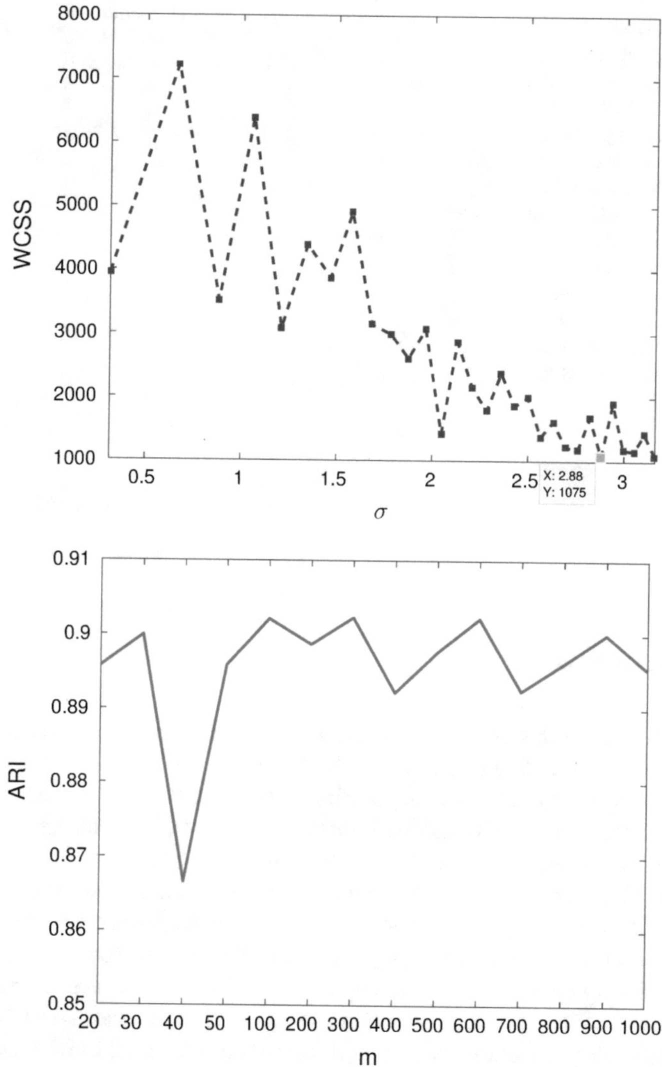

Fig. 4 *FSKSC parameters selection. (Top)* Tuning of the Gaussian kernel bandwidth σ *(Bottom)* Change of the cluster performance (median ARI over 30 runs) with respect to the Nyström subset size m. The simulations refer to the S1 dataset

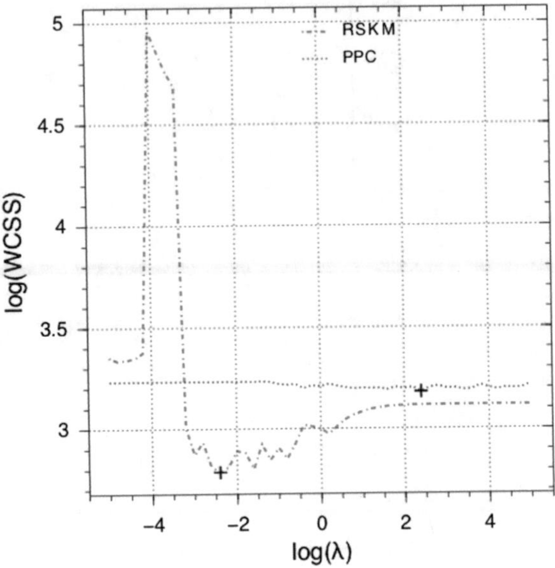

Fig. 5 RSKM and PPC parameters selection. Tuning of the regularization parameter for RSKM and PPC approaches by means of the WCSS criterion, concerning the toy dataset shown in Fig. 3. In this case RSKM is insensitive to the outliers and allows to perfectly detect the three Gaussians (ARI = 0.99), while the best performance reached by the PPC method is ARI = 0.60

6 Experiments

In this section a number of large-scale clustering algorithms are compared in terms of accuracy and execution time. The methods that are analyzed are: fixed-size kernel spectral clustering (FSKSC), regularized stochastic k-means (RSKM), parallel plane clustering [56] (PPC), parallel k-means [9] (PKM). The datasets used in the experiments are listed in Table 1 and mainly comprise databases available at the UCI repository [57]. Although they relate to classification problems, in view of the cluster assumption [58] [7] they can also be used to evaluate the performance of clustering algorithms (in this case the labels play the role of the ground-truth).

The clustering quality is measured by means of two quality metrics, namely the Davies-Bouldin (DB) [59] criterion and the adjusted rand index (ARI [60]). The first quantifies the separation between each pair of clusters in terms of between cluster scatter (how far the clusters are) and within cluster scatter (how tightly grouped the data in each cluster are). The ARI index measures the agreement between two partitions and is used to assess the correlation between the outcome of a clustering algorithm and the available ground-truth.

All the simulations are performed on an eight cores desktop PC in *Julia*,[8] which is a high-level dynamic programming language that provides a sophisticated compiler and an intuitive distributed parallel execution.

[7]The cluster assumption states that if points are in the same cluster they are likely to be of the same class.

[8]http://julialang.org/.

Table 2 Clustering performance. Comparison of the RSKM and FSKSC approaches against k-means (KM) and PPC algorithms in terms of ARI (the higher the better) and DB (the lower the better). The results indicate the best performance over 20 runs

Dataset	RSKM-L1		RSKM-L2		PKM		PPC		FSKSC	
	DB	ARI	DB	ARI	DB	ARI	DB	ARI	DB	ARI
Iris	0.834	0.653	0.834	0.641	0.663	0.645	0.872	**0.758**	**0.000**	0.744
Ecoli	**0.865**	0.722	0.960	**0.727**	1.032	0.704	3.377	0.430	1.063	0.679
Dermatology	1.188	0.891	1.557	**0.903**	1.479	0.891	1.619	0.669	**1.252**	0.829
Vowel	1.553	0.121	1.512	0.127	1.544	**0.133**	4.846	0.060	**0.000**	0.109
Libras	**1.231**	**0.357**	1.234	0.338	1.288	0.327	4.900	0.109	1.352	0.298
Pen digits	1.381	**0.638**	1.262	0.624	1.352	0.634	4.024	0.257	**0.000**	0.546
Opt digits	1.932	0.559	1.991	0.571	1.967	**0.593**	6.315	0.139	**1.140**	0.512
Spambase	3.527	**0.549**	0.650	0.532	**0.613**	0.517	1.244	0.288	3.871	0.382
S1	0.367	0.981	0.509	0.903	**0.365**	**0.986**	1.581	0.551	0.366	0.984
S2	0.478	0.905	0.571	0.867	**0.465**	**0.938**	2.051	0.420	0.518	0.876
S3	0.686	0.681	0.689	0.679	0.644	**0.725**	3.360	0.395	**0.000**	0.648
S4	0.663	0.602	0.665	0.598	0.657	**0.632**	4.158	0.305	**0.000**	0.607
Magic	1.430	0.005	1.435	0.008	1.439	0.006	1.424	0.002	**0.313**	**0.089**
Shuttle	0.775	0.395	0.733	0.396	0.730	0.395	1.295	**0.468**	**0.000**	0.440
Skin	0.695	-0.020	0.696	–0.028	0.699	-0.030	0.811	**–0.005**	**0.687**	-0.027
Gzoo	0.989	**0.302**	0.988	0.297	**0.981**	0.295	1.420	0.253	1.693	0.255
Covertype	2.399	0.107	2.450	0.099	1.825	**0.120**	5.600	0.059	**0.000**	0.095
Poker	2.193	**0.001**	2.190	0.000	2.204	0.000	17.605	0.000	**0.000**	0.000
Susy	2.08	**0.12**	2.10	**0.12**	**2.07**	0.11	3.46	0.09	2.17	**0.12**
Higgs	**2.68**	0.007	2.84	0.007	**2.68**	0.006	3.50	0.002	3.34	**0.008**

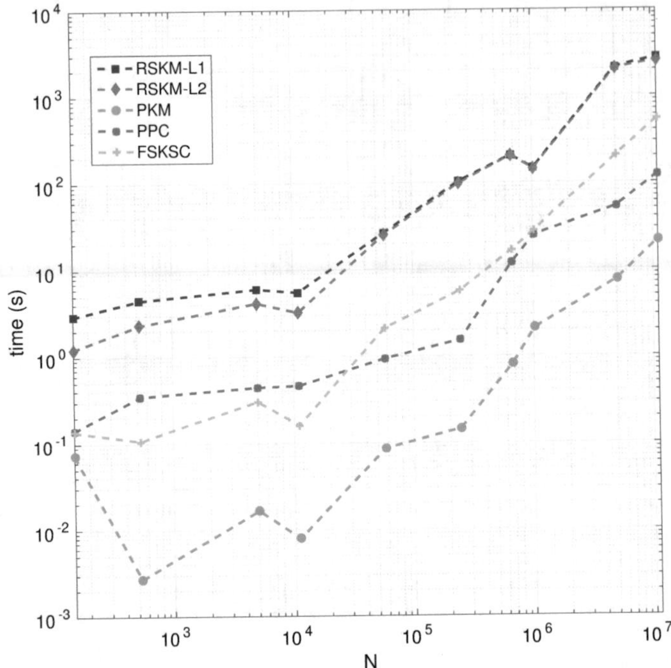

Fig. 6 Efficiency evaluation. Runtime of FSKSC (train + test), RSKM with l1 and l2 regularization, parallel k-means and PPC algorithms related to the following datasets: Iris, Vowel, S1, Pen Digits, Shuttle, Skin, Gzoo, Poker, Susy, Higgs described in Table 2

The selection of the tuning parameters has been done as follows. For all the methods the number of clusters k has been set equal to the number of classes and the tuning parameters are selected by means of the within cluster sum of squares or WCSS criterion [61]. WCSS quantifies the compactness of the clusters in terms of sum of squared distances of each point in a cluster to the cluster center, averaged over all the clusters: the lower the index, the better (i.e. the higher the compactness). Concerning the FSKSC algorithm, the Gaussian kernel defined as $k(\mathbf{x}_i, \mathbf{x}_j) = \exp\left(-\frac{||\mathbf{x}_i - \mathbf{x}_j||^2}{\sigma^2}\right)$ is used to induce the nonlinear mapping. In this case, WCSS allows to select an optimal bandwidth σ as shown at the top side of Fig. 4 for the S1 dataset. Furthermore, the Nyström subset size has been set to $m = 100$ in case of the small datasets and $m = 150$ for the medium and large databases. This setting has been empirically found to represent a good choice, as illustrated at the bottom of Fig. 4 for the S1 dataset. Also in case of RSKM and PPC the regularization parameter C is found as the value yielding the minimum WCSS. An example of such a tuning procedure is depicted in Fig. 5 in case of a toy dataset consisting of a Gaussian mixture with three components surrounded by outliers.

Table 2 reports the results of the simulations, where the best performance over 20 runs is indicated. While the regularized stochastic k-means and the parallel k-means approaches perform better in terms of adjusted rand index, the fixed-size kernel spectral clustering achieves the best results as measured by the Davies-Bouldin criterion. The computational efficiency of the methods is compared in Fig. 6, from which it is evident that parallel k-means has the lowest runtime.

7 Conclusions

In this chapter we have revised two large-scale clustering algorithms, namely regularized stochastic k-means (RSKM) and fixed-size kernel spectral clustering (FSKSC). The first learns in parallel the cluster prototypes by means of stochastic optimization schemes implemented through Map-Reduce, while the second relies on the Nyström method to speed-up a kernel-based formulation of spectral clustering known as kernel spectral clustering. These approaches are benchmarked on real-life datasets of different sizes. The experimental results show their competitiveness both in terms of runtime and cluster quality compared to other state-of-the-art clustering algorithms such as parallel k-means and parallel plane clustering.

Acknowledgements EU: The research leading to these results has received funding from the European Research Council under the European Union's Seventh Framework Programme (FP7/2007–2013) / ERC AdG A-DATADRIVE-B (290923). This chapter reflects only the authors' views, the Union is not liable for any use that may be made of the contained information. Research Council KUL: GOA/10/09 MaNet, CoE PFV/10/002 (OPTEC), BIL12/11T; PhD/Postdoc grants. Flemish Government: FWO: projects: G.0377.12 (Structured systems), G.088114N (Tensor based data similarity); PhD/Postdoc grants. IWT: projects: SBO POM (100031); PhD/Postdoc grants. iMinds Medical Information Technologies SBO 2014. Belgian Federal Science Policy Office: IUAP P7/19 (DYSCO, Dynamical systems, control and optimization, 2012–2017).

References

1. H. Tong and U. Kang, "Big data clustering," in *Data Clustering: Algorithms and Applications*, 2013, pp. 259–276.
2. R. T. Ng and J. Han, "Clarans: A method for clustering objects for spatial data mining," *IEEE Trans. on Knowl. and Data Eng.*, vol. 14, no. 5, pp. 1003–1016, 2002.
3. T. Zhang, R. Ramakrishnan, and M. Livny, "Birch: An efficient data clustering method for very large databases," in *Proceedings of the 1996 ACM SIGMOD International Conference on Management of Data*, 1996, pp. 103–114.
4. S. Guha, R. Rastogi, and K. Shim, "Cure: An efficient clustering algorithm for large databases," *SIGMOD Rec.*, vol. 27, no. 2, pp. 73–84, 1998.
5. C. Boutsidis, A. Zouzias, and P. Drineas, "Random projections for k-means clustering," in *Advances in Neural Information Processing Systems* 23, 2010, pp. 298–306.
6. H. Tong, S. Papadimitriou, J. Sun, P. S. Yu, and C. Faloutsos, "Colibri: Fast mining of large static and dynamic graphs," in *Proceedings of the 14th ACM SIGKDD International Conference on Knowledge Discovery and Data Mining*, 2008, pp. 686–694.
7. J. Dean and S. Ghemawat, "Mapreduce: Simplified data processing on large clusters," *Commun. ACM*, vol. 51, no. 1, pp. 107–113, 2008.

8. G. Karypis and V. Kumar, "Multilevel k-way partitioning scheme for irregular graphs," *J. Parallel Distrib. Comput.*, vol. 48, no. 1, pp. 96–129, 1998.
9. W. Zhao, H. Ma, and Q. He, "Parallel k-means clustering based on mapreduce," in *Proceedings of the 1st International Conference on Cloud Computing*, 2009, pp. 674–679.
10. S. Papadimitriou and J. Sun, "Disco: Distributed co-clustering with map-reduce: A case study towards petabyte-scale end-to-end mining," in *Proceedings of the 2008 Eighth IEEE International Conference on Data Mining*, 2008, pp. 512–521.
11. A. F. et al., "A survey of clustering algorithms for big data: Taxonomy and empirical analysis," *IEEE Transactions on Emerging Topics In Computing*, vol. 2, no. 3, pp. 267–279, 2014.
12. A. M. et al., "Iterative big data clustering algorithms: a review," *Journal of Software: practice and experience*, vol. 46, no. 1, pp. 107–129, 2016.
13. F. R. K. Chung, *Spectral Graph Theory*, 1997.
14. A. Y. Ng, M. I. Jordan, and Y. Weiss, "On spectral clustering: Analysis and an algorithm," in *NIPS*, T. G. Dietterich, S. Becker, and Z. Ghahramani, Eds., Cambridge, MA, 2002, pp. 849–856.
15. U. von Luxburg, "A tutorial on spectral clustering," *Statistics and Computing*, vol. 17, no. 4, pp. 395–416, 2007.
16. H. Jia, S. Ding, X. Xu, and R. Nie, "The latest research progress on spectral clustering," *Neural Computing and Applications*, vol. 24, no. 7–8, pp. 1477–1486, 2014.
17. J. A. K. Suykens, T. Van Gestel, J. De Brabanter, B. De Moor, and J. Vandewalle, *Least Squares Support Vector Machines*. World Scientific, Singapore, 2002.
18. R. Langone, R. Mall, C. Alzate, and J. A. K. Suykens, *Unsupervised Learning Algorithms*. Springer International Publishing, 2016, ch. Kernel Spectral Clustering and Applications, pp. 135–161.
19. C. Alzate and J. A. K. Suykens, "Multiway spectral clustering with out-of-sample extensions through weighted kernel PCA," *IEEE Transactions on Pattern Analysis and Machine Intelligence*, vol. 32, no. 2, pp. 335–347, February 2010.
20. C. Baker, *The numerical treatment of integral equations*. Clarendon Press, Oxford, 1977.
21. L. Bottou, "Large-Scale Machine Learning with Stochastic Gradient Descent," *in Proceedings of the 19th International Conference on Computational Statistics (COMPSTAT 2010)*, Y. Lechevallier and G. Saporta, Eds. Paris, France: Springer, Aug. 2010, pp. 177–187.
22. Y. Nesterov, "Primal-dual subgradient methods for convex problems," Mathematical Programming, vol. 120, no. 1, pp. 221–259, 2009.
23. M. Meila and J. Shi, "A random walks view of spectral segmentation," in *Artificial Intelligence and Statistics AISTATS*, 2001.
24. J. B. MacQueen, "Some methods for classification and analysis of multivariate observations," in *Proceedings of 5th Berkeley Symposium on Mathematical Statistics and Probability*, 1985, pp. 193–218.
25. R. Langone, R. Mall, V. Jumutc, and J. A. K. Suykens, "Fast in-memory spectral clustering using a fixed-size approach," in *Proceedings of the European Symposium on Artitficial Neural Networks (ESANN)*, 2016, pp. 557–562.
26. F. Lin and W. W. Cohen, "Power iteration clustering," in *International Conference on Machine Learning*, 2010, pp. 655–662.
27. C. Fowlkes, S. Belongie, F. Chung, and J. Malik, "Spectral grouping using the Nyström method," *IEEE Transactions on Pattern Analysis and Machine Intelligence*, vol. 26, no. 2, pp. 214–225, Feb. 2004.
28. H. Ning, W. Xu, Y. Chi, Y. Gong, and T. S. Huang, "Incremental spectral clustering with application to monitoring of evolving blog communities." in *SIAM International Conference on Data Mining*, 2007, pp. 261–272.
29. A. M. Bagirov, B. Ordin, G. Ozturk, and A. E. Xavier, "An incremental clustering algorithm based on hyperbolic smoothing," *Computational Optimization and Applications*, vol. 61, no. 1, pp. 219–241, 2014.
30. R. Langone, O. M. Agudelo, B. De Moor, and J. A. K. Suykens, "Incremental kernel spectral clustering for online learning of non-stationary data," *Neurocomputing*, vol. 139, no. 0, pp. 246–260, September 2014.

31. W.-Y. Chen, Y. Song, H. Bai, C.-J. Lin, and E. Chang, "Parallel spectral clustering in distributed systems," IEEE Transactions on Pattern Analysis and Machine Intelligence, vol. 33, no. 3, pp. 568–586, March 2011.
32. C. Alzate and J. A. K. Suykens, "Sparse kernel models for spectral clustering using the incomplete Cholesky decomposition," in *Proc. of the 2008 International Joint Conference on Neural Networks (IJCNN 2008)*, 2008, pp. 3555–3562.
33. K. Frederix and M. Van Barel, "Sparse spectral clustering method based on the incomplete cholesky decomposition," *J. Comput. Appl. Math.*, vol. 237, no. 1, pp. 145–161, Jan. 2013.
34. M. Novak, C. Alzate, R. Langone, and J. A. K. Suykens, "Fast kernel spectral clustering based on incomplete Cholesky factorization for large scale data analysis," *Internal Report 14–119, ESAT-SISTA, KU Leuven (Leuven, Belgium)*, pp. 1–44, 2014.
35. X. Chen and D. Cai, "Large scale spectral clustering with landmark-based representation," in *AAAI Conference on Artificial Intelligence*, 2011.
36. D. Luo, C. Ding, H. Huang, and F. Nie, "Consensus spectral clustering in near-linear time," in *International Conference on Data Engineering*, 2011, pp. 1079–1090.
37. K. Taşdemir, "Vector quantization based approximate spectral clustering of large datasets," *Pattern Recognition*, vol. 45, no. 8, pp. 3034–3044, 2012.
38. L. Wang, C. Leckie, R. Kotagiri, and J. Bezdek, "Approximate pairwise clustering for large data sets via sampling plus extension," *Pattern Recognition*, vol. 44, no. 2, pp. 222–235, 2011.
39. J. A. K. Suykens, T. Van Gestel, J. Vandewalle, and B. De Moor, "A support vector machine formulation to PCA analysis and its kernel version," *IEEE Transactions on Neural Networks*, vol. 14, no. 2, pp. 447–450, Mar. 2003.
40. B. Schölkopf, A. J. Smola, and K. R. Müller, "Nonlinear component analysis as a kernel eigenvalue problem," *Neural Computation*, vol. 10, pp. 1299–1319, 1998.
41. S. Mika, B. Schölkopf, A. J. Smola, K. R. Müller, M. Scholz, and G. Rätsch, "Kernel PCA and de-noising in feature spaces," in *Advances in Neural Information Processing Systems 11*, M. S. Kearns, S. A. Solla, and D. A. Cohn, Eds. MIT Press, 1999.
42. M. Meila and J. Shi, "Learning segmentation by random walks," in *Advances in Neural Information Processing Systems 13*, T. K. Leen, T. G. Dietterich, and V. Tresp, Eds. MIT Press, 2001.
43. J. C. Delvenne, S. N. Yaliraki, and M. Barahona, "Stability of graph communities across time scales," *Proceedings of the National Academy of Sciences*, vol. 107, no. 29, pp. 12 755–12 760, Jul. 2010.
44. C. K. I. Williams and M. Seeger, "Using the Nyström method to speed up kernel machines," in *Advances in Neural Information Processing Systems*, 2001.
45. B. Kvesi, J.-M. Boucher, and S. Saoudi, "Stochastic k-means algorithm for vector quantization." *Pattern Recognition Letters*, vol. 22, no. 6/7, pp. 603–610, 2001.
46. W. Sun and J. Wang, "Regularized k-means clustering of high-dimensional data and its asymptotic consistency," *Electronic Journal of Statistics*, vol. 6, pp. 148–167, 2012.
47. D. M. Witten and R. Tibshirani, "A framework for feature selection in clustering," *Journal of the American Statistical Association*, vol. 105, no. 490, pp. 713–726, Jun. 2010.
48. F. Bach, R. Jenatton, and J. Mairal, *Optimization with Sparsity-Inducing Penalties (Foundations and Trends in Machine Learning)*. Hanover, MA, USA: Now Publishers Inc., 2011.
49. J. Whang, I. S. Dhillon, and D. Gleich, "Non-exhaustive, overlapping k-means," in *SIAM International Conference on Data Mining (SDM)*, 2015, pp. 936–944.
50. S. Boyd and L. Vandenberghe, *Convex Optimization*. New York, NY, USA: Cambridge University Press, 2004.
51. Y. Nesterov, *Introductory Lectures on Convex Optimization: A Basic Course (Applied Optimization)*, 1st ed. Springer Netherlands.
52. A. Rakhlin, O. Shamir, and K. Sridharan, "Making gradient descent optimal for strongly convex stochastic optimization." in *ICML*. icml.cc / Omnipress, 2012.
53. L. Xiao, "Dual averaging methods for regularized stochastic learning and online optimization," *J. Mach. Learn. Res.*, vol. 11, pp. 2543–2596, Dec. 2010.

54. J. Duchi, E. Hazan, and Y. Singer, "Adaptive subgradient methods for online learning and stochastic optimization," *J. Mach. Learn. Res.*, vol. 12, pp. 2121–2159, Jul. 2011.
55. V. Jumutc, R. Langone, and J. A. K. Suykens, "Regularized and sparse stochastic k-means for distributed large-scale clustering," in *IEEE International Conference on Big Data*, 2015, pp. 2535–2540.
56. Y.-H. Shao, L. Bai, Z. Wang, X.-Y. Hua, and N.-Y. Deng, "Proximal plane clustering via eigenvalues," ser. Procedia Computer Science, vol. 17, 2013, pp. 41–47.
57. A. Frank and A. Asuncion, "UCI machine learning repository, http://archive.ics.uci.edu/ml," 2010.
58. O. Chapelle, B. Schölkopf, and A. Zien, Eds., *Semi-Supervised Learning*. Cambridge, MA: MIT Press, 2006.
59. D. L. Davies and D. W. Bouldin, "A cluster separation measure," *IEEE Transactions on Pattern Analysis and Machine Intelligence*, vol. 1, no. 2, pp. 224–227, April 1979.
60. L. Hubert and P. Arabie, "Comparing partitions," *Journal of Classification*, pp. 193–218, 1985.
61. M. Halkidi, Y. Batistakis, and M. Vazirgiannis, "On clustering validation techniques," *Journal of Intelligent Information Systems*, vol. 17, pp. 107–145, 2001.

On High Dimensional Searching Spaces and Learning Methods

Hossein Yazdani, Daniel Ortiz-Arroyo, Kazimierz Choroś and Halina Kwasnicka

Abstract In data science, there are important parameters that affect the accuracy of the algorithms used. Some of these parameters are: the type of data objects, the membership assignments, and distance or similarity functions. In this chapter we describe different data types, membership functions, and similarity functions and discuss the pros and cons of using each of them. Conventional similarity functions evaluate objects in the vector space. Contrarily, Weighted Feature Distance (WFD) functions compare data objects in both feature and vector spaces, preventing the system from being affected by some dominant features. Traditional membership functions assign membership values to data objects but impose some restrictions. Bounded Fuzzy Possibilistic Method (BFPM) makes possible for data objects to participate fully or partially in several clusters or even in all clusters. BFPM introduces intervals for the upper and lower boundaries for data objects with respect to each cluster. BFPM facilitates algorithms to converge and also inherits the abilities of conventional fuzzy and possibilistic methods. In Big Data applications knowing the exact type of data objects and selecting the most accurate similarity [1] and membership assignments is crucial in decreasing computing costs and obtaining the best performance. This chapter provides data types taxonomies to assist data miners in selecting the right

H. Yazdani (✉) · D. Ortiz-Arroyo
Department of Energy Technology, Aalborg University Esbjerg, Esbjerg, Denmark
e-mail: yazdanihossein@yahoo.com

D. Ortiz-Arroyo
e-mail: doa@et.aau.dk

H. Yazdani
Faculty of Electronics, Wroclaw University of Science and Technology, Wroclaw, Poland

H. Yazdani · K. Choroś
Faculty of Computer Science and Management, Wroclaw University of Science and Technology, Wroclaw, Poland
e-mail: kazimierz.choros@pwr.edu.pl

H. Yazdani · H. Kwasnicka
Department of Computational Intelligence, Wroclaw University of Science and Technology, Wroclaw, Poland
e-mail: halina.kwasnicka@pwr.wroc.pl

W. Pedrycz and S.-M. Chen (eds.), *Data Science and Big Data: An Environment of Computational Intelligence*, Studies in Big Data 24,
DOI 10.1007/978-3-319-53474-9_2

learning method on each selected data set. Examples illustrate how to evaluate the accuracy and performance of the proposed algorithms. Experimental results show why these parameters are important.

Keywords Bounded fuzzy-possibilistic method · Membership function · Distance function · Supervised learning · Unsupervised learning · Clustering · Data type · Critical objects · Outstanding objects · Weighted feature distance

1 Introduction

The growth of data in recent years has created the need for the use of more sophisticated algorithms in data science. Most of these algorithms make use of well known techniques such as sampling, data condensation, density-based approaches, grid-based approaches, divide and conquer, incremental learning, and distributed computing to process big data [2, 3]. In spite of the availability of new frameworks for Big Data such as Spark or Hadoop, working with large amounts of data is still a challenge that requires new approaches.

1.1 Classification and Clustering

Classification is a form of supervised learning that is performed in a two-step process [4, 5]. In the training step, a classifier is built from a training data set with class labels. In the second step, the classifier is used to classify the rest of the data objects in the testing data set.

Clustering is a form of unsupervised learning that splits data into different groups or clusters by calculating the similarity between the objects contained in a data set [6–8]. More formally, assume that we have a set of n objects represented by $O = \{o_1, o_2, \dots, o_n\}$ in which each object is typically described by numerical *feature − vector* data that has the form $X = \{x_1, \dots, x_m\} \subset R^d$, where d is the dimension of the search space or the number of features. In classification, the data set is divided into two parts: learning set $O_L = \{o_1, o_2, \dots, o_l\}$ and testing set $O_T = \{o_{l+1}, o_{l+2}, \dots, o_n\}$. In these kinds of problems, classes are classified based on a class label x_l. A cluster or a class is a set of c values $\{u_{ij}\}$, where u represents a membership value, i is the ith object in the data set and j is the jth class. A partition matrix is often represented as a $c \times n$ matrix $U = [u_{ij}]$ [6, 7]. The procedure for membership assignment in classification and clustering problems is very similar [9], and for convenience in the rest of the paper we will refer only to clustering.

The rest of the chapter is organized as follow. Section 2 describes the conventional membership functions. The issues with learning methods in membership assignments are discussed in this section. Similarity functions and the challenges on conventional distance functions are described in Sect. 3. Data types and their behaviour are analysed in Sect. 4. Outstanding and critical objects and areas are discussed in this section. Experimental results on several data sets are presented in Sect. 5. Discussion and conclusion are presented in Sect. 6.

2 Membership Function

A partition or membership matrix is often represented as a $c \times n$ matrix $U = [u_{ij}]$, where u represents a membership value, i is the ith object in the data set and j is the jth class. Crisp, fuzzy or probability, possibilistic, bounded fuzzy possibilistic are different types of partitioning methods [6, 10–15]. Crisp clusters are non-empty, mutually-disjoint subsets of O:

$$M_{hcn} = \left\{ U \in \Re^{c \times n} | \; u_{ij} \in \{0, 1\}, \;\; \forall j, i; \right.$$

$$\left. 0 < \sum_{i=1}^{n} u_{ij} < n, \;\; \forall j; \;\; \sum_{j=1}^{c} u_{ij} = 1, \;\; \forall i \right\} \tag{1}$$

where u_{ij} is the membership of the object o_i in cluster j. If the object o_i is a member of cluster j, then $u_{ij} = 1$; otherwise, $u_{ij} = 0$. Fuzzy clustering is similar to crisp clustering, but each object can have partial membership in more than one cluster [16–20]. This condition is stated in (2), where data objects may have partial nonzero membership in several clusters, but only full membership in one cluster.

$$M_{fcn} = \left\{ U \in \Re^{c \times n} | \; u_{ij} \in [0, 1], \;\; \forall j, i; \right.$$

$$\left. 0 < \sum_{i=1}^{n} u_{ij} < n, \;\; \forall j; \;\; \sum_{j=1}^{c} u_{ij} = 1, \;\; \forall i \right\} \tag{2}$$

An alternative partitioning approach is *possibilistic clustering* [8, 18, 21]. In (3) the condition $\sum_{j=1}^{c} u_{ij} = 1$ is relaxed by substituting it with $\sum_{j=1}^{c} u_{ij} > 0$.

$$M_{pcn} = \left\{ U \in \Re^{c \times n} | \; u_{ij} \in [0, 1], \;\; \forall j, i; \right.$$

$$\left. 0 < \sum_{i=1}^{n} u_{ij} < n, \; \forall j; \;\; \sum_{j=1}^{c} u_{ij} > 0, \;\; \forall i \right\} \tag{3}$$

Based on (1), (2) and (3), it is easy to see that all crisp partitions are subsets of fuzzy partitions, and a fuzzy partition is a subset of a possibilistic partition, i.e., $M_{hcn} \subset M_{fcn} \subset M_{pcn}$ [8].

2.1 Challenges on Learning Methods

Regarding the membership functions presented above we look at the pros and cons of using each of these functions. In crisp memberships, if the object o_i is a member of cluster j, then $u_{ij} = 1$; otherwise, $u_{ij} = 0$. In such a membership function, members are not able to participate in other clusters and therefore it cannot be used in some applications such as in applying hierarchical algorithms [22]. In fuzzy methods (2), each column of the partition matrix must sum to 1 ($\sum_{j=1}^{c} u_{ij} = 1$) [6]. Thus, a property of fuzzy clustering is that, as c becomes larger, the u_{ij} values must become smaller.

Possibilistic methods have also some drawbacks such as offering trivial null solutions [8, 23] and lack of upper and lower boundaries with respect to each cluster [24]. Possibilistic methods do not have this constraint that fuzzy method have, but fuzzy methods are restricted by the constraint ($\sum_{j=1}^{c} u_{ij} = 1$).

2.2 Bounded Fuzzy Possibilistic Method (BFPM)

Bounded Fuzzy Possibilistic Method (BFPM) makes it possible for data objects to have full membership in several or even in all clusters. This method also does not have the drawbacks of fuzzy and possibilistic clustering methods. BFPM in (4), has the normalizing condition $1/c \sum_{j=1}^{c} u_{ij}$. Unlike Possibilistic method ($u_{ij} > 0$) there is no boundary in the membership functions. BFPM employs defined intervals [0, 1] for each data object with respect to each cluster. Another advantage of BFPM is that its implementation is relatively easy and that it tends to converge quickly.

$$M_{bfpm} = \Bigg\{ U \in \Re^{c \times n} |\ u_{ij} \in [0,1],\ \ \forall j, i;$$

$$0 < \sum_{i=1}^{n} u_{ij} < n,\ \ \forall j;\ \ 0 < 1/c \sum_{j=1}^{c} u_{ij} \leq 1,\ \ \forall i \Bigg\} \tag{4}$$

BFPM avoids the problem of decreasing the membership degrees of objects, as the number of clusters increases [25, 26].

2.3 Numerical Example

Assume $U = \{u_{ij}(x) | x_i \in L_j\}$ is a function that assigns a membership degree for each point x_i to a line L_j, where a line represents a cluster. Now consider the following equation which describes n lines crossing at the origin:

$$AX = 0 \tag{5}$$

where matrix A is a $n \times m$ coefficient matrix, and X is an $m \times 1$ matrix, in which n is the number of lines and m is the number of dimensions. From a geometrical point of view, each line containing the origin is a subspace of R^m. Equation (5) describes n with its different lines as a subspace. Without the origin, each of those lines is not a subspace, since the definition of a subspace comprises the existence of the null vector as a condition, in addition to other properties [27].

When trying to design a probability/fuzzy-based clustering method that could create clusters using all the points in all lines, it should be noted that removing or even decreasing the membership value of the origin ruins the subspace. For instance, $x = 0$, $y = 0$, $x = y$, and $x = -y$ are equations representing some of those lines with some data objects (points) on them as shown in the following equation. Note that all lines contain point (0, 0).

$$\begin{bmatrix} 1 & 0 \\ 0 & 1 \\ 1 & 1 \\ 1 & -1 \end{bmatrix} \times \begin{bmatrix} X \\ Y \end{bmatrix} = \begin{bmatrix} 0 \\ 0 \\ 0 \\ 0 \end{bmatrix}$$

Assume that we have two of those lines $L_1 : \{y = 0\}$ and $L_2 : \{x = 0\}$ with five points on each, including the origin, as shown in the following definitions:

$$L_1 = \{p_{11}, p_{12}, p_{13}, p_{14}, p_{15}\} = \{(-1, 0), (-2, 0), (0, 0), (1, 0), (2, 0)\}$$

$$L_2 = \{p_{21}, p_{22}, p_{23}, p_{24}, p_{25}\} = \{(0, -1), (0, -2), (0, 0), (0, 1), (0, 2)\}$$

where $p_{ij} = (x, y)$. As mentioned, the origin is part of all lines, but for convenience, we have given it different names such as p_{13} and p_{23} in each line above.

The point distances with respect to each line and Euclidean $||X||_2$ norm $\left(d_k(x, y) = \left(\sum_{i=1}^{d} \mid x_i - y_i \mid^2 \right)^{(1/2)} \right)$ are shown in the (2×5) matrices below, where 2 is the number of clusters and 5 is the number of objects.

$$D_1 = \begin{bmatrix} 0.0 \,, 0.0 \,, 0.0 \,, 0.0 \,, 0.0 \\ 2.0 \,, 1.0 \,, 0.0 \,, 1.0 \,, 2.0 \end{bmatrix} \qquad D_2 = \begin{bmatrix} 2.0 \,, 1.0 \,, 0.0 \,, 1.0 \,, 2.0 \\ 0.0 \,, 0.0 \,, 0.0 \,, 0.0 \,, 0.0 \end{bmatrix}$$

A zero value in the first matrix in the first row indicates that the object is on the first line. For example in D_1, the first row shows that all the members of set X_1 are on the first line. The second row shows how far each one of the points on the line are from the second cluster. Likewise the matrix $D2$ shows the data points on the second line. We assigned membership values to each point, using crisp and fuzzy logic as shown in the matrices below by using the following membership function (6) for crisp and fuzzy methods and also the conditions for these methods described in (1) and (2).

$$U_{ij} = \begin{cases} 1 & \text{if } d_{p_{ij}} = 0 \\ 1 - \frac{d_{p_{ij}}}{d_\delta} & \text{if } 0 < d_{p_{ij}} \leqslant d_\delta \\ 0 & \text{if } d_{p_{ij}} > d_\delta \end{cases} \tag{6}$$

where $d_{p_{ij}}$ is the Euclidean distance of object x_i from cluster j, and d_δ is a constant that we use to normalize the values. In our example we used $d_\delta = 2$.

$$U_{crisp}(L_1) = \begin{bmatrix} 1.0 , 1.0 , \mathbf{1.0} , 1.0 , 1.0 \\ 0.0 , 0.0 , \mathbf{0.0} , 0.0 , 0.0 \end{bmatrix} \; U_{crisp}(L_2) = \begin{bmatrix} 0.0 , 0.0 , \mathbf{0.0} , 0.0 , 0.0 \\ 1.0 , 1.0 , \mathbf{0.0} , 1.0 , 1.0 \end{bmatrix}$$

or

$$U_{crisp}(L_1) = \begin{bmatrix} 1.0 , 1.0 , \mathbf{0.0} , 1.0 , 1.0 \\ 0.0 , 0.0 , \mathbf{0.0} , 0.0 , 0.0 \end{bmatrix} \; U_{crisp}(L_2) = \begin{bmatrix} 0.0 , 0.0 , \mathbf{0.0} , 0.0 , 0.0 \\ 1.0 , 1.0 , \mathbf{1.0} , 1.0 , 1.0 \end{bmatrix}$$

$$U_{Fuzzy}(L_1) = \begin{bmatrix} 1.0 , 0.5 , \mathbf{0.5} , 0.5 , 1.0 \\ 0.0 , 0.5 , \mathbf{0.5} , 0.5 , 0.0 \end{bmatrix} \; U_{Fuzzy}(L_2) = \begin{bmatrix} 0.0 , 0.5 , \mathbf{0.5} , 0.5 , 0.0 \\ 1.0 , 0.5 , \mathbf{0.5} , 0.5 , 1.0 \end{bmatrix}$$

In crisp methods, the origin can be a member of just one line or cluster. Therefore, the other lines without the origin can not be subspaces [27]. In other words, the example "crossing lines at origin" can not be represented by crisp methods.

Given the properties of the membership functions in fuzzy methods, if the number of clusters increases, the membership value assigned to each object will decrease proportionally.

Methods such as PCM, allow data objects to obtain larger values in membership assignments [8, 21]. But PCM needs a good initialization to perform clustering [23]. According to PCM condition ($u_{ij} \geq 0$), the trivial null solutions should be handled by modifying the membership assignments [8, 21, 23]. The authors in [8] did not change the membership function to solve this problem, instead they introduce an algorithm to overcome the issue of trivial null solutions by changing the objective function as:

$$J_m(U, V) = \sum_{j=1}^{c} \sum_{i=1}^{n} u_{ij}^m \, ||X_i - V_j||_A^2 + \sum_{i=1}^{C} \eta_i \sum_{j=1}^{n} (1 - u_{ij})^m \tag{7}$$

where η_i are suitable positive numbers. The authors of [23] discuss more details about (7), without considering membership functions. Implementation of such algorithm needs proper constraints and also requires good initializations, otherwise the accuracy and the results will not be reasonable [23]. U_{pcm} can obtain different values, since the implementation of PCM can be different because the boundaries for membership assignments with respect to each cluster are not completely defined.

In conclusion, crisp membership functions are not able to assign membership values to data objects participating in more than one cluster. Fuzzy membership functions reduce the membership values assigned to data objects with respect to each cluster, and possibilistic membership function is not well defined with respect to clusters. BFPM avoids the problem of reducing the membership degrees of objects when the number of clusters increases.

$$U_{bfpm}(L_1) = \begin{bmatrix} 1.0 , 1.0 , \mathbf{1.0} , 1.0 , 1.0 \\ 0.0 , 0.5 , \mathbf{1.0} , 0.5 , 0.0 \end{bmatrix} \; U_{bfpm}(L_2) = \begin{bmatrix} 0.0 , 0.5 , \mathbf{1.0} , 0.5 , 0.0 \\ 1.0 , 1.0 , \mathbf{1.0} , 1.0 , 1.0 \end{bmatrix}$$

BFPM allows data objects (such as the origin in the lines presented by previous example) to be members of all clusters with full membership. Additionally, BFPM may show which members can affect the algorithm if moved to other clusters. In critical systems, identifying these types of objects is a big advantage, because we may see how to encourage or prevent objects from contributing to other clusters. The method also includes those data objects that participate in just one cluster. Some of the issues on membership functions are described in [6, 24]. In [24] some other examples on different membership methods are discussed.

3 Similarity Functions

Similarity function is a fundamental part in learning algorithms [6, 28–32], as any agent, classifier, or method make use of these functions. Most of the learning methods compare a given problem with other problems to find the most suitable solution. This methodology indicates that the solution for the most similar problem can be the desired solution for the given problem [33].

Distance functions are based on the similarity between data objects or use probability measures. Tables 1, 2 and 3 show some well-known similarity functions (Eqs. 8–26) in L_1, L_2, and L_n norms [38, 39]. The taxonomy is divided into two categories: *vector* and *probabilistic* approaches. P and Q represent data objects or probability measures, in d dimensional search space, and $D(P, Q)$ presents a distance function between P and Q. Equation (13) is introduced to normalize the search space

Table 1 Distance functions or probability measures on Minkowski family

Minkowski family	Euclidean (L_2)	$D_E = \sqrt{\sum_{i=1}^{d} \lvert P_i - Q_i \rvert^2}$	(8)
	City block (L_1) [34]	$D_{CB} = \sum_{i=1}^{d} \lvert P_i - Q_i \rvert$	(9)
	Minkowski (L_p) [34]	$D_{MK} = \sqrt[p]{\sum_{i=1}^{d} \lvert P_i - Q_i \rvert^p}$	(10)
	Chebyshev (L_∞) [35]	$D_{Checb} = max_i \lvert P_i - Q_i \rvert$	(11)

Table 2 Distance functions or probability measures on Lvovich Chebyshev (L_1) family

Lvovich family	Sorensen [28]	$D_{Sor} = \frac{\sum_{i=1}^{d} \lvert P_i - Q_i \rvert}{\sum_{i=1}^{d} (P_i + Q_i)}$	(12)
	Gower [36]	$D_{Gov} = \frac{1}{d} d \sum_{i=1}^{d} \frac{\lvert P_i - Q_i \rvert}{R_i}$	(13)
		$D_{Gov} = \frac{1}{d} \sum_{i=1}^{d} \lvert P_i - Q_i \rvert$	(14)
L_1 family	Soergel [29]	$D_{Sg} = \frac{\sum_{i=1}^{d} \lvert P_i - Q_i \rvert}{\sum_{i=1}^{d} max(P_i, Q_i)}$	(15)
	Kulczynski [30]	$D_{Sg} = \frac{\sum_{i=1}^{d} \lvert P_i - Q_i \rvert}{\sum_{i=1}^{d} min(P_i, Q_i)}$	(16)
	Canberra [30]	$D_{Can} = \sum_{i=1}^{d} \frac{\lvert P_i - Q_i \rvert}{P_i + Q_i}$	(17)
	Lorentzian [30]	$D_{Lor} = \sum_{i=1}^{d} ln(1 + \lvert P_i - Q_i \rvert)$	(18)

Table 3 Distance functions or probability measures on x^2 (L_2) family

x^2 family	Squared euclidean	$D_{SE} = \sum_{i=1}^{d} (P_i - Q_i)^2$	(19)
	Pearson x^2 [1]	$D_P = \sum_{i=1}^{d} \frac{(P_i - Q_i)^2}{Q_i}$	(20)
L_2 family	Neyman x^2 [1]	$D_N = \sum_{i=1}^{d} \frac{(P_i - Q_i)^2}{P_i}$	(21)
	Squared x^2 [1]	$D_{SQ} = \sum_{i=1}^{d} \frac{(P_i - Q_i)^2}{P_i + Q_i}$	(22)
	Probabilistic x^2 [37]	$D_{PSQ} = 2 \sum_{i=1}^{d} \frac{(P_i - Q_i)^2}{P_i + Q_i}$	(23)
	Divergence [37]	$D_{Div} = 2 \sum_{i=1}^{d} \frac{(P_i - Q_i)^2}{(P_i + Q_i)^2}$	(24)
	Clark [30]	$D_{Clk} = \sqrt{\sum_{i=1}^{d} \left(\frac{(P_i - Q_i)}{(P_i + Q_i)} \right)^2}$	(25)
	Additive x^2 [30]	$D_{Ad} = \sum_{i=1}^{d} \frac{(P_i - Q_i)^2 (P_i + Q_i)}{(P_i Q_i)}$	(26)

boundaries by dividing the equation by R, the range of the population in the data set. The method scales down the search space by dividing the equation by d, the number of dimensions [36]. Asymmetric distance functions (Pearson (20), Neyman (21)) and symmetric versions of those functions (squared x^2 (22)) have been proposed, additionally to probabilistic symmetric x^2 (23) functions. There are other useful distance functions such as distance functions based on histograms, signatures, and probability density [40, 41] that we do not discuss in this paper.

3.1 Challenges on Similarity Functions

Assume there are two objects in a three dimensional search space, such as $O_1 = (2, 2, 5)$ and $O_2 = (1, 1, 1)$, and a prototype $P = (2, 2, 2)$. Now if we use a distance function such as Euclidean distance, object O_2 seems overall more similar to the prototype, but from a features' perspective, O_1 is more similar to the prototype when compared with O_2 given that they share two out of three features. This example motivates the following distance functions. These functions can be applied in high dimensional search spaces ($d' \gg d$) typical of big data applications [16, 42, 43] where d' is a very large number. Let us consider:

$$O_1' = (2, 2, 2, ..., x), O_2' = (1, 1, 1, ..., 1), P' = (2, 2, 2, ..., 2)$$

where

$$O_{1,1}' = O_{1,2}' = O_{1,3}' = \cdots = O_{1,d'-1}' = 2 \quad and \quad O_{1,d'}' = x = \sqrt{d'} + 2$$

$$O_{2,1}' = O_{2,2}' = O_{2,3}' = \cdots = O_{2,d'}' = 1$$

$$P_1' = P_2' = P_3' = \cdots = P_{d'}' = 2$$

According to all similarity functions presented in Tables 1, 2 and 3, we see how these functions may have some dominant features ($x > \sqrt{d'} + 2$) that may cause algorithms to misclassify data objects.

We should evaluate the data objects' features from different perspectives, not just using the same scale. This is because each feature has its own effect on the similarity function and a single feature should not have a large impact on the final result.

3.2 Weighted Feature Distances

Assume a set of n objects represented by $O = \{o_1, o_2, \ldots, o_n\}$ in which each object is typically represented by numerical *feature* – *vector* data, with the same priority in features, that has the form $X = \{x_1, \ldots, x_m\} \subset R^d$, where d is the dimension of the search space or the number of features. We introduce Weighted Feature Distance (*WFD*) that overcome some of the issues with distance function that we have described.

$\boldsymbol{WFD_{(L_1)}}$: Weighted feature distance (WFD_{L_1}) for L_1 norm is:

$$WFD_{(L_1)} = \left(|W_iO_i - W_jO_j|\right) =$$

$$= \sum_{k=1}^{d} \left(|w_k x_{ik} - w_k^{'} x_{jk}|\right) \tag{27}$$

$\boldsymbol{WFD_{(L_2)}}$: Weighted feature distance (WFD_{L_2}) for L_2 norm is:

$$WFD_{L_2} = \sqrt{\left(W_iO_i - W_jO_j\right)^2} =$$

$$= \left(\sum_{k=1}^{d} \left(|w_k x_{ik} - w_k^{'} x_{jk}|^2\right) \right)^{(\frac{1}{2})} \tag{28}$$

where d is the number of variables, or dimensions for numerical data objects. w_k and $w_k^{'}$ are the weights assigned to features of the first and the second objects respectively. We make ($w_k = w_k^{'}$) if both objects are in the same scale.
We can also obtain the Euclidean distance function from (28) by assigning the same values to w_k as:

$$w_1 = w_2 = \cdots = w_d = 1$$

$\boldsymbol{WFD_{(L_p)}}$: Weighted feature distance (WFD_{L_p}) for L_p norm is:

$$WFD_{(L_1)} = \left(|W_iO_i - W_jO_j|^p\right)^{(\frac{1}{r})} =$$

$$= \sum_{k=1}^{d} \left(|w_k x_{ik} - w_k^{'} x_{jk}|^p\right)^{(\frac{1}{r})} \tag{29}$$

where d is the number of variables, or dimensions for numerical data objects. p and r are coefficients that allow us to use different metrics but p and r can be equal. w_k and $w_k^{'}$ are the weights assigned to features of the first and the second objects respectively. ($w_k = w_k^{'}$), if both objects are in the same scale.

4 Data Types

Data mining techniques extract knowledge from data objects. To obtain the most accurate results, we need to consider the data types in our mining algorithms. Each type of object has its own characteristic and behaviour in data sets. The type of objects discussed in this paper help to avoid the cost of redoing mining techniques caused by treating objects in a wrong way.

4.1 Data Objects Taxonomies

Data mining methods evaluate data objects based on their (*descriptive* and *predictive*) patterns. The type of data objects should be considered, as each type of data object has different effect on the final results [44]. For instance, data objects known as *outlier(s)* are interesting objects in anomaly detection. On the other hand, outliers do not play any role in other applications since they are considered noise. Since each type of data object has different effects on the final result of an algorithm, we aim to look at different types of data from different perspectives. We start with the simplest definition of data objects and categorize them into single variable or with two or more variables [45].

- ***Univariate Data Object***:
  Observations on a single variable on data sets $X = \{x_1, x_2, ..., x_n\}$, where n is the number of single variable observations (x_i). Univariate Data Object can be categorized into two groups:

  1. Categorical or qualitative [31], that can be represented by *frequency distributions* and *bar charts*.
  2. Numerical or quantitative, which can be discrete or continuous data. *Dotplots* can be used to represent this type of variables.

- ***Multivariate Data Object***:
  Observations on a set of variables on data sets or populations presented as $X = \{X_1, X_2, ..., X_n\}$, where $X_i = \{x_1, x_2, ..., x_d\}$, n is the number of observations, and d is the number of variables or dimensions. Each variable can be a member of the above mentioned categories.

4.2 Complex and Advanced Objects

The growth of data in various types prevents data taxonomies for classifying data objects into above mentioned categories. Methods dealing with data objects need to distinguish their type to create more efficient methodologies and data mining algorithms. *Complex* and *Advanced* categories are two main topics for sophisticated data objects.

These objects have sophisticated structures, and also need advanced techniques for storage, representation, retrieval and analysis. Table 4 shows these data objects without the details. Further information can be found in [24]. An advantage of sophisticated objects is in allowing miners to reduce the cost of using similarity functions on these type of objects instead of comparing the data objects individually. For example two networks can be compared at once instead of being compared individually.

Table 4 Data types (Complex and advanced data objects)

Complex objects	Advanced objects
Structured data object [46]	Sequential patterns [47, 48]
Semi-structured data object [49]	Graph and sub-graph patterns [50, 51]
Unstructured data object [52]	Objects in interconnected networks [53, 54]
Spatial data object [55]	Data stream or stream data [56]
Hypertext [57]	Time series [58]
Multimedia [59]	

4.3 Outlier and Outstanding Objects

Data objects from each of the categories previously presented, can be considered as normal data objects that do fit the data model and obey the discovered data patterns. Now we introduce some data objects known as *Outlier* and *Outstanding*, that cannot be considered as normal data objects. These data objects affect the results obtained from knowledge extracted from data sets. Data objects from these categories can be any data object from above mentioned categorizes (complex, advanced, univariate, and multivariate data objects). Outliers and outstanding objects are important since they have potential ability to change the results produced by the learning algorithms.
Outlier: A data set may contain objects that do not fit the model of the data, and do not obey the discovered patterns [60, 61]. These data objects are called '*outliers*' [17, 24]. Outliers are important because they might change the behaviour of the model, as they are far from the discovered patterns and are mostly known as noise or exceptions. Outliers are useful in some applications such as fraud and anomaly detection [62], as these rare cases are more interesting than the normal cases. Outlier analysis is used in a data mining technique known as *outlier mining*.
Outstanding Objects Unlike outliers, a data set may contain objects that do fit the model of the data and obey the discovered patterns fully, even in all models or clusters. These data objects are important because they do not change the behaviour of the model, as they are in the discovered patterns and are known as full members. These critical objects named as "*outstanding*" objects cannot be removed from any cluster that they participate in [25]. The another important property of outstanding objects is that they may easily move from one cluster to another by small changes in even one dimension [24]. The crossing lines at origin example describes the behaviour and properties of outstanding objects. Origin should be a member of each line with full membership degree, otherwise each line without the origin can not be considered as a subspace. In such cases, we can see the importance of outstanding objects, in having full membership in several or in all objective functions [63].

In next section we describe some experimental results on clustering methods to illustrate how mining methods deal with outstanding objects, and how these data objects can affect the final results.

5 Experimental Results

Experiments are based on three scenarios. The first scenario compares the accuracy of membership functions on clustering and classification methods on some data sets shown in Table 5. In the second scenario, we check the effect of dominant features on similarity functions and consequently on final results of clustering methods. Data sets are selected based on a different number of features as we aim to check how proposed methods can be influenced by dominant features. Finally, the third scenario provides an environment to evaluate the behaviour of critical areas and objects that we have called *Outstanding*. In all scenarios in our experiments, we compare the accuracy of different fuzzy and possibilistic methods with BFPM and BFPM-WFD algorithms presented in Algorithms (1) and (2).

BFPM Algorithm This algorithm uses the conventional distance functions for membership assignments. Equations (30) and (31) show how the algorithm calculates (u_{ij}) and how the prototypes (v_j) will be updated in each iteration. The algorithm runs until the condition is false:

$$max_{1 \leq k \leq c}\{||V_{k,new} - V_{k,old}||^2\} < \varepsilon$$

The value assigned to ε is a predetermined constant that varies based on the type of objects and clustering problems.

U is the $(n \times c)$ partition matrix, $V = v_1, v_2, ..., v_c$ is the vector of c cluster centers in $\mathfrak{R}^d$, m is the fuzzification constant, and $||.||_A$ is any inner product A-induced norm [6, 64], and Euclidean distance function presented by (32).

$$D_E = \sqrt{\sum_{i=1}^{d} |X_i - Y_i|^2}$$

$$= \sqrt{(X_1 - Y_1)^2 + (X_2 - Y_2)^2 + \cdots + (X_d - Y_d)^2} \quad (32)$$

Table 5 Multi dimensional data sets

Dataset	Attributes	No. objects	Clusters
Iris	4	150	3
Pima Indians	8	768	2
Yeast	8	1299	4
MAGIC	11	19200	2
Dermatology	34	358	6
Libras	90	360	15

Algorithm 1 BFPM Algorithm

Input: **X,** c, m
Output: **U, V**
Initialize V;
while $max_{1\leq k\leq c}\{||V_{k,new} - V_{k,old}||^2\} > \varepsilon$ **do**

$$u_{ij} = \Big[\sum_{k=1}^{c}\Big(\frac{||X_i - v_j||}{||X_i - v_k||}\Big)^{\frac{2}{m-1}}\Big]^{\frac{1}{m}}, \quad \forall i,j \tag{30}$$

$$V_j = \frac{\sum_{i=1}^{n}(u_{ij})^m x_i}{\sum_{i=1}^{n}(u_{ij})^m}, \quad \forall j\,; \qquad (0 < \frac{1}{c}\sum_{j=1}^{c} u_{ij} \leq 1). \tag{31}$$

end while

where d is the number of features or dimensions, and X and Y are two different objects in d dimensional search space.

BFPM-WFD Since BFPM algorithm assigns (u_{ij}) based only on the total distance shown by (32), we implement algorithm BFPM-WFD (BFPM Weighted Feature Distance) not only to compare the objects based on their similarity using the distance function, but also to check the similarity between features of objects and similar features of prototypes individually.

Algorithm 2 BFPM-WFD

Input: **X,** c, m
Output: **U, V**
Initialize V;
while $max_{1\leq k\leq c}\{||V_{k,new} - V_{k,old}||^2\} > \varepsilon$ **do**

$$\Big\{u_{ij} = \Big[\sum_{k=1}^{c}\Big(\frac{||X_i - v_j||}{||X_i - v_k||}\Big)^{\frac{2}{m-1}}\Big]^{\frac{1}{m}}, \quad \forall i,j\,;$$

$$||X_i - X_j|| = \Big(\sum_{f=1}^{d}\big(|w_f.x_{if} - w_f'.x_{jf}|^2\big)\Big)^{(\frac{1}{2})}\Big\} \tag{33}$$

$$V_j = \frac{\sum_{i=1}^{n}(u_{ij})^m x_i}{\sum_{i=1}^{n}(u_{ij})^m}, \quad \forall j\,; \qquad (0 < \frac{1}{c}\sum_{j=1}^{c} u_{ij} \leq 1). \tag{34}$$

end while

w_f and w_f' are weights assigned to features (x_{if} and x_{jf}) of objects X_i and X_j respectively, presented by (28). Table 6 illustrates the compared results between BFPM and other fuzzy and possibilistic methods: Type-1 fuzzy sets (T1), Interval Type-2 fuzzy sets (IT2), General Type-2 (GT2), Quasi-T2 (QT2), FCM and PCM on four data sets

Table 6 Compared accuracy between conventional fuzzy, possibilistic, and BFPM methods

Methods	Iris	Pima Indian	Yeast	MAGIC
T1 (fuzzy) [65]	95.15	73.59	60.03	77.26
IT2 (fuzzy) [65]	94.18	74.38	54.81	75.63
QT2 (fuzzy) [65]	94.28	75.05	55.97	77.44
GT2 (fuzzy) [65]	94.76	74.40	58.22	78.14
FCM (fuzzy) [12]	88.6	74	67.4	54
PCM (possibilistic) [12]	89.4	59.8	32.8	62
BFPM	97.33	99.9	67.71	100.0
BFPM-WFD	100.0	100.0	82.3	100.0

Table 7 Compared accuracy based on distance functions

Dataset ↓ | *Dis.Func.* ⟶	Euclidean (L_2)	$WFD_{L_2}(w = \frac{1}{2})$	WFD (L_2) $w = \frac{1}{3}$	$WFD_{L_2}(w = \frac{1}{d})$
Irish	97.33	100	100	100
Pima	99.9	100	100	100
Yeast	67.71	77.2	77.3	**82.03**
MAGIC	100.0	100.0	100.0	100.0
Dermatology	77.4	89.5	83.0	**92.4**
Libras	57.0	**69.0**	62.5	61.4

"Iris", "Pima", "Yeast" and "MAGIC". This comparison is based on the first scenario, and as results show, BFPM performs better than the conventional fuzzy and possibilistic methods.

Table 7 compares the accuracy between WFD with different weights ($w = 1/2$, $w = 1/3, w = 1/d$) and Euclidean distance function on different data sets "Iris", "Pima", "Yeast", "Magic", "Dermatology" and "Libras", where d is the number of dimensions. According to the table, dominant features has less impact on weighted features distance functions. The table also shows that in some data sets such as "Libras" larger values for assigned weights are most desirable and in some other such as "Yeast" lower values are most suitable. The comparison between conventional similarity function and WFD was implemented with respect to the second scenario.

Table 8 Outstanding objects with ability to move from one cluster to another

Dataset	No. objects	>90%	>80%	>70%
Irish	150	25	99	99
Pima	768	677	751	751
Yeast	1299	868	1135	1264
MAGIC	19200	0	34	424
Dermatology	358	286	331	355
Libras	360	238	317	336

According to the last scenario, we aim to check the ability of outstanding objects to participating in other clusters. Table 8 demonstrates the potential ability of data objects to get membership values from the closest cluster, besides their own clusters. For example, the first row of the table shows that 25 data objects from Iris data set have the potential ability of more than 90% to participate in the closest cluster. As the table presents, some data objects are able to move to another cluster with small changes. These kinds of behavior can be beneficial or produce errors.

In some safety critical systems such as cancerous human cell detection, or fraudulent banking transactions, we need to prevent data objects to move to other clusters.

Figures 1, 2, 3 and 4 plot data objects on data sets "Iris", and "Libras" [66] obtained by Fuzzy and BFPM methods with respect to two closest clusters. By comparing the plots for BFPM and fuzzy methods [24], critical areas and objects are being shown. This comparison is being highlighted when we look at the accuracy of Fuzzy, Possibilistic, and BFPM as well as considering the ability of outstanding objects to affect performance. In fuzzy methods, data objects are mostly separated. In this situation the critical areas are not shown, but instead in BFPM method, critical areas, and also outstanding objects may be identified.

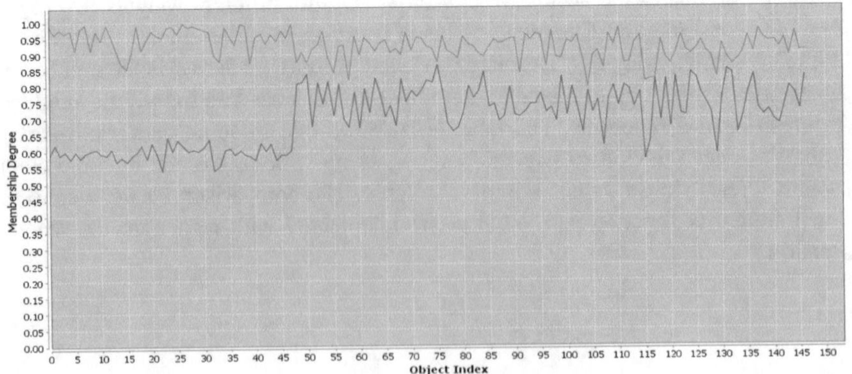

Fig. 1 Mutation plot for Iris data set, prepared by BFPM method

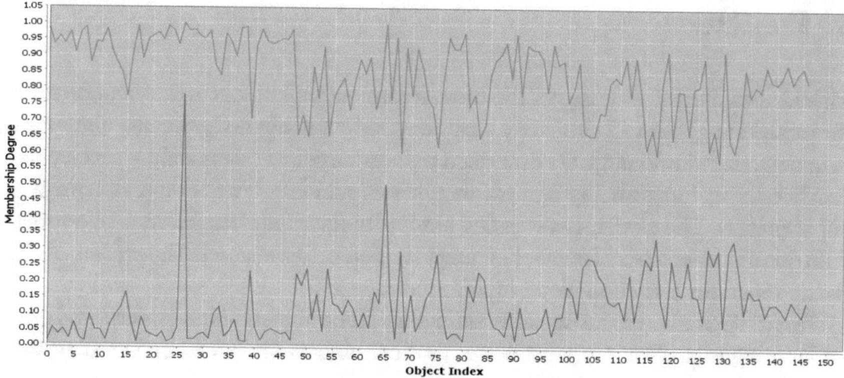

Fig. 2 Mutation plot for Iris data set, prepared by Fuzzy method

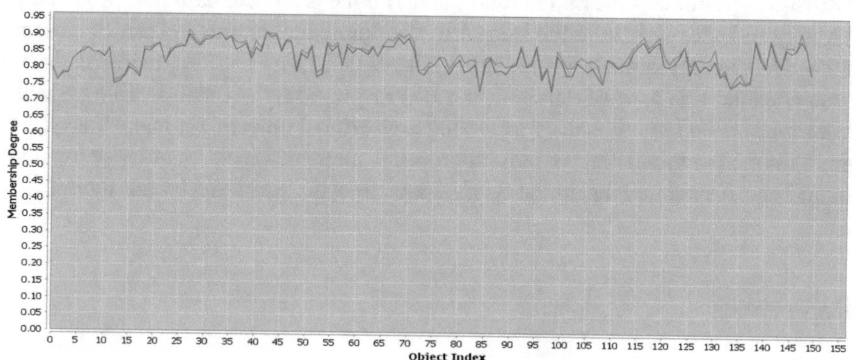

Fig. 3 Mutation plot for Libras data set, prepared by BFPM method

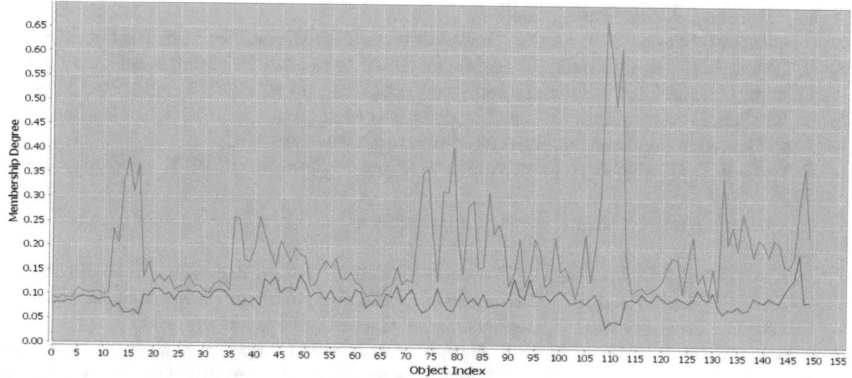

Fig. 4 Mutation plot for Libras data set, prepared by Fuzzy method

6 Conclusion

This chapter describes some of the most important parameters in learning methods for partitioning, such as similarity functions, membership assignments, and type of data objects. Additionally, we described the most used and well known membership functions. The functionality of these membership functions was compared on different scenarios. The challenges in using similarity functions that could deal correctly with dominant features is another concept studied in this chapter. The presented similarity functions were compared in different aspects.

This chapter also discusses different types of data objects and their potential effect on learning algorithm's performance. Critical objects known as outstanding were described in the context of several examples.

Our results show that BFPM performs better than other conventional fuzzy and possibilistic algorithms discussed in this chapter on the presented data sets. WFD helps learning methods to handle the impact of the dominant features in their processing steps. Outstanding objects are the most critical and many learning methods do not even consider this type of data objects. BFPM provides the most flexible environment for outstanding objects by analysing how critical objects can make the system more stable. We found that the most appropriate membership and similarity function should be selected, regarding the type of data objects considered by our model.

References

1. S.-H. Cha, "Comprehensive Survey On Distance/Similarity Measures Between Probability Density Functions," Int. J. Math. Models Methods Appl. Sci., vol. 1, no. 4, pp. 300–307, 2007.
2. P. N. Tan, M. Steinbach, V. Kumar, "Introduction to Data Mining Instructor's Solution Manual," Pearson Addison-Wesley, 2006.
3. P. N. Tan, M. Steinbach, V. Kumar, "Introduction to Data Mining," Pearson Wesley, 2006.
4. B. Taskar, E. Segal, D. Koller, "Probabilistic classification and clustering in relational data," In Proc. Int. Joint Conf. Artificial Intelligence (IJCAI01), pp. 870–878, Seattle, WA, 2001.
5. H. Yazdani, H. Kwasnicka, "Fuzzy Classiffication Method in Credit Risk," in Springer Int. Conf. Computer and Computational Intelligence, pp. 495–505, 2012.
6. H. Yazdani, D. O. Arroyo, H. Kwasnick, "New Similarity Functions", IEEE, AIPR, pp. 47–52, 2016.
7. R. Xu, D. C. Wunsch, "Recent advances in cluster analysis," Intelligent Computing and Cybernetics, 2008.
8. D. T. Anderson, J. C. Bezdek, M. Popescu, J. M. Keller. "Comparing Fuzzy, Probabilistic, and Posibilistic Partitions," IEEE Transactions On Fuzzy Systems, Vol. 18, No. 5, 2010.
9. C. Borgelt, "Prototype-based Classification and Clustering." Magdeburg, 2005.
10. L. F. S. Coletta, L. Vendramin, E. R. Hruschka, R. J. G. B. Campello, W. Pedrycz, "Collaborative Fuzzy Clustering Algorithms: Some Refinements and Design Guidelines," IEEE Transactions On Fuzzy Systems, Vol. 20, No. 3, pp. 444–462, 2012.
11. X. Wu, V. Kumar, J. R. Quinlan, J. Ghosh, Q. Yang, H. Motoda, G. J. McLachlan, A. Ng, B. Liu, P. S. Yu, Z. H. Zhou, M. Steinbach, D. J. Hand, D. Steinberg. "Top 10 algorithms in data mining," Springer-Verlag London, 2007.
12. N. R. Pal, K. Pal, J. M. Keller, J. C. Bezdek. "A Possibilistic Fuzzy c-Means Clustering Algorithm," IEEE Transactions On Fuzzy Systems, Vol. 13, No. 4. 2005.

13. S. Singh, A.K. Solanki, N. Trivedi, M. Kumar, "Data Mining Challenges and Knowledge Discovery in Real Life Applications," IEEE, 978-1-4244-8679-3/11/, 2011.
14. T. Hastie, R. Tibshirani, J. Friedman, "The Elements of Statistical Learning Data Mining, Inference, and Prediction," Springer Series in Statistics, 2005.
15. L.A.Zadeh. "Fuzzy Sets," Information and Control, 338–353, 1965.
16. R. J. Hathawaya, J. C. Bezdekb. "Extending fuzzy and probabilistic clustering to very large data sets," Elsevier, 2006.
17. J. Han, M. Kamber, "Data Mining Concepts and Techniques," Elsevier, Morgan Kaufmann series, 2006.
18. L.A. Zadeh, "Fuzzy Sets As A Basis For A Theory Of Possibility" North-Holland Publishing Company Fuzzy Sets and Systems 1, 1978.
19. L.A. Zadeh. "Toward Extended Fuzzy Logic- A First Step," Elsevier, Fuzzy Sets and Systems, 3175–3181, 2009.
20. H. C. Huang, Y. Y. Chuang, C. S. Chen, "Multiple Kernel Fuzzy Clustering," IEEE Transactions On Fuzzy Systems, 2011.
21. R. Krishnapuram, J. M. Keller, "A Possibilistic Approach to Clustering," IEEE, Transaction On Fuzzy Systems, Vol. 1, No. 2. 1993.
22. F. Memoli, G. Carlsson, "Characterization, Stability and Convergence of Hierarchical Clustering Methods", Journal of Machine Learning Research, pp. 1425–1470, 2010.
23. M. Barni, V. Cappellini, A. Mecocci, "Comments on A Possibilistic Approach to Clustering," IEEE, Transactions On Fuzzy Systems, Vol. 4, No. 3. 1996.
24. H. Yazdani, H. Kwasnicka, "Issues on Critical Objects in Mining Algorithms", IEEE, AIPR, pp. 53–58, 2016.
25. H. Yazdani, D. Ortiz-Arroyo, K. Choros, H. Kwasnicka, "Applying Bounded Fuzzy Possibilistic Method on Critical Objects", IEEE, CINTI, 2016.
26. H. Yazdani, "Bounded Fuzzy Possibilistic On Different Search Spaces", IEEE, CINTI, 2016.
27. G. Strang, "Introduction to Linear Algebra", Wellesley-Cambridge Press, 2016.
28. J. Looman, J.B. Campbell, "Adaptation of Sorensen's K For Estimating Unit Affinities In Prairie Vegetation," Ecology, Vol. 41, No. 3, 1960.
29. V. Monev, "Introduction to Similarity Searching in Chemistry," Communications in Mathematical And Computer Chemistry, No. 51, 2004.
30. P. Kumar, A. Johnson, "On A Symmetric Divergence Measure And Information Inequalities," Journal of Inequalities in Pure And Applied Mathematics, Vol. 6, Issue 3, Article 65, 2005.
31. S. Boriah, V. Chandola, V. Kumar, "Similarity Measures For Categorical Data: A Comparative Evaluation," SIAM, 2008.
32. M. Minor, A. Tartakovski, R. Bergmann, "Representation and structure-based similarity assessment for agile workflows," Springer, 7th International Conf. on Case-Based Reasoning and Development, pp. 224–238, 2007.
33. X. Chen, X. Li, B. Ma, P. M.B. Vitanyi, "The Similarity Metric," IEEE Transactions On Information Theory, Vol. 50, No. 12, 2004.
34. E.F. Krause, "Taxicab Geometry An Adventure in Non-Euclidean Geometry," Dover, 1987.
35. D. M. J. Tax, R. Duin, D. De Ridder, "Classification, Parameter Estimation and State Estimation: An Engineering Approach Using MATLAB," John Wiley and Sons, 2004.
36. J.C. Gower, "A General Coefficient Of Similarity And Some Of Its Properties," Biometrics, Vol. 27, No. 4, pp. 857–871, 1971.
37. M. Deza and E. Deza, "Encyclopedia of Distances," Springer-Verlag, 2014.
38. Y. Rubner, C. Tomasi, L. J. Guibas, "A Metric For Distributions With Applications to Image Databases," IEEE International Conference on Computer Vision, 1998.
39. D.G. Gavin, W.W. Oswald, E.R. Wahl, J.W. Williams, "A Statistical Approach To Evaluating Distance Metrics And Analog Assignments For Pollen Records," Quaternary Research, Vol. 60, Issue 3, pp. 356–367, 2003.
40. F.D. Jou, K.C. Fan, Y.L. Chang, "Efficient Matching of Large-Size Histograms," Elsevier, Pattern Recognition, pp. 277–286, 2004.

41. S.H. Cha, "Taxonomy of Nominal Type Histogram Distance Measures," American Conference On Applied Mathematics, Harvard, Massachusetts, USA, pp. 24–26, 2008.
42. L. Parsons, E. Haque, H. Liu, "Subspace Clustering for High Dimensional Data: A Review," ACM. Sigkdd Explorations, Vol. 6, pp. 90–105, 2004.
43. T. C. Havens, J. C. Bezdek, C. Leckie, L. O. Hall, M. Palaniswami, "Fuzzy c−Mean Algorithms for Very Large Data," in IEEE Transactions on Fuzzy Information and Engineering (ICFIE), pp. 865–874, 2007.
44. R. Peck, J. L. Devore, "Statistics The Exploration and Analysis of Data," Cengage Learning, 2010.
45. W. Hardle, L. Simar, "Applied Multivariate Statistical Analysis," Springer, 2003.
46. H. Bunke, B.T.Messmer, "Similarity Measures for Structured Representations," Springer, Vol. 837, pp. 106–118, 1993.
47. J. Pei, J. Han, B. Mortazavi-Asl, J. Wang, H. Pinto, Q. Chen, U. Dayal, M.C. Hsu, "Mining Sequential Patterns by Pattern-Growth: The Prefix Span Approach," IEEE Transactions on Knowledge and Data Engineering, pp. 1424–1440, 2004.
48. S. Cong, J. Han, D. Padua, "Parallel Mining of Closed Sequential Patterns," Knowledge Discovery in Databases, pp. 562–567, 2005.
49. S. Chakrabarti, "Mining the Web: Statistical Analysis of Hypertext and Semi-Structured Data," Morgan Kaufmann, 2002.
50. P. Kefalas, P. Symeonidis, Y. Manolopoulos, "A Graph-Based Taxonomy of Recommendation Algorithms and Systems in LBSNs," IEEE Transactions On Knowledge And Data Engineering, Vol. 28, No. 3, pp. 604–622, 2016.
51. M. Kuramochi, G. Karypis, "Frequent Sub-graph Discovery," Data Mining, pp. 313–320, 2001.
52. S. Weiss, N. Indurkhya, T. Zhang, F. Damerau, "Text Mining: Predictive Methods for Analysing Unstructured Information," Springer, 2004.
53. P. J. Carrington, J. Scott, S.Wasserman, "Models and Methods in Social Network Analysis," Cambridge University Press, 2005.
54. P. Domingos, "Mining Social Networks for Viral Marketing," IEEE Intelligent Systems, pp. 80–82, 2005.
55. K. Koperski, J. Han, "Discovery of Spatial Association Rules in Geographic Information Databases," Large Spatial Databases, pp. 47–66, 1995.
56. J. Gehrke, F. Korn, D. Srivastava, "On Computing Correlated Aggregates Over Continuous Data Streams," Conf. Management of Data, pp. 13–24, 2001.
57. H. J. Oh, S. H. Myaeng, M. H. Lee, "A Practical Hypertext Categorization Method Using Links and Incrementally Available Class Information," Research and Development in Information Retrieval, pp. 264–271, 2000.
58. A. Bagnall, J. Lines, J. Hills, A. Bostrom "Time-Series Classification with COTE: The Collective of Transformation-Based Ensembles", IEEE Transactions On Knowledge And Data Engineering, Vol. 27, No. 9, pp. 2522–2535, 2015.
59. A. Hinneburg, D. A. Keim, "An Efficient Approach to Clustering in Large Multimedia Databases with Noise," Knowledge Discovery and Data Mining, pp. 58–65, 1998.
60. C. C. Aggarwal, "Outlier Analysis," Springer, 2013.
61. V. Barnett, T. Lewis, "Outliers in Statistical Data," John Wiley and Sons, 1994.
62. D. J. Weller-Fahy, B. J. Borghetti, A. A. Sodemann, "A Survey of Distance and Similarity Measures Used Within Network Intrusion Anomaly Detection", IEEE Communication Surveys and Tutorials, Vol. 17, No. 1, 2015.
63. H. Yazdani, H. Kwasnicka, D. Ortiz-Arroyo, "Multi Objective Particle Swarm Optimization Using Fuzzy Logic," in Springer Int. Conf. Computer and Computational Intelligence, pp. 224–233, 2011.
64. R. Xu, D. Wunsch, "Clustering," IEEE Press Series on Computational Intelligence, 2009.
65. O. Linda, Milos Manic, "General Type-2 Fuzzy C-Means Algorithm for Uncertain Fuzzy Clustering," in IEEE Transactions on Fuzzy Systems, Vol 20, pp. 883–897, 2012.
66. J. Zhou, C. L. P. Chen, L. Chen, H. X. Li, "A Collaborative Fuzzy Clustering Algorithm in Distributed Network Environments", IEEE, Transactions On Fuzzy Systems, Vol. 22, No. 6, pp. 1443–1456, 2014.

Enhanced Over_Sampling Techniques for Imbalanced Big Data Set Classification

Sachin Subhash Patil and Shefali Pratap Sonavane

Abstract Facing hundreds of gigabytes of data has triggered a need to reconsider data management options. There is a tremendous requirement to study data sets beyond the capability of commonly used software tools to capture, curate and manage within a tolerable elapsed time and also beyond the processing feasibility of the single machine architecture. In addition to the traditional structured data, the new avenue of NoSQL Big Data has urged a call to experimental techniques and technologies that require ventures to re-integrate. It helps to discover large hidden values from huge datasets that are complex, diverse and of a massive scale. In many of the real world applications, classification of imbalanced datasets is the point of priority concern. The standard classifier learning algorithms assume balanced class distribution and equal misclassification costs; as a result, the classification of datasets having imbalanced class distribution has produced a notable drawback in performance obtained by the most standard classifier learning algorithms. Most of the classification methods focus on two-class imbalance problem inspite of multi-class imbalance problem, which exist in real-world domains. A methodology is introduced for single-class/multi-class imbalanced data sets (Lowest vs. Highest —LVH) with enhanced over_sampling (O.S.) techniques (MEre Mean Minority Over_Sampling Technique—MEMMOT, Majority Minority Mix mean—MMMm, Nearest Farthest Neighbor_Mid—NFN-M, Clustering Minority Examples—CME, Majority Minority Cluster Based Under_Over Sampling Technique—MMCBUOS, Updated Class Purity Maximization—UCPM) to improve classification. The study is based on broadly two views: either to compare the enhanced non-cluster techniques to prior work or to have a clustering based approach for advance O.S. techniques.

S.S. Patil (✉)
Faculty of Computer Science and Engineering,
Rajarambapu Institute of Technology Rajaramnagar, Post-Rajaramnagar,
Islampur 415414, Maharashtra, India
e-mail: sachin.patil@ritindia.edu

S.P. Sonavane
Faculty of Information Technology, Walchand College of Engineering Vishrambag,
Post-Vishrambag, Sangli 416415, Maharashtra, India
e-mail: shefali.sonavane@walchandsangli.ac.in

W. Pedrycz and S.-M. Chen (eds.), *Data Science and Big Data: An Environment of Computational Intelligence*, Studies in Big Data 24,
DOI 10.1007/978-3-319-53474-9_3

Finally, this balanced data is to be applied to form Random Forest (R.F.) tree for classification. O.S. techniques are projected to apply on imbalanced Big Data using mapreduce environment. Experiments are suggested to perform on Apache Hadoop and Apache Spark, using different datasets from UCI/KEEL repository. Geometric mean, F-measures, Area under curve (AUC), Average accuracy, Brier scores are used to measure the performance of this classification.

Keywords Imbalanced datasets · Big data · Over_sampling techniques · Data level approach · Minority class · Multi-class · Mapreduce · Clustering · Streaming inputs · Reduct

1 Introduction

Big Data is a watchword of today's research which is basically dependent on huge digital data generated in exabytes per year. There is no dearth of data in today's enterprise, but the spotlight is to focus on integration, exploitation and analysis of information. The study of some performance techniques is needed to harness the efficient handling of Big Data streams.

Scientific research has a wash in a flood of data today, which has set a revolution by this Big Data. The new notion of decision-making through the promise of Big Data is impeded with various problems like heterogeneity, scale, timelines, complexity and privacy. Addressing the same at all phases will help to create value from data. Traditional data management principles are unable to address the full spectrum of enterprise requirements with less structured data. As per resources [1], the size of digital data in 2011 is roughly 1.8 Zettabytes (1.8 trillion gigabytes) which is estimated to be supported by networking infrastructure having to manage 50 times more information by year 2020 [2]. Concentric considerations of efficiency, economics and privacy should carefully be planned. Big Data challenges induced by traditional data generation, consumption and analytics are handled efficiently. But, recently in sighted characteristics of Big Data has shown vital trends of access, mobility, utilization as well as ecosystem capabilities [3, 5, 7]. Now-a-days, mining knowledge from huge varied data for better decision making is a challenge [4].

1.1 Basics of Data Mining

Data mining is the extraction of predictive information from large datasets. The various data mining tools are used to predict useful information from available datasets, which helps organization to make proactive business-driven decisions [6].

Data mining is an application of the algorithm to find patterns and relationships that may exist in databases. The databases may contain a large number of records and machine learning is used to perform a search over this space. Now, machine learning is defined as "algorithmic part of data mining process". With the aim of extracting useful knowledge from complex real-world problems, machine learning techniques has also been useful in the last few years. Many real-world problems contain few examples of the concept to be described due to either the rarity or the cost to obtain them. This situation consequences for rare classes or rare cases in datasets, leading to confront new avenues in mining and learning [8–10].

1.2 Classification

Data mining is a data driven approach in the sense that it automatically extracts the patterns from large datasets. Various data mining techniques are present, including association rules, decision tree classification, clustering, sequence mining and so on.

Classification is the most popular application area of data mining. Classification is one kind of predictive modelling in data mining. It is a process of assigning new objects to predefined categories or classes. In classification, building of the model is a supervised learning process and this model can be represented in various forms, like neural networks, mathematical formulae, decision trees and classification rules [11].

Training examples are described in two terms: (a) attributes (b) class labels. The attributes can be categorical or numeric and class labels are called as predictive or output attributes. The training examples are processed using some machine learning algorithms to build decision functions and these functions are used to predict class labels for new datasets. Numbers of classification techniques are available, some of them are as follows:

1. Decision Tree based Methods
2. Rule-based Methods
3. Memory based reasoning
4. Neural Networks
5. Naïve Bayes and Bayesian Belief Networks
6. Support Vector Machines

The classification plays an important role and is an essential tool for several organizations and in many individuals' lives. Classification makes easier to locate, retrieve and maintain things and information. Without classification it is impossible to organize things in a specific order. Hence, it becomes difficult to find things since there is no index to refer. For this reason, classification comes into existence to make our life better and easier.

1.3 Clustering

Clustering is considered as the unsupervised learning problem. Every problem of this type deals with finding structure or similarities in a collection of unlabeled data or objects. A cluster is a collection of objects or data which are "similar" between the particular clusters and "dissimilar" to the object or data belongs to the other clusters [11]. The most common similarity measure criterion is 'distance'. If cluster is formed on the basis of distance, then it is called as 'distance based clustering'; in which two or more objects or data will belong to the same cluster if they are "too close to each other" using a given distance. There are various types of distance measures: Euclidian distance, Manhattan distance, Minikowski distance, Correlation distance. Different distance measures, gives different resulting clusters. Clustering is of the two types:

i. Hierarchical Clustering:
   Algorithms of hierarchical clustering create hierarchical decomposition of the objects or the data using similarity criteria with the agglomerative approach and divisive approach. By this algorithm, cluster is formed from dendrogram which helps for visualization.
ii. Partitional Clustering:
   Partitioning algorithms construct the partitions of the data or objects and then evaluate them as per the similarity criteria.

Hierarchical algorithms are desirable for the small datasets and partitioning algorithms are desirable for the larger datasets [13]. The performance of various clustering algorithms which are supported by Mahout on different cloud runtime such as Hadoop and Granule are evaluated in [11, 14]. Granule (stream processing system) gives better result than hadoop, but doesn't support for failure recovery. K-means requires a user to specify a K value. K-means clustering is the best choice when all points belong to distinct groups and when it initially approaching a new dataset. K-means runs quickly and can find large distinctions within the data. Fuzzy C-means operates in the same manner as k-means, with the modification that instead of each data point belongs to a single cluster; each data point is assigned a probability of belonging to every cluster. Dirichlet and LDA take long time to run than K-means and fuzzy C-means due to its complexity. [15] compares k-means and fuzzy c-means for clustering a noisy realistic and big Wikipedia's dataset using apache mahout. The variance of both algorithms is according to initial seeding. Generally, k-means is slower than the fuzzy version. However, with random seeding one cannot assume which method will be faster. In this research, it is found that in a noisy dataset, fuzzy c-means can lead to inferior cluster quality than k-means. [13] has proven Partitional Clustering algorithm is best suited for large datasets by studying and comparing the results of Partitional K-means, Hierarchical algorithm and Expectation Maximization (EM) Clustering Algorithm.

Osama [16] has proposed an idea for text categorization in the vector space model using Term Frequency-Inverse Document Frequency (TF-IDF). TF-IDF is very sensitive for feature selection settings and less stable. It is the most common weighting function for information retrieval problem and text mining. It assigns weight to each vector component of the document and categorizes those [12].

Another challenge for data mining algorithms is Big Data. The techniques used to deal with Big Data are focused on fast, scalable and parallel implementations. To reach this goal, mapreduce framework is followed. The mapreduce framework is one of the most popular procedures used for handling Big Data. The working of this framework is based on—divide and conquer strategy, where the dataset is divided into subsets that are easy to manage and partial results from earlier stages are combined for further needful processing.

Furthermore, many real world applications present classes which have an insignificant number of samples as compared to other classes. This situation is called as a class imbalance problem. Usually, the insignificant samples are the main focus of study; hence it is necessary to classify them correctly. In machine learning research, learning from imbalanced datasets is an issue that has attracted a lot of attention. The statistical learning methods are suited for balanced data sets and may ignore the negligible samples which are important. That's why it is required to consider the features of the problem and solve it correctly.

In this chapter, the various enhanced O.S. techniques used to deal with single-class/multi-class imbalanced data problem are presented. These approaches are evaluated on the basis of potency in correct classification of each instance of each class and time required to build the classification model. In order to perform classification, R.F. which is a popular and well-known decision tree ensemble method can be used. It is proven that R.F. is scalable, robust and gives better performance [6]. In experimental studies, the mapreduce based implementation of non-cluster or clustering based O.S. is required to carry out. In an imbalanced data domain, the efficacy in classification can basically be evaluated using two measures: Geometric Mean for true rates and β-F- Measure [11].

While the benefits of Big Data are real and significant, there are many technical challenges that must be addressed to fully realize the potential. The challenges are:

- New data management and analysis platforms
- Techniques for addressing imbalance data sets
- Handling huge streaming data
- Feature selection for reduced data set
- Enhancement for traditional mining iteration based techniques over new frameworks
- Soft computing and machine learning techniques for efficient processing
- Privacy management

2 Mapreduce Framework and Classification of Imbalanced Data Sets

2.1 MapReduce Framework

Real world areas such as Telecommunication, Financial Businesses, Health Care, Pharmaceuticals, etc. are generating enormous amounts of data. It is the need of an hour to get useful business insights from this data and has become a challenge to standard data mining techniques. Traditionally, large amount of data is managed by data warehouses. But, data warehouses have some drawbacks as they don't cope-up with management of massive data that grow day by day. Consequently, the Big Data concept came into existence. Big Data is data that overshoot the processing capacity of traditional database systems. The data is too large, moves too fast or doesn't fit into a predefined structure of database architecture [11]. To process this massive size, high velocity structured/unstructured streaming data, there is a need of advance processing framework to deal with distributed computing using commodity hardware. Hadoop is the Apache framework, has capabilities as a data platform. Hadoop and its surrounding ecosystem solution vendors provide the enterprise requirements. It also helps to integrate together the Data Warehouse and other enterprise data systems as the part of modern data architecture. It is a step on the journey toward delivering an enterprise 'Data Lake' [8].

Since 2012 Big Data has become a hot IT buzzword and it is defined in terms of 3 V's such as Volume, Velocity and Variety. [3, 19].

- Volume: Deals with the amount of data that is generated per day and grows from MB's to GB's to PB's.
- Velocity: Deals with how fast data is coming and how fast it should be analysed.
- Variety: Deals with different forms of data, structured and unstructured [18].

Initially in 2004, the mapreduce programming framework was proposed by Google. It is a platform designed for processing tremendous amount of data in an extremely parallel manner. It provides an environment to easily develop scalable and fault tolerant applications. The mapreduce programming model carries out the calculation process in two phases and are as follows:

- Map: Master node divides the input dataset into independent sub-problems and distributes them to slave nodes, the slave nodes process these sub problems in a parallel way and pass the result back to its master node.
- Reduce: Master node takes the results from all the sub-problems and combines them to form the output.

In the mapreduce model all the computation is organized around (key, value) pairs. First stage deals with Map function, taking a single (key, value) pair as input and produce lists of intermediate (key; value) pairs as the output. It can be represented as [19]:

$$\text{Map}(\text{key};\ \text{value}) \rightarrow \text{list}(\text{key}';\ \text{value}')$$

Then, the system merges and groups these intermediate pairs by keys and passes them to the Reduce function. Finally, the Reduce function takes a key and an associated value list as input and generates a new list of values as output, which can be represented as follows in [20, 21]:

$$\text{Reduce}(\text{key}',\ \text{list}(\text{value}') \rightarrow (\text{key}',\ \text{value1})$$

The Map-Reduce techniques are used to perform the tasks in parallel fashion over the framework provided. The performance of executing is the major concern of improvement, which can be increased without increasing the hardware cost, but is achieved by just tuning some parameters like input size, processing intricacy and cluster conditions. Further, tuning factors in architecture design schemes helps to improve the overall performance of multi-GPU mapreduce [9, 22]. Data analytics are becoming a good practice in most domains. The economy and everyday life are full of guiding examples, which provide the basis for updating or refining our understanding of Big Data Analytics and for exploring new ground [19]. The focus of research should further apply the Knowledge Discovery process (KDD) to create data backbones for decision support systems to aid in ergonomic evaluations [21].

In [22], the author has proposed and implemented a non-parametric extension of Hadoop. It allows for early results for arbitrary workflows in an incremental manner, besides providing accuracy in the computation on reliable on-line estimates. Early results for simple aggregates are experimented, which have laid a foundation in [22] to achieve similar objectives. These estimates are based on a technique called bootstrapping [25, 26], used widely in statistics applicable to arbitrary functions and data distributions. The technique is selected based on its generality and accuracy. While data sampling over memory-resident, disk resident has been widely studied, sampling over a distributed file system has not been fully addressed [21]. Summary of various files (database files) sampling techniques is studied which are closely related to random sampling over Hadoop Distributed File System (HDFS). As the initial samples are drawn from the original dataset, they are further re-sampled based on replacement drawn from it. They are additionally used to generate results, which derive the final result distribution. The sample result distribution is used for estimating the accuracy. The error for arbitrary analytical functions can be estimated via the bootstrapping technique described in [23]. There are other resampling techniques, such as the jackknife, which performs resampling without replacement. It is known that jackknife does not work for many functions such as the median [23].

The Apache Hadoop is an open source project and provides the mapreduce framework as in Fig. 1 for processing Big Data. Hadoop distributed file system (HDFS) is used to store the data which makes data available to all computing nodes. The Hadoop usage has three steps: 1. Load data into HDFS 2. mapreduce operations 3. Retrieve data from HDFS.

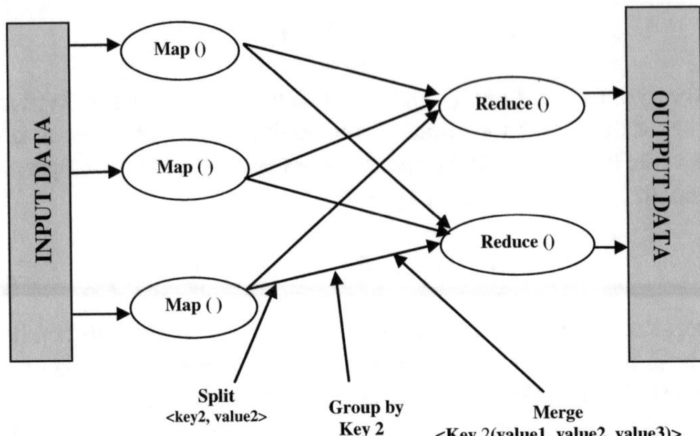

Fig. 1 Working of mapreduce framework [18]

By nature this process is of batch operation type, well suited for analytics and non-interactive applications. The Hadoop is not a database or data warehouse, but it can act as an analytical tool. The role of the users in this scenario is only to define what should be computed in the Map and Reduce functions while the system automatically distributes the data processing over a distributed cluster of machines.

Apache Mahout is an open source project and runs on top of the hadoop platform [6]. It provides a set of machine learning algorithms for clustering, recommendation and classification problems. Mahout contains various implementations of classification models such as Logistic Regression, Bayesian models, Support Vector Machines, and Random Forest. Like Hadoop, Apache Spark is an open-source cluster computing framework and was developed in the AMPLab at UC Berkeley. It is an in-memory based mapreduce paradigm and performs 100 times faster than Hadoop for certain applications [6].

2.2 Classification of Imbalanced Datasets

Many real-world problems contain some concepts only with very few examples in a huge set. They are described either by their rarity or the cost to obtain them. The results from these rarities have been identified as one of the main challenges in data mining. In mining, the categorization of input into pre-labeled classes is done on certain similarity aspects. The categories may either be of multi-class or two classes. The analysis of various techniques for multi-class imbalanced data problem is required to be focused as many real world problems such as medical diagnosis [24], fraud detection, finances, risk management, network intrusion, E-mail foldering, Software Defect Detection [25]. The classification of imbalanced datasets poses problems where class distributions having number of examples in one class are outnumbered by other classes [10]. A class having an abundant number of examples is called as majority or negative class and a class having a significant number of examples called as minority or positive class. Additionally, the positive (minority) class is the class of interest from the learning point of view and has a great impact when misclassified [6]. Learning from imbalanced datasets is a difficult task for machine learning algorithm due to global search measure which does not take into account the difference between the numbers of instances of each class. The specific rules are used for identification of minority class instances. But during model construction, these specific rules are ignored in the presence of more general rules, which are used to identify the instances of majority class.

Several techniques are available to address the classification of imbalanced data [18]. These techniques are categorized into various groups [6]:

1. Data Level Approach:
   An original imbalanced dataset is modified to get a balanced dataset and further analysed by standard machine learning algorithms to get the required results.
2. Algorithm Level Approach:
   An existing algorithm is modified to launch procedures that can deal with imbalanced data.
3. Cost-Sensitive Approach:
   Both, the data level and the algorithm level approaches are combined to get accuracy and reduce misclassification costs.

The techniques discussed, deals with the data level approach in detail. Furthermore, data level approaches are divided into various groups: O.S., Undersampling and Hybrid technique. In O.S. technique, new data from minority classes are added to the original dataset in order to obtain a balanced dataset. In Undersampling technique, data from majority classes are removed in order to balance the datasets. With hybrid technique, previous techniques are combined to achieve the goal of balanced dataset. Usually in hybrid approach, first O.S. technique is used to create new samples for minority class and then Undersampling technique is applied to delete samples from the majority class [9]. The O.S. and Undersampling techniques have some drawbacks. The noisy data may get replicated in O.S. techniques. Undersampling may lead to loss of important data due to random selection scheme [6].

Synthetic Minority Oversampling Technique (SMOTE) algorithm is used as a powerful solution to solve imbalanced dataset problem, which has shown success in various application domains [26]. The SMOTE algorithm is an O.S. technique. This technique adds synthetic minority class samples to original dataset to achieve the balanced dataset [6].

In SMOTE algorithm, minority class is oversampled by duplicating samples from minority class. Depending on the O.S. required, numbers of the nearest neighbors are chosen randomly [6]. The synthetic data is then generated based on feature space likeliness prevails between existing samples of a minority class. For subset $S_{min} \epsilon S$, consider k nearest neighbors for each sample $x_i \in X$. The

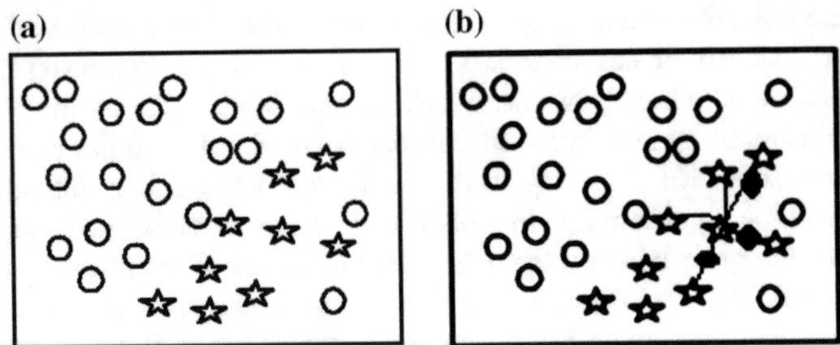

Fig. 2 **a** Imbalanced class distribution, **b** generation of synthetic data, k = 5

K-nearest neighbors are the K elements whose Euclidean distance between itself and x_i has smallest weight along the n-dimensional feature space of X. The samples are generated as simply as, randomly selecting one of the K-nearest neighbors and multiplying the corresponding Euclidean distance with random numbers between [0, 1]. Finally, this value is added to x_i.

$$x_{i\_new} = x_{i\_old} + (x_{i\_KNN} - x_{i\_old})^{*}\delta \quad (1)$$

where, $x_i \in S_{min}$ is a sample from minority class used to generate synthetic data. x_{i_KNN} is the nearest neighbor for x_i, and δ is a random number between [0, 1]. The generated synthetic data is a point on line fragment between x_i under consideration and x_{i_KNN} k-nearest neighbors for x_i.

The following Fig. 2a, shows imbalanced class distribution, where the circles represent the majority class and stars represent the minority class. The K-nearest neighbor is set to K = 5. The Fig. 2b, illustrates synthetic data generation on the line segment joining x_i and x_{i_KNN} and it is spotlighted by diamond. Finally, synthetic data are added to the original dataset in order to balance it.

Though SMOTE is a popular technique in imbalanced domain, it has some drawbacks including over-generalization, only applicable for binary class problem, over-sampling rate varying with the dataset. To avoid these drawbacks, some approaches are defined such as Borderline-SMOTE and Adaptive Synthetic Sampling for generalization. Evolutionary algorithms and sampling methods are used to deal with the class imbalanced problem [27]. SMOTE + GLMBoost [28] and NRBoundary-SMOTE are based on the neighborhood Rough Set Model [29] and are used to solve the class imbalance problem. The ensemble methods like AdaBoost, RUSBoost and SMOTEBoost are coupled with SMOTE to solve imbalanced data problems [24]. All these approaches are focused on two-class problem. In [30], the author proposed a solution for multi-class problem based on fuzzy rule classification. Ensembles of decision trees (R.F.) have been the most successful general-purpose classification algorithm in modern times [6]. R.F. was proposed by Leo Breiman in 2002 to improve the classification of a dataset having a small training dataset and large testing dataset i.e. R.F. is suited for large number attributes and a small number of observations [18]. R.F. is scalable, fast and durable approach for classification of high-dimensional data and can handle continuous data, categorical data and time-to-event data [31].

Huge attribute sets with categorical data values lead to inaccurate results of imbalanced data sets. Attribute reduction of large featured data sets has moved attention for handling imbalanced data sets in alignment to reduce storage cum efficient retrieval. It aims to delete superfluous condition attributes from a decision system. However, most of the existing techniques for selecting an informative attribute set are developed for a single type of attributes. It is challenging to deduce some effective attribute reduction techniques for heterogeneous data [32]. There are two categories of techniques which deal with heterogeneous conditional attribute reduction viz. pre-processed into single-type data and secondly, measure diverse types of attributes using dissimilar criteria [33]. Both can be combined to provide an overall evaluation of mixed attributes. Pre-processing into a single type deals with conversion from one form to another, i.e. symbolic to real-valued and vice versa, but may lead to loss of mutual information. Also measuring and combining unlike attributes may sometimes direct to problematic undesirable results [20]. Both categories further fail to comprehend the substitution ability among heterogeneous attributes, their classification ability towards decision attributes and inconsistency tending to decision labels. Rough set helps to address inconsistencies between conditional and decision attributes. Rough set deals attribute reduction with in-hand available data itself to process, which is different than all other methods and obtains reducts keeping original inconsistencies [34]. Classical rough sets only deal with symbolic/integer valued attributes. As a generalization of rough set model, fuzzy rough sets were developed to handle real-valued attributes. Attribute reduction with fuzzy rough sets was developed and then improved in [17, 35, 36] with several heuristic algorithms to find reducts. By discerning two samples to a certain degree related to decision labels, fuzzy rough set methods help to tackle the inconsistency between condition attributes and decision labels [36]. In terms of transforming attribute types, both classical and fuzzy rough sets are not able to exactly handle heterogeneous data [37].

3 Methodology

3.1 Architecture

The projected research is based on experimental analysis, which comprises a range of exploratory statistical and quantitative methods. Quantitative research is generally associated with the positivist paradigm. It involves collecting and converting data into numerical form for statistical calculations and drawing out conclusions. Hypotheses address predictions about possible relationships between the things they want to investigate. In order to find answers, the researchers need to avail certain techniques and tools with a clearly defined plan of action. Quantitative research mainly tends towards deductive reasoning directing to shift from the general to the specific things i.e. top down approach. The conclusions can be validated based on earlier findings.

The study involves an overview of data science and the challenge of working with Big Data using statistical methods. It requires integrating the insights from data analytics into knowledge generation and decision-making based on both structured and unstructured data. The study incorporates the process of how to acquire data, process it, store it and convert it into a format suitable for analysis, model the experimental design and interpret the final results.

The theoretical concept deals with the development of some techniques to enhance the traditional approaches to handle streaming high velocity data mining aspects in parallel execution environment. Moreover, the imbalanced classification of minority data versus majority samples is to be studied to provide a well-balanced data.

The analysis of Big Data involves multiple distinct phases as shown in the Fig. 3. The huge streams of data are required to handle, accept, record and store. The 'storage and retrieval' is a major concern of performance. Further, the system attempts to implement an optional clustering model that clusters the data parallely based on the similarities or theme of the text data. It helps to find the characteristic similarity of data under pre-processing for further needful clustering. Moreover, streaming inputs with heterogeneity are to be addressed with approximation techniques to provide some useful early insight. The inputs are further processed with enhanced O.S techniques for creating balanced dataset. The outputs are finally analysed for improved accuracy.

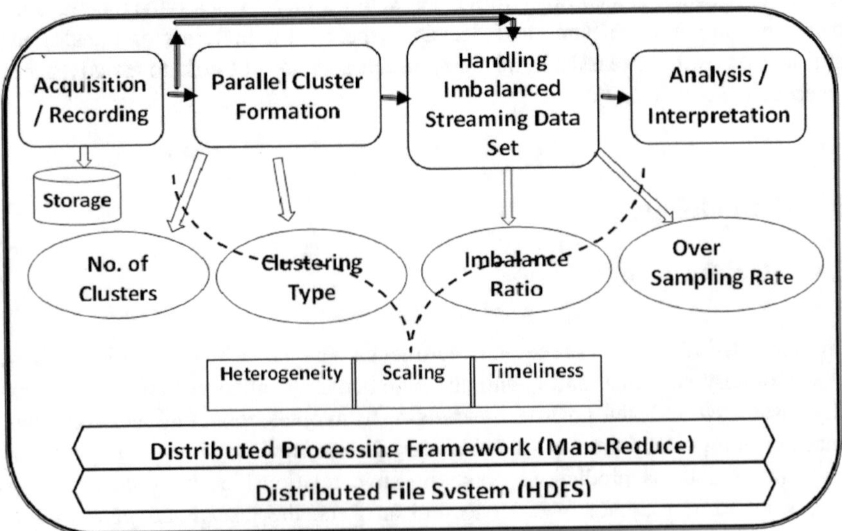

Fig. 3 Overall system architecture

3.1.1 Input Pre-processing and Similarity Based Parallel Clustering of Streaming Data

The designed system consists of following modules:

- *Data acceptance*—input for data preparation model.
- *Data preparation* of (D) docs (Using Porter Stemmer)—having n terms of data (d) as d_i and d_j (i, j = 0, 1, 2,…, n). D_i and D_j are description similarities and F_i and F_j are functional similarities
  - Tokenization
  - Stop word removal
  - Stemming—to give output as D_i' and D_j'
- *Similarity Based Analysis*
  - Description (S_D) and Functional similarity (S_F) using Cosine/Jaccard similarity coefficient
  - Characteristic similarity (S_C) by weighted sum of S_D and S_F
  - Table of S_C for data under study
- *Clustering algorithms*—Clustering data (considering either majority and minority instances or especially minority instances) can be done either according to their S_C in D using a parallel clustering algorithm or using various clustering algorithms as like DBSCAN, BIRCH, K-Means.

The output of the system is a document clustered in the form of hierarchical hyperlinks, pure/hybrid clusters for the further O.S. process.

Further, the work may be extended for analysis showing that the datasets containing a number of features may contain some superfluous attributes which are intended to be reduced (fuzzy based techniques). This may help to provide the reduced dataset characterizing the same notion as the original set for efficient storage and needful analysis. The proposed major steps of the analysis are as follows:

- *Decision information table with the discernibility relation of mixed attributes is to be formed from given input data set*:
  An information system is a pair (U, A), where U = $\{x_1, x_2,\ldots, x_n\}$ is a nonempty finite universe of discourses and A = $\{a_1, a_2,\ldots, a_m\}$ is a nonempty finite set of heterogeneous attributes. With every subset of attributes B $\subseteq$ A, associate a binary relation IND(B), called the B-indiscernibility relation, and defined as IND(B) = $\{(x, y) \in U \times U: a(x) = a(y), a \in B\}$; then IND(B) is an equivalence relation and IND(B) = $\cap a \in B$ IND({a}). By [x]B the equivalence class of IND(B) including x is denoted.
  A decision information table (DT), is an information system A* = (A $\cup$ D) where A $\cap$ D = φ. A is a set of heterogeneous conditional attributes, while D is the decision attribute and U/IND(D) = $\{D_1, D_2,\ldots, D_r\}$.

The following approach of discernibility matrix is used as the mathematical foundation of finding reducts.
Let $A^* = (U, A \cup D)$ be a DT, where $U = \{x_1, x_2, \ldots, x_n\}$ is a nonempty finite universe of discourses, A is a set of symbolic conditional attributes, and D is the decision attribute. $M(A^*)$ denotes an $n \times n$ matrix (C_{ij}), called the discernibility matrix of A^*, such that (s.t.) $C_{ij} = \{a_k \in A: a_k(x_i) \neq a_k(x_j)\}$, for x_i and x_j satisfying one of the following conditions:

i. $x_i \in pos_A(D)$ and $x_j \in pos_A(D)$;
ii. $x_i \in pos_A(D)$ and $x_j \in pos_A(D)$; and
iii. $x_i, x_j \in pos_A(D)$ and $D(x_i) \neq D(x_j)$.

Suppose U is a nonempty universe, $A^* = (U, A \cup D)$ is a decision system, $A \cap D = \varphi$ and $A = A_S \cup A_R$, where A_S and A_R denote the families of symbolic and real valued condition attributes, respectively. Suppose $R_D(A) = R_D(A_S) \cup R_D(A_R)$; then, $R_D(A)$ is also a fuzzy relation, where the relation $R_D(A_S)$ is considered as a fuzzy relation, i.e., $R_D(A_S)(x, y) = 1$ if $(x, y) \in R_D(A_S)$ Or $= 0$ if $(x, y) \in R_D(A_S)$.

- *The discernible ability of conditional attributes to decision labels is to be modeled using dependence function:*
  A Boolean function denoted by $f_U(A \cup D) = \wedge(\vee C_{ij})$, $C_{ij} \neq \varphi$ is referred to as the discernibility function for A^*. A discernible relation of $a \in A$ is defined as $R_D(a) = \{(x, y): a(x) \neq a(y)\}$ for $D(x) \neq D(y)$.
- *Attributes are to be selected using reduct algorithm:*

$$\mathrm{RedX}(A \cup D) = \{A_1, \ldots, A_t\}.$$

- *Formation of rules based on selected attributes in conjunctive/disjunctive forms:*
  Let $g_U(A \cup D)$ be the reduced disjunctive form of $f_U(A \cup D)$ by applying the distribution and absorption laws as many times as possible; then there exist t and $A_i \subseteq A$ for $i = 1, \ldots, t$ s.t. $g_U(A \cup D) = (\wedge A_1) \vee \cdots \vee (\wedge A_t)$.
- *Reducing the data set as per reducts.*

3.1.2 Enhanced Over_Sampling Techniques for Imbalanced Dataset

A methodology for O.S. of two-class or multi-class imbalanced data is addressed. The O.S. techniques can be applied to the non-cluster imbalanced Big Data (I.B.D.) sets or clustered based.

The O.S. techniques are based on linear interpolation methods. These can be further classified on the basis of variables within viz. bi-linear, tri-linear interpolation methods. The Linear interpolation methods have disadvantages of low precision, less differentiability of interpolated points and error proportional to the square of the distance between points. Subsequently, Polynomial or Spline interpolation can be used to reduce error, smoothing interpolation, estimating local

Maxima-Minima and providing infinite differentiability of the interpolated points. But these methods are computationally more complex and might exhibit oscillatory artifacts at the end points.

I. **Un-Clustered Simplistic Imbalanced Data Sets**

i. Two Class Data Sets (Assuming N—No. of instances):

- Technique 1—**Mere Mean Minority Over_Sampling** Technique **(MEMMOT)**:
  This technique is enhanced SMOTE methodology. A SMOTE provides a disadvantage of duplicating majority/minority samples. In MEMMOT, following procedure avoids almost a duplication of interpolated instances. Let the training dataset be D, D_{maj}—majority class instances z_m (m = 1, 2, …., m) and D_{min}—minority class instances x_n (n = 1, 2,…, n).
  Find safe levels of all instances before processing [12].
  Further, for each minority instance x_n (for 100% O.S. rate):

  1. Find K-NN instances comprising the whole training data set.
  2. Find SMOTE based interpolated instances of all the K-NN instances.
  3. Take the average of all interpolated instances to get a new interpolated instance.
  4. Check for duplication—if Yes, reduce the lowest safe level nearest neighbor instance from the current K-NN. Reduce the interpolated instance of the respective lowest safe level nearest neighbor instance from the current interpolated instances. Repeat step 3.

  For O.S. rate above 100%:
  Reduce the lowest safe level nearest neighbor instance from the current K-NN. Reduce the interpolated instance of the respective lowest safe level nearest neighbor instance from the current interpolated instances. Repeat step 3 to comply the O.S. rate. The value of K should satisfy the condition as-

$$K > \%\text{over\_sampling rate}/100$$

  OR
  Repeatedly use the current over sampled set for further over sampling based on MEMMOT, till the satisfaction of O.S. rate.
  OR

  i. Based on safe levels or random basis—select 50% samples out of the first 100% over sampled instances and remaining 50% from the original set. Use this combined set for next O.S. generations with MEMMOT.
  ii. For more O.S. rate, based on safe levels or random basis, select 33% samples from each Original, First 100% and Second 100% over

sampled sets. Use this combined set for next O.S. generations with MEMMOT.

iii. Continue step ii. with reduced selection ratios of 25, 12.5, 6.25% and so on from original and over sampled sets….till the O.S. rate is satisfied.

For O.S. rate below 100%, select the interpolated samples either randomly or on the basis of high safe level—which comply the required O.S. rate. In either case if the technique fails to comply I.R., than under-Sampling Based on Clustering [39] can be used to reduce majority classes to match the number of minority instances.

The above method helps to provide more generalized synthetic minority instances with low repetition rates and improves classification accuracy.

- Technique 2—**M**inority **M**ajority **M**ix mean **(MMMm)**:

  This technique is a unique advancement of SMOTE methodology. SMOTE, only considers K-NN minority samples for creating synthetic samples. As stated above, it leads to duplication and as well a less generalized interpolated sample.

  The technique (MMMm)—considers K-NN minority as well as majority samples for further O.S. of interpolated instance. It helps to provide a more generalized interpolated sample, less duplication faults with overcoming boundary-line samples.

  Find safe levels of all instances before processing [12].

  Further, for each minority instance x_n (n = 1, 2,…,n and for 100% O.S. rate):

  1. Find K-NN instances comprising the whole training data set.
  2. Check for K-NN instances set—either all instances are minority or majority or minority-majority mix.
  3. If all instances are minority class—follow step 6 to 7.
  4. If all instances are of majority class—follow step 8 to 9.
  5. Else—follow step 10 to 11.
  6. Select an instance on the highest safe level basis from available K-NN. Find a SMOTE based interpolated instance.
  7. Check for duplication—if Yes, reduce the highest safe level instance from the current K-NN. Repeat step 6.
  8. Select an instance from K-NN on a random basis. Find a nearest minority sample to the current majority sample under consideration. Find a SMOTE based interpolated instance of both—majority and nearest minority instance. Take the average of both interpolated instances to get the new interpolated instance.
  9. Check for duplication—if Yes, select the next most nearest minority sample to the current majority sample under consideration. Repeat step 8.
  10. Select an instance from K-NN on a random basis.

i. If the selected instance is of minority class—find a SMOTE based interpolated instance.
ii. If the selected instance is of majority class—find a highest safe level minority sample from in hand minority samples in K-NN set. Find a SMOTE based interpolated instance for both—majority and nearest minority instance. Take the average of both interpolated instances to get the new interpolated instance.

11. Check for duplication—if Yes, remove the instance under consideration. Repeat step 10.

For O.S. rate above 100%:
Reduce the lowest safe level instance from the current k-NN set. Further, step 2 to 10 can repeatedly be used for all remaining instances to comply the O.S. rate.
The value of K should satisfy the condition as-
$K \geq$ % over_ sampling rate/100
OR
Repeatedly use the current over sampled set for further over sampling based on MMMm, till the satisfaction of O.S. rate.
OR

i. Based on safe levels or random basis—select 50% samples out of the first 100% over sampled instances and remaining 50% from the original set. Use this combined set for next O.S. generations with MMMm.
ii. For more O.S. rate, based on safe levels or random basis, select 33% samples from each Original, First 100% and Second 100% over sampled sets. Use this combined set for next O.S. generations with MMMm.
iii. Continue step ii with reduced selection ratios of 25, 12.5, 6.25% and so on from original and over sampled sets....till the O.S. rate is satisfied.

For O.S. rate below 100%, select the interpolated samples either randomly or on the basis of high safe level—which comply the required O.S. rate. In either case if the technique fails to comply I.R., than under-Sampling Based on Clustering [39] can be used to reduce majority classes to match the number of minority instances.

- Technique 3—**N**earest **F**arthest **N**eighbor_**M**id—**(NFN-M)**:
This technique proposes to consider K - nearest as well as farthest instances for interpolation with an add-on middle element.
Find safe levels of all instances before processing [39].
Further, for each minority instance x_n (n = 1, 2,..., n and for 100% O.S. rate):

1. Find KNN dataset such that—K/2 nearest, K/2 farthest and mid-point element from the whole data set, where k $\leq$ N − 2.
   - For even number data sets above 2, either lower-end/higher-end mid-point is considered
   - For odd number data sets—K > 1
2. Check for K-NN instances set—either all instances are minority or majority or minority-majority mix.
3. If all instances are minority class—follow step 6 to 7.
4. If all instances are of majority class—follow step 8 to 9.
5. Else—follow step 10 to 11.
6. Select an instance on the highest safe level basis from available K-NN. Find a SMOTE based interpolated instance.
7. Check for duplication—if Yes, reduce the highest safe level instance from the current K-NN. Repeat step 6.
8. Select an instance from K-NN on a random basis. Find a nearest minority sample to the current majority sample under consideration. Find a SMOTE based interpolated instance of both—majority and nearest minority instance. Take the average of both interpolated instances to get the new interpolated instance.
9. Check for duplication—if Yes, select the next most nearest minority sample to the current majority sample under consideration. Repeat step 8.
10. Select an instance from K-NN on a random basis.
    i. If the selected instance is of minority class—find a SMOTE based interpolated instance.
    ii. If the selected instance is of majority class—find a highest safe level minority sample from in hand minority samples in K-NN set. Find a SMOTE based interpolated instance for both—majority and nearest minority instance. Take the average of both interpolated instances to get the new interpolated instance.
11. Check for duplication—if Yes, remove the instance under consideration. Repeat step 10.

For O.S. rate above 100%:
Reduce the lowest safe level instance from the current k-NN set. Further, step 2 to 10 can repeatedly be used for all remaining instances to comply the O.S. rate.
The value of K should satisfy the condition as-
K $\geq$ % over_ sampling rate/100
OR
Repeatedly use the current over sampled set for further over sampling based on NFN-M, till the satisfaction of O.S. rate.
OR

i. Based on safe levels or random basis—select 50% samples out of the first 100% over sampled instances and remaining 50% from the original set. Use this combined set for next O.S. generations with NFN-M.
ii. For more O.S. rate, based on safe levels or random basis, select 33% samples from each Original, First 100% and Second 100% over sampled sets. Use this combined set for next O.S. generations with NFN-M.
iii. Continue step ii with reduced selection ratios of 25, 12.5, 6.25% and so on from original and over sampled sets….till the O.S. rate is satisfied.

For O.S. rate below 100%, select the interpolated samples either randomly or on the basis of high safe level—which comply the required O.S. rate. In either case if the technique fails to comply I.R., than under-Sampling Based on Clustering [39] can be used to reduce majority classes to match the number of minority instances.
This proposed method helps to consider a wide range of inputs for creating synthetic samples and avoids repetition.

- Technique 4—Clustering **M**inority **E**xamples **(CME)**:
This technique is a pure cluster based technique. The technique involves only the instances of minority classes for synthetic samples generation. The means of clusters basically seem to synthetic instances. The technique helps to provide the same objective as like DBSMOTE of enriching centroids based O.S.
Find safe levels of all instances before processing [12].
Further, for each minority instance x_n (n = 1, 2,…, n and for 100% O.S. rate):

1. Cluster only minority data set using any clustering algorithm (basically K-Means).
2. Each mean (Centroid) seems to be a new interpolated synthetic sample.
3. Check for duplication—if Yes, remove that respective Centroid.
For achieving the O.S. rate:

   - Add the centroids obtained in iteration to previous minority data set forming new set and continue Step 1 to 3 (Keeping cluster no. and initial seeds same as pervious iteration)
   OR
   - Either repeat the Step 1 to 3 by reducing an element within the original data set based on the lowest safe level, till dataset_size >2 (Keeping cluster no. same as pervious iteration but the initial seeds will be different)

The above methods provide generalized synthetic instances with low repetition rates and helps improving classification.

In either case if the technique fails to comply I.R., than under-Sampling Based on Clustering [39] can be used to reduce majority classes to match the number of minority instances.

- Technique 5—**M**ajority **M**inority **C**luster **B**ased **U**nder_**O**ver **S**ampling Technique—**(MMCBUOS)**:
  The idea is to consider both between-class and with-in class imbalances to address them simultaneously. Technique proposes to cluster individual class beforehand to under-over sampling.
  Find safe levels of all instances before processing [12].

  1. Cluster individual classes from training dataset. (e.g. using K-means)
  2. Check for majority class cluster (Mj_l) which is immediate large compared to largest minority class cluster.
  3. Under sample all the majority clusters above Mj_l to the level of Mj_l.
  4. Calculate the total 't' of all majority clusters after under sampling.
  5. Find the over sampling 'o' requirement based on 't' (considering I.R. $\leq$ 1.5)
  6. Find individual minority cluster O.S. rate (equalizing each cluster sample instance) based on 'o' to meet I.R.
  7. O.S. is carried out in each individual minority cluster using MEMMOT/MMMm/NFN-M. (while O.S., the neighbor is selected from the same cluster from minority class)
  8. The induced balanced data set is used for further classification.

- Technique 6—**U**pdated **C**lass **P**urity **M**aximization—**(UCPM)**:
  This technique focuses on over-sampling under observation of under sampling technique [40]. Clustering is used to solve the class imbalance issue through O.S. This idea is to find as many clusters of majority class instances which are pure. Remaining impure clusters are considered for further needful O.S. It also considers both between-class and with-in class imbalances to address them simultaneously. The technique leverages the drawback of complying I.R. using better O.S. techniques, improving classification results. Technique proposes to cluster individual class beforehand to under-over sampling [40, 44].

  1. Select a pair of minority and majority instances as centers from the training data set.
  2. The remaining instances are partitioned into two subsets according to their nearest centers, with at least one subset having a high class purity.
  3. Step 2 is repeated recursively for each of the two subsets until we can no longer form two clusters, with at least one yielding higher class purity than its parent cluster.
  4. Eliminate the pure clusters and remaining impure clusters are seen for further O.S.
  5. Calculate the total 't' of all majority class instances in all impure clusters after under sampling.

6. Find the over sampling 'o' requirement based on 't' (considering I.R. $\leq$ 1.5)
7. Find individual minority cluster O.S. rate (equalizing each cluster sample instance) based on 'o' to meet I.R.
8. O.S. is carried out in each individual minority cluster using MEMMOT/MMMm/NFN-M. (while O.S., the neighbor is selected from the same cluster from minority class)
9. The induced balanced data set is used for further classification.

ii. Multi-Class Imbalanced Data Sets:-

Profound techniques in multi-class imbalanced data set handling are One versus One (OVO) and One versus All (OVA). In the proposed technique, Lowest versus Highest (LVH) is a unique method sufficing all disadvantages of OVO and OVA technique. The technique helps to reduce computation in addition to improve classification performance. The details of techniques with respective advantages and disadvantages as:

OVA—helps to reduce computation for balanced data set formations, but it doesn't comply with the realistic need of classification cum O.S. as compared to second lowest minority set underlying "All" label. In certain cases, minority class may exceed some majority sub-classes after over sampling has led to excess synthetic sample creation. This may lead to poor classification results.

OVO—helps to address an individual class of data sets compared to each set within. It basically coverts a multi-class issue into a binary model, applying all other binarization models of the induced sets. With the same, it inherits the disadvantages of binary based techniques. It incurs a heavy computation overhead compared to OVA and also marginal accuracy is achieved. In addition, the balancing of each individual class leads to more over_sampled instances affecting the classification results. The test sample has to needlessly proceed to more binary models for final classification output.

LVH—The idea quoted is more robust compared to the above two techniques for multi-class data set balancing.

- Compares the individual lowest minority class (all classes below a certain threshold, i.e. 35–40%) versus one highest majority class only.
- O.S. of minority classes is done one by one, forming the final training data set.
- Avoid duplication cum reduced computation.
- Reduces synthetic samples generations and provides more realistic interpolated samples compared to all other minority set underlying.
- Evade overshooting of other majority sub-classes after O.S.
- Reduces computation of test samples for final classification.

– Comply the highest majority class indirectly, conforming all remaining classes within.

The Fig. 4 depict the in-sight of LVH method. Figure 4a states the whole data set. Figure 4b states the comparison of the lowest (a) versus highest (A) class. Figure 4c states the comparison of second-last lowest (b) versus highest (A) class. Figure 4d states the comparison of third-last lowest (c) versus highest (A) class. This method will surely help to enhance the accuracy of the model.
The above six proposed methods can be used similarly with LVH for balancing of multi-class data sets to improve classification results.

II. **Clustered Based Over_Sampling Techniques for Imbalanced Data Sets**
In contrast to above non-clustered O.S. techniques for two-class/multi-class, clustering based techniques work with unsupervised approach and further over sample the minority class data sets for required balanced objective. For two/multi-class datasets (LVH based), all three clustering techniques (CME/MMCBUOS/UCPM) can be used.

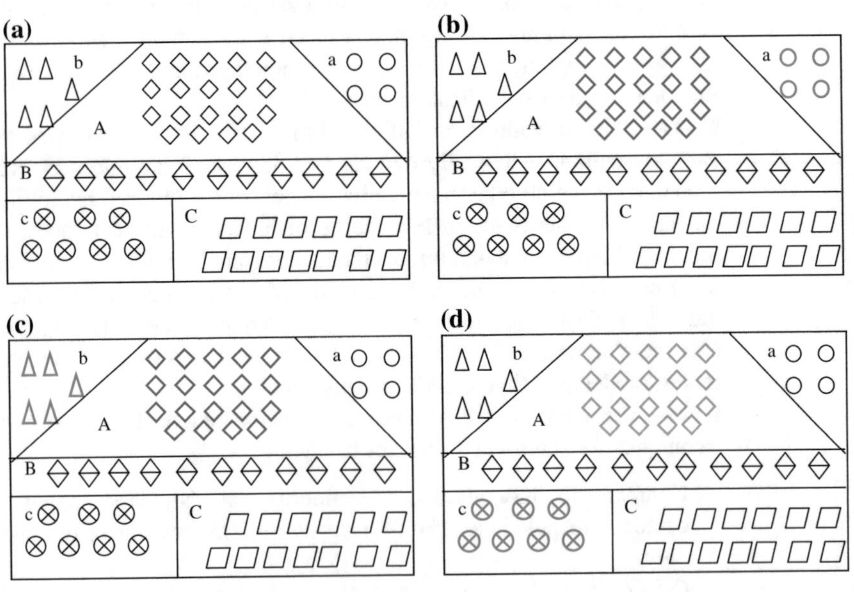

Fig. 4 **a** Whole dataset, **b** LVH method, **c** LVH method, **d** LVH method

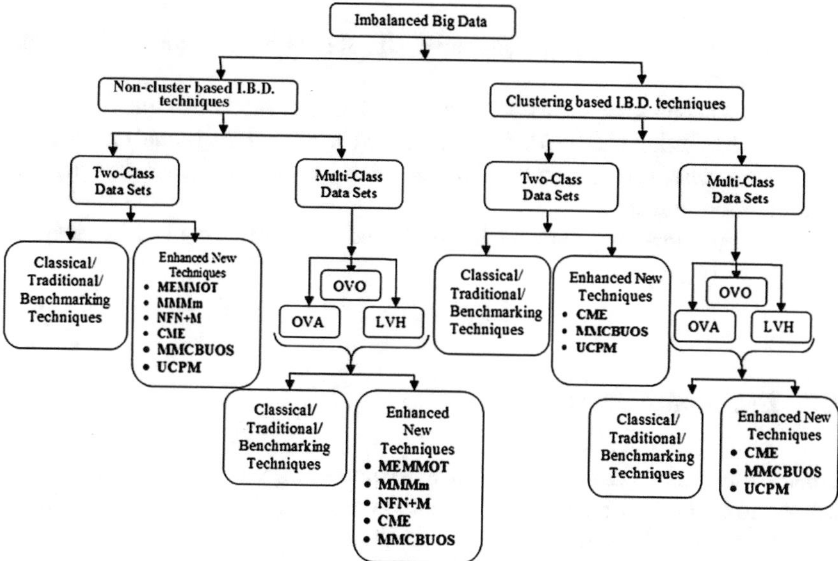

Fig. 5 Structure of I.B.D. techniques

Various clustering algorithms can be implemented with the three enhanced techniques. In [41], an online fault detection algorithm based on incremental clustering is studied. [42], which addresses the issue of ordinal classification methods in imbalanced data sets. A novel class detection in data streams with efficient string based methodology is deliberated in [43].

The overall structure of I.B.D. handling techniques is shown in Fig. 5. O.S. techniques are used to handle large imbalanced datasets.

Further, the work can be extended for analysis showing that the datasets containing a number of features may contain some superfluous attributes which are intended to be reduced. This may help to provide the reduced dataset characterizing the same notion as the original set for efficient storage and needful analysis.

3.2 *Sampling Design and Assumptions*

i. Sampling:

   In some cases, it is highly inefficient and impossible to test the entire population—thus the focus of study may lead toward samples. Samples help to test the entire population, which produces error or a cross section of it. Sampling does come with several problems as like missing elements, foreign elements, duplicate elements etc.

ii. Assumptions:
   Support for incremental sample input_size should be incorporated in Hadoop framework.
   Heterogeneous data set may be converted into a symbolic/numeric attribute values for further O.S. procedures and analysis. High dimensional data set leads to bias classification results towards minority classes, hence the need for dimension reduction.
   The input streaming data set is considered to be of known class labels for empirical analysis.

3.3 Evaluation Parameters

Experiments can be performed on Apache Hadoop and Apache Spark using diverse datasets from the UCI repository. Geometric mean and F-measures can be planned to measure the performance of this classification.

The proposed major steps for implementation are as follows:

1. To convert the dataset from CSV format into Hadoop Sequence File format consisting of key/value pairs.
2. To map the dataset in OVO/OVA/LVH model.
3. To implement O.S. algorithms, to convert imbalanced dataset to balanced form.
4. To implement R.F. forest algorithm for training set and analyse.
5. To analyse the results of performance metrics as a Geometric Mean (GM) and F-measure for varying data size, number of data nodes.

The effectiveness is to be evaluated using above two measures that are able to efficaciously rate the success in imbalanced classification.

- **Geometric Mean (GM)**: gives a balanced accuracy of true rates. It attempts to maximize the accuracy of each one of the two classes with balancing good link between both objectives.

$$GM = \sqrt{\text{sensitivity} * \text{specificity}} = \sqrt[k]{\pi \sum_{i=1}^{k} R_i} \tag{2}$$

k no. of classes
R_i the recall of ith call

- Sensitivity (recall) = True Positive/(True Positive + False Negative)
- Specificity = True Negative/(False Positive + True Negative)

– **F-Measure**: Another metric used to assess quality of classifiers in imbalanced domain is F-Measure and it is given as:

$$\text{F} - \text{Measure} = \frac{2\, R_i \text{P}_i}{R_i + P_i} \; for\, all\, i = 1, 2, \ldots k \tag{3}$$

P_i—precision of ith class.

- Precision = True Positive/(True Positive + False Positive) – predicted positive cases that were correctly classified.

4 Conceptual Framework

4.1 *Pre-processing and Efficient Parallel Clustering Architecture*

Architecture in Fig. 6 is planned for similarity based clustering model that clusters the Big Data based on the similarities.

– The similarities between data/documents are calculated based on description similarity. Further functional similarity is used as a measure to analyse clusters.
– The clustering would be implemented using M-R techniques based on weighted means of characteristic similarity.

As shown in Fig. 7, the input data may be partitioned and provided to mappers, thereafter to combiner for computing the relative clusters, including their centroids and merge the local centroids for global mean to achieve required clusters.

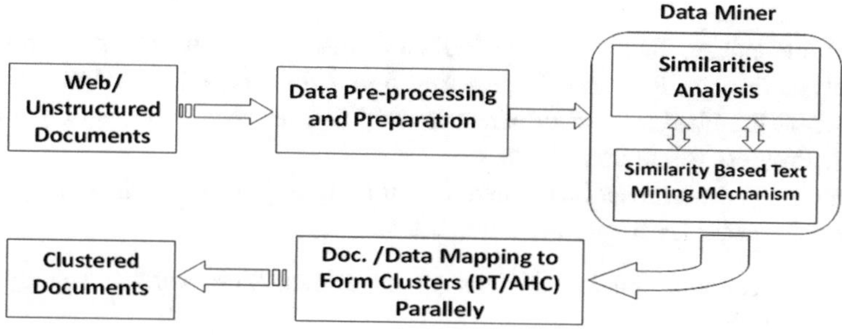

Fig. 6 Similarity based mining architecture

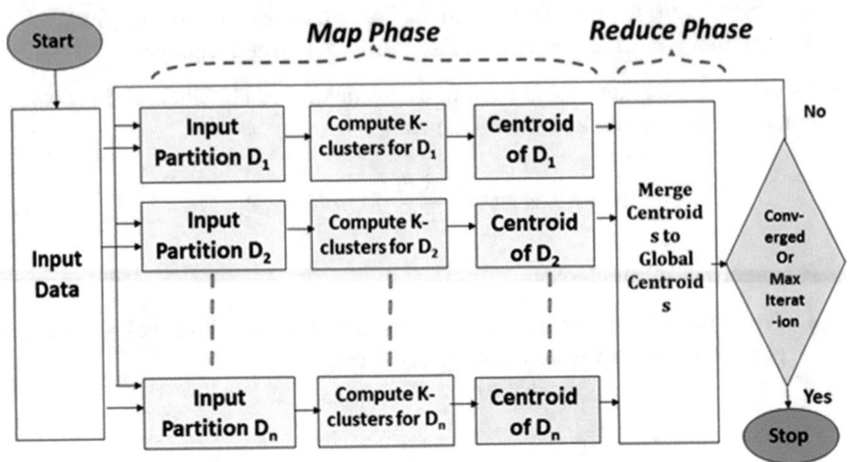

Fig. 7 Parallel clustering architecture

4.2 Conceptual Flow of Experimentation

The conceptual flow of analysis consists of following steps:

1. Identify the imbalanced training set and form a R.F. tree of it. Use testing set to check for initial bare results for comparison at last. Parallely, this data set can be clustered and checked for cluster cohesiveness.
2. Carry out O.S. of two-class or multi-class imbalanced training data sets, either using un-clustered simplistic term or on clustering basis.
3. Perform—R.F. on same.
4. Use Testing set—for model prediction and accuracy testing.
5. Update R.F., if the I.R. goes above 1.5 (+10%) OR error rate goes above a certain threshold (recall). <If the error rate goes above a certain threshold—then re-correct the data set with prior known classes>
6. The new corrected data can further be used to improve R.F. tree using Step 2 and 3. Parallely this data can be analysed for cluster cohesiveness and similarity cohesiveness.
7. Continuously, collect real time incoming data set for further prediction and analysis through R.F. tree. Repeat step 5 to 7 for this real time data set.
8. Analyse the classifier performance on said measurements by respective methods. Progress to the conclusion.
9. Optionally—can repeat Step 7 and 8 for infinite sequence of time to improve the classifier accuracy as per the requirement.

Figure 8 explains the logical flow of experimentation work and analyse for necessary results.

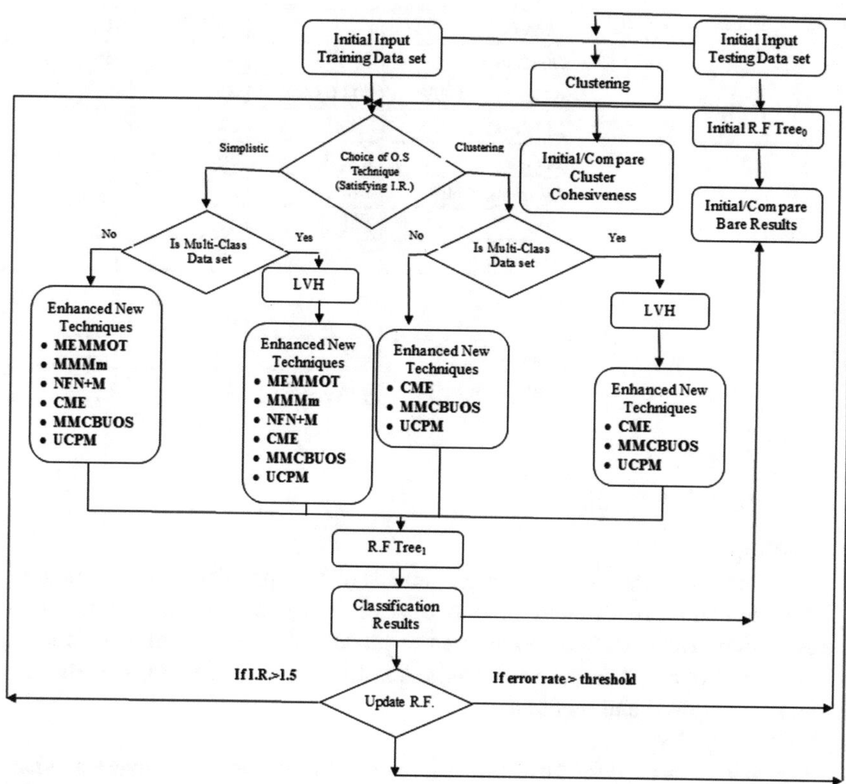

Fig. 8 Conceptual flow of experimentation

4.3 Data Sets

HBase is an open source, distributed, column oriented database modelled after Google's BigTable. HBase is a distributed Key/Value store built on top of Hadoop. HBase shines with large amounts of data and read/write concurrency. The data are acquired for further analysis in HBase.

Basically, the standard datasets are to be collected from UCI (UC Irvine Machine Learning Repository) library and KEEL (Knowledge Extraction based on Evolutionary Learning) database repository for experimentation as below:

In order to analyse the quality of solutions provided for the anticipated problem, datasets from the UCI Database Repository is to be used. As selected datasets have multiple classes, diverse solutions are to be experimented to address this issue separately. Table 1 summarizes the details of selected datasets which includes documented name of the data set, number of instances (#EX), number of attributes (#ATTR) and number of classes (#CL).

Table 1 Characteristics of dataset [38]

Dataset	#EX	#ATTR	#CL
Mashup	14,193	06	Multi-class
KDD Cup 1999	40,00,000	42	2
Mushroom	8,124	22	2
Landsat	6,435	36	6
Lymphography	148	18	4
Zoo	101	17	7
Segment	2,310	19	7
Iris	150	4	3
Car	1,728	6	4
Vehicle	946	19	4
Waveform	5,000	40	3

Data sets referred [38]:

- **Mashup-**
  The data are collected from ProgrammableWeb, a popular online community built around user-generated mashup services. It provides the most characteristic collection. The extracted data is used to produce datasets for the population of mashup services. The dataset included mashup service name, tags, APIs used, technology used and its class.
- **KDD Cup 1999-**
  This data set is used for The Third International Knowledge Discovery and Data Mining Tools Competition, which was held in conjunction with KDD-99_The Fifth International Conference on Knowledge Discovery and Data Mining. The competition task was to build a network intrusion detector, a predictive model capable of distinguishing between "bad" connections called intrusions or attacks and "good" normal connections. This database contains a standard set of data to be audited, which includes a wide variety of intrusions simulated in a military network environment.
- **Mushroom-**
  This data set includes descriptions of hypothetical samples corresponding to 23 species of gilled mushrooms in the Agaricus and Lepiota Family. Each species is identified as definitely edible, definitely poisonous, or of unknown edibility and not recommended. This latter class was combined with the poisonous one. The Guide clearly states that there is no simple rule for determining the edibility of a mushroom.
- **Landsat-**
  The database consists of the multi-spectral values of pixels in 3 × 3 neighborhoods in a satellite image and the classification associated with the central pixel in each neighborhood. The aim is to predict this classification, given the multi-spectral values. In the sample database, the class of a pixel is coded as a number.

- **Lymphography-**
  This is one of three domains provided by the Oncology Institute that has repeatedly appeared in the machine learning literature.
- **Zoo-**
  A simple database containing 17 Boolean-valued attributes. The "type" attribute appears to be the class attribute. The breakdown of animal types is provided.
- **Segment-**
  The instances were drawn randomly from a database of 7 outdoor images. The images were hand segmented to create a classification for every pixel. Each instance is a 3×3 region.
- **Iris-**
  This is perhaps the best known database to be found in the pattern recognition literature. Fisher's paper is a classic in the field and is referenced frequently to this day. The data set contains 3 classes of 50 instances each, where each class refers to a type of iris plant. One class is linearly separable from the other 2; the latter are NOT linearly separable from each other.
- **Car-**
  Car Evaluation Database was derived from a simple hierarchical decision model originally developed for the demonstration of DEX, M. Bohanec, V. Rajkovic: Expert system for decision making.
- **Vehicle-**
  The purpose is to classify a given silhouette as one of four types of vehicle, using a set of features extracted from the silhouette. The vehicle may be viewed from one of many different angles.
- **Waveform-**
  There are 3 classes of waves with 40 attributes, all of which include noise. The latter 19 attributes are all noise attributes with mean 0 and variance 1 and contains 5000 instances.

4.4 Methods of Data Computations

Methods of data computations are as:

- For similarity based parallel clustering following methods are used-
  - *Step 1:*
    Morphological similar words are clubbed together under the assumption that they are also semantically similar.
  - *Step 2:* **Compute Description Similarity (D_S)–**
    For e.g.: using Jaccard Similarity Coefficients for D_i *and* D_j as cardinality of their intersection divide by the cardinality of their union.

$$D_s(d_i, d_j) = \frac{\left|D_i' \cap D_j'\right|}{\left|D_i' \cup D_j'\right|} \tag{4}$$

Larger the numerator tends to more similarity and the denominator is scaling factor to ensure similarity between 0 and 1.

- ***Step 3:*** **Compute Functional Similarity (D_F)–**
  For e.g.: using Jaccard Similarity Coefficients for F_i *and* F_j as cardinality of their intersection divide by the cardinality of their union.

$$D_F(d_i, d_j) = \frac{\left|F_i' \cap F_j'\right|}{\left|F_i' \cup F_j'\right|} \tag{5}$$

- ***Step 4:*** **Compute Characteristic Similarity (D_C)–**
  Characteristic similarity between d_i *and* d_j *is computed* by a weighted sum of description similarity and functionality similarity.

$$D_C(d_i, d_j) = \alpha^* D_S(d_i, d_j) + \beta^* D_F(d_i, d_j) \tag{6}$$

Where $\alpha \in [0, 1]$—weight of D_S and $\beta \in [0, 1]$—weight of D_F and $\alpha + \beta = 1$ Weight expresses relative importance between two

- ***Step 5:***
  A: Partition Clustering (K-means)-

Step 1.

map (k1, v1)
do clustering for each sample by K-means
Output (C_i, N_i, P_i)

Step 2.

reduce (k2, v2)
merge sequence (C_i, N_i, P_i)
output (c1,c2,…,ck)

Where

C_i—collection of ith sample's cluster centres
N_i—collection of points—number of C_i
P_i—collection of points for each centre of C_i

B: Hierarchical Algometric Clustering (HAC)-
Assume 'n' data items, each initialized to be a cluster of its own. At each reduction step, the two most similar clusters are merged until only K (K < n) clusters remains. Reduction during cluster formation as:

Step1. Search for the pair of maximum similarity in table of S_C and merge them.
Step2. Create a new similarity table by calculating the average values of similarities between clusters.
Step3. Save and reuse these similarities and cluster partitions for later visualization.
Step4. Repeat step 1–3 until the K clusters remains on the table

5 Conclusion

In this study, the enhanced data pre-processing techniques for two-class and multi-class imbalanced data has been presented using non-cluster/cluster based O.S. techniques. R.F. algorithm is used as a base classifier, which is decision tree ensemble and known for its good performance. Traditional data mining techniques are unable to survive with requirements urged by Big Data; hence, the mapreduce framework under Hadoop environment is used to deal with it.

Experimental analysis can be carried out using various datasets of UCI repository. The system quality testing benchmark may be indexed in terms of the parameters like accuracy, AUC, G-Mean and F-measure. These can be applied to three methods, namely MEMMOT, MMMm and NFN-M in non-cluster mode as well as CME, MMCBUOS and UCPM in clustering mode using OVO, OVA and LVH method. It is believed that, the proposed techniques will outperform the existing methods i.e. SMOTE, BorderlineSMOTE, SafeLevel SMOTE, DBSMOTE. At the same time, concerns raised out of the intrinsic data characteristics like small disjuncts, lack of density, overlapping and impact of borderline instances will be addressed. The issues related to dataset shift and changing O.S. rate, needs to be further addressed in detail.

References

1. A. Gandomi and M. Haider, “Beyond the hype: big data concepts, methods, and analytics,” *International Journal of Information Management*, vol. 35, no. 2, pp. 137–144, 2015.
2. W. Zhao, H. Ma, and Q. He., “Parallel k-means clustering based on mapreduce,” *CloudCom*, pp. 674–679, 2009.
3. D. Agrawal et al., “Challenges and Opportunity with Big Data,” *Community White Paper*, pp. 01–16, 2012.
4. X. Wu et al., “Data Mining with Big Data,” *IEEE Trans. Knowledge Data Engg*, vol. 26, no. 1, pp. 97–107, 2014.
5. X.-W. Chen et al., “Big data deep learning: Challenges and perspectives,” *IEEE Access Practical Innovations: open solutions*, vol. 2, pp. 514–525, 2014.

6. M. A. Nadaf, S. S. Patil, "Performance Evaluation of Categorizing Technical Support Requests Using Advanced K-Means Algorithm," *IEEE International Advance Computing Conference (IACC)*, pp. 409–414, 2015.
7. "Big Data: Challenges and Opportunities, Infosys Labs Briefings - Infosys Labs," http://www.infosys. com/infosys-labs/publications/Documents/bigdata-challenges-opportunities.pdf.
8. "A Modern Data Architecture with Apache Hadoop," *A White paper developed by Hortonworks*, pp. 01–18, 2014.
9. H. Jiang, Y. Chen, and Z. Qiao, "Scaling up mapreduce-based big data processing on multi-GPU systems," *SpingerLink Clust. Comput*, vol. 18, no. 1, pp. 369–383, 2015.
10. N. Chawla, L. Aleksandar, L. Hall, and K. Bowyer, "SMOTEBoost: Improving prediction of the minority class in boosting," *PKDD Springer Berlin Heidelberg*, pp. 107–119, 2003.
11. R. C. Bhagat, S. S. Patil, "Enhanced SMOTE algorithm for classification of imbalanced bigdata using Random Forest," *IEEE International Advance Computing Conference (IACC)*, pp. 403–408, 2015.
12. W. A. Rivera, O. Asparouhov, "Safe Level OUPS for Improving Target Concept Learning in Imbalanced Data Sets," *Proceedings of the IEEE Southeast Con,* pp. 1–8, 2015.
13. K. Shvachko, H. Kuang, S. Radia, R. Chansler, "The Hadoop Distributed File System," *IEEE 26th Symposium on Mass Storage Systems and Technologies*, pp. 1–10, 2010.
14. *F. Khan, "An initial seed selection algorithm for k-means clustering of georeferenced data to improve replicability of cluster assignments for mapping application," Elsevier publication-Journal of Applied Soft Computing,* vol. 12, pp. 3698– 3700, *2012.*
15. R. Esteves, C. Rong, "Using Mahout for Clustering Wikipedia's Latest Articles: A Comparison between K-means and Fuzzy C-means in the Cloud," *IEEE Third International Conference on Cloud Computing Technology and Science*, pp. 565–569, 2011.
16. A. Osama, "Comparisons between data clustering algorithms," *The international Arab Journal of Information Technology*, vol. 5, pp. 320–325, 2008.
17. R. Jensen and Q. Shen, "New approaches to fuzzy-rough feature selection," *IEEE Trans. Fuzzy Syst*, vol. 17, no. 4, pp. 824–838, 2009.
18. R. Sara, V. Lopez, J. Benitez, and F. Herrera, "On the use of MapReduce for imbalanced big data using Random Forest," *Elsevier: Journal of Information Sciences*, pp. 112–137, 2014.
19. B. Japp et al., "No More Secrets with Big Data Analytics," *White paper developed by Sogeti Trend Lab VINT*, pp. 01–210, 2013.
20. E. Tsang, D. Chen, D. Yeung, and X. Wang, "Attributes reduction using fuzzy rough sets," *IEEE Trans. Fuzzy Syst*, vol. 16, no. 5, pp. 1130–1141, 2008.
21. R. Renu, G. Mocko, and A. Koneru, "Use of big data and knowledge discovery to create data backbones for decision support systems," *Elsevier Procedia Comput. Sci*, vol. 20, pp. 446–453, 2013.
22. L. Nikolay, Z. Kai, and Z. Carlo, "Early accurate results for advanced analytics on MapReduce," *Proceedings of the VLDB Endowment*, vol. 5, no.10, pp. 1028–1039, 2012.
23. J. Shao, and D. Tu., "The jackknife and bootstrap," *Springer series in statistics Springer Verlag*, pp. 01–13, 2013.
24. B. Chumphol, K. Sinapiromsaran, and C. Lursinsap, "Safe-level-smote: Safelevel- synthetic minority over-sampling technique for handling the class imbalanced problem," *AKDD Springer Berlin Heidelberg*, pp. 475–482, 2009.
25. P. Byoung-Jun, S. Oh, and W. Pedrycz, "The design of polynomial function-based neural network predictors for detection of software defects," *Elsevier: Journal of Information Sciences*, pp. 40–57, 2013.
26. N. Chawla, K. W. Bowyer, L. O. Hall, and W. P. Kegelmeyer, "SMOTE: synthetic minority over-sampling technique," *Journal of Artificial Intelligence Research*, vol. 16, pp. 321–357, 2002.
27. S. Garcia et al., "Evolutionary-based selection of generalized instances for imbalanced classification," *Elsevier: Journal of Knowl. Based Syst*, pp. 3–12, 2012.

28. H. Xiang, Y. Yang, and S. Zhao, "Local clustering ensemble learning method based on improved AdaBoost for rare class analysis," *Journal of Computational Information Systems*, Vol. 8, no. 4, pp. 1783–1790, 2012.
29. H. Feng, and L. Hang, "A Novel Boundary Oversampling Algorithm Based on Neighborhood Rough Set Model: NRSBoundary-SMOTE," *Hindawi Mathematical Problems in Engineering*, 2013.
30. F. Alberto, M. Jesus, and F. Herrera, "Multi-class imbalanced data-sets with linguistic fuzzy rule based classification systems based on pairwise learning," *Springer IPMU*, pp. 89–98, 2010.
31. J. Hanl, Y. Liul, and X. Sunl, "A Scalable Random Forest Algorithm Based on MapReduce," *IEEE*, pp. 849–852, 2013.
32. D. Chen, Y. Yang, "Attribute reduction for heterogeneous data based on the combination of classical and Fuzzy rough set models," *IEEE Trans. Fuzzy Syst*, vol. 22, no. 5, pp. 1325–1334, 2014.
33. J. Ji, W. Pang, C. Zhou, X. Han, and Z. Wang, "A fuzzy k-prototype clustering algorithm for mixed numeric and categorical data," *Elsevier Knowl. Based Syst*, vol. 30, pp. 129–135, 2012.
34. D. Chen, S. Zhao, L. Zhang, Y. Yang, and X. Zhang, "Sample pair selection for attribute reduction with rough set," *IEEE Trans. Knowl. Data Engg*, vol. 24, no. 11, pp. 2080–2093, 2012.
35. D. G. Chen and S. Y. Zhao, "Local reduction of decision system with fuzzy rough sets," *Elsevier Fuzzy Sets Syst*, vol. 161, no. 13, pp. 1871–1883, 2010.
36. D. Chen, L. Zhang, S. Zhao, Q. Hu, and P. Zhu, "A novel algorithm for finding reducts with fuzzy rough sets," *IEEE Trans. Fuzzy Syst*, vol. 20, no. 2, pp. 385–389, 2012.
37. Q. Hu, D. Yu, J. Liu, and C. Wu, "Neighborhood rough set based heterogeneous feature subset selection," *Elsevier Information Sci*, vol. 178, pp. 3577–3594, 2008.
38. UCI Machine Learning Repository, http://archieve.ics.uci.edu/ml/
39. S. Yen and Y. Lee, "Under-Sampling Approaches for Improving Prediction of the Minority Class in an Imbalanced Dataset," *ICIC 2006, LNCIS 344*, pp. 731–740, 2006.
40. K.Yoon, S. Kwek, "An Unsupervised Learning Approach to Resolving the Data Imbalanced Issue in Supervised Learning Problems in Functional Genomics," *International Conference on Hybrid Intelligent Systems*, pp. 1–6, 2005.
41. J. Kwak, T. Lee, C. Kim, "An Incremental Clustering-Based Fault Detection Algorithm for Class-Imbalanced Process Data," *IEEE Transactions on Semiconductor Manufacturing*, pp. 318–328, 2015.
42. S. Kim, H. Kim, Y. Namkoong, "Ordinal Classification of Imbalanced Data with Application in Emergency and Disaster Information Services," *IEEE Intelligent Systems*, pp. 50–56, 2016.
43. M. Chandak, "Role of big-data in classification and novel class detection in data streams," *Springer Journal of Big Data*, pp. 1–9, 2016.
44. M. Bach, A. Werner, J. Żywiec, W. Pluskiewicz, "The study of under- and over-sampling methods' utility in analysis of highly imbalanced data on osteoporosis," *Elsevier Information Sciences*(In Press, Corrected Proof), 2016.

Online Anomaly Detection in Big Data: The First Line of Defense Against Intruders

Balakumar Balasingam, Pujitha Mannaru, David Sidoti, Krishna Pattipati and Peter Willett

Abstract We live in a world of abundance of information, but lack the ability to fully benefit from it, as succinctly described by John Naisbitt in his 1982 book, "we are drowning in information, but starved for knowledge". The information, collected by various sensors and humans, is corrupted by noise, ambiguity and distortions and suffers from the *data deluge problem*. Combining the noisy, ambiguous and distorted information that comes from a variety of sources scattered around the globe in order to synthesize accurate and actionable knowledge is a challenging problem. To make things even more complex, there are intentionally developed intrusive mechanisms that aim to disturb accurate information fusion and knowledge extraction; these mechanisms include cyber attacks, cyber espionage and cyber crime, to name a few. Intrusion detection has become a major research focus over the past two decades and several intrusion detection approaches, such as rule-based, signature-based and computer intelligence based approaches were developed. Out of these, computational intelligence based anomaly detection mechanisms show the ability to handle hitherto unknown intrusions and attacks. However, these approaches suffer from two different issues: (i) they are not designed to detect similar attacks on a large number of devices, and (ii) they are not designed for quickest detection. In this chapter, we describe an approach that helps to scale-up existing computational intelligence approaches to implement quickest anomaly detection in millions of devices at the same time.

B. Balasingam (✉) · P. Mannaru · D. Sidoti · K. Pattipati · P. Willett
Department of Electrical and Computer Engineering, University of Connecticut,
371 Fairfield Way, U-4157, Storrs, CT 06269, USA
e-mail: bala@engr.uconn.edu; balakumar.balasingam@uconn.edu

P. Mannaru
e-mail: pujitha.mannaru@uconn.edu

D. Sidoti
e-mail: david.sidoti@uconn.edu

K. Pattipati
e-mail: krishna.pattipati@uconn.edu

P. Willett
e-mail: peter.willett@uconn.edu

W. Pedrycz and S.-M. Chen (eds.), *Data Science and Big Data: An Environment of Computational Intelligence*, Studies in Big Data 24,
DOI 10.1007/978-3-319-53474-9_4

Keywords Cyber-physical-human systems · Cyber threat detection · Malware · Intrusion detection · Online anomaly detection · Cyber-security · Big data · Computational intelligence · Quickest detection of changes · Likelihood ratio test

1 Introduction

Advances in miniature computing have led to the development of physical systems that perform very complex tasks and their need to communicate defines cyber-physical systems. In addition, cyber and physical systems interact with humans, to accomplish tasks that require cognitive inputs that can only be provided by humans, or to enhance the comfort and quality of experience of humans, or to achieve both. These cyber-physical-human systems (CPHS), as illustrated in Fig. 1, are expected to become more and more commonplace in the future.

Revolutions in computing and connectivity have also exposed the CPHS infrastructure to premeditated attacks with potentially catastrophic consequences by those within as well as outside enterprises and geographic boundaries, armed only with a computer and the knowledge needed to identify and exploit vulnerabilities.

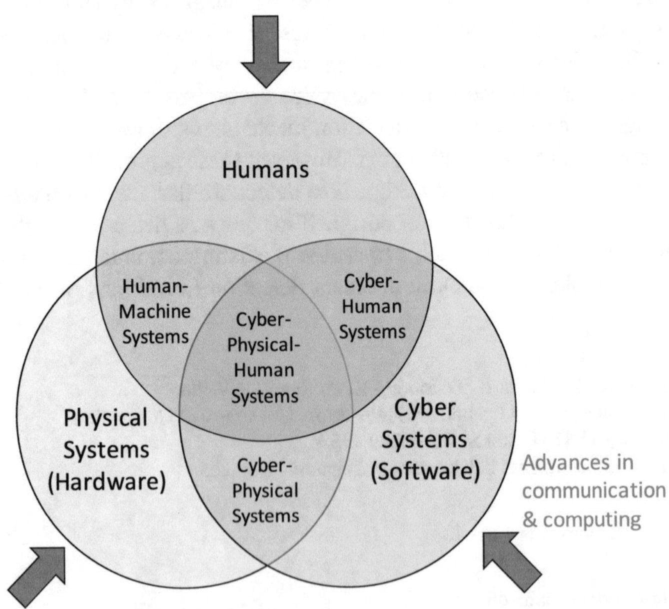

Fig. 1 Cyber-physical-human systems (CPHS). Due to the nature of the increasingly connected world, the standalone nature of physical systems, cyber systems (e.g., a web page) and humans continue to shrink (i.e., all three of the *circles* are getting closer and closer with time and the area occupied by cyber-physical-human systems is increasing). The *arrows* indicate that each of the circles is moving closer and closer due to advances in communication and computing technology

The past two decades have seen tremendous research activity in developing intrusion detection mechanisms. These approaches can be generally categorized into the following: [1], *signature-based detection, rule-based detection,* and *computer intelligence based detection.* The signature and rule based approaches are designed based largely on known attacks or intrusive behaviors, whereas the computational intelligence based approaches [2–5] promise defense against unknown attacks as well. Even though the term "intrusion detection" is more explanatory in nature, most of the intrusion detection mechanisms are designed to detect "contextual anomalies". Hence, we will use the more general term, anomaly detection [6–9], in the rest of this chapter.

Advanced persistent threats (APT) are insidious cyber-attacks executed by sophisticated and well-resourced adversaries targeting specific information in high-profile companies and governments, usually in a long term campaign involving different phases and multiple attack vectors (e.g., cyber, physical, and deception, automated tools, social engineering) [10]. Some examples of APTs are: Stuxnet worm [11], Aurora Trojan [12], Diginotar hack [13], RAS Breach [14], Operation Ke3chang [15], Operation Snowman [16].[1] The distinguishing characteristics of APTs are:

1. *Specific digital assets with clear objectives* that bring competitive advantage in a conflict or strategic benefits, such as national security data, intellectual property, trade secrets, etc.
2. *Highly organized and well-resourced attackers* having access to zero-day vulnerabilities and distributed denial of service (DDoS) attack tools working in a coordinated way.
3. *A long-term campaign that adapts to defensive mechanisms* to stay undetected in the target's network for several months or years.
4. *Stealthy and evasive attack techniques* by concealing themselves within enterprise network traffic. For example, APT actors may use zero-day exploits to avoid signature-based detection, and encryption to obfuscate network traffic.

Expanding the APT attack models in [18, 19], we view an APT as being comprised of the following eight phases: (1) reconnoitre; (2) gain access; (3) internal reconnaissance; (4) expand access; (5) gather information; (6) extract information; (7) malware control (command and control); and (8) erase tracks. Cyber defense against APTs, like conventional warfare, necessitates situational awareness and rapid decision-making (faster "OODA" loop) to eliminate or mitigate the effects of a specific attack. Approaches to detecting network attacks must resolve issues with false positives and missed detections, achieve high scalability, facilitate the detection of highly complex attacks, and adapt to new types of attacks by piecing together how individual attack steps are configured to enable their detection. It is not uncommon

[1]Malware is an umbrella term used to refer to a variety of software intrusions, including viruses, worms, Trojan horses, ransomware, spyware, scareware, adware and so on. These can take the form of executable code, scripts, active content and other software. The majority of recent active malware threats were worms or Trojans rather than viruses [17]. When malware is used in a deliberate and concerted manner, as in APTs, one needs sophisticated monitoring and mitigating strategies to address them.

to have thousands of false positive alerts per day. Indeed, a threat analysis system must scan and fuse context-specific network and host activities from multiple sources to *reliably* detect changes that could signal a developing attack, identify the attack sources, assess its impact, and respond before it affects network and combat system operations (e.g., data corruption, functional degradation, system latencies). The typical attack methods, features of the attack that can be used to detect the anomalies, locations where anomalies can be detected, the data used for detection and the type of analysis for anomaly detection are broadly described in Table 1 [18, 19].

Anomaly detection [20–26] forms the first line of defense against unauthorized users and deliberate intruders. This can be done by continuously monitoring user activities and by detecting patterns of behavior that are potentially harmful to the network. Recent developments in big data technology [27] allow the ability to store large amounts of historical user activity data so that one can visit past instances in order to analyze and learn about threat patterns and behaviors. Further, due to the relatively recent and dramatic changes in the field, it is often hard to decide exactly what kind of information needs to be stored. As a result, any available information regarding user activity, such as ethernet source address, ethernet destination address, internet protocol (IP) source address, IP destination address, geographical information of the subscriber, etc. are collected and their relevance to security threats needs to be accessed online. Hence, the first challenge in anomaly detection is *feature selection*, i.e., deciding which features give the most information for effective detection.

Feature selection and data reduction have been the focus of extensive research in the past [28–31]. Studies of data reduction techniques broadly categorize them into *linear* and *nonlinear methods* [31–33]. Linear data reduction techniques include principal component analysis (PCA) [34], partial least squares (PLS) [35], independent component analysis [36, 37], factor analysis [29, 33], and random projections [38]. Nonlinear data reduction methods include locally linear embedding [39], principal curves [40], multifactor dimensionality reduction [41, 42], multidimensional scaling [43], and self-organizing maps [44, 45].

The second challenge in anomaly detection comes from the transactions that are event triggered and hence are asynchronous in nature. Also, the observed features are *asymmetrical*, i.e., not all the features are observed during the same sampling time. Two possible approaches for online monitoring of asynchronous transactions are *time windowing* and *time interval modeling*. In time windowing, the incoming data from a constant time window is processed to make one snapshot of data corresponding to that time window; the time window is defined in a way that all the features can be captured. This approach is relatively easy to implement and is incorporated in this chapter. In time interval modeling, the anomaly detection scheme needs to be equipped to handle the time difference between two subsequent observations through time series modeling of each individual features [46].

The third challenge for online anomaly detection is the *anomaly detection* process itself: How can one perform anomaly detection in millions of devices by continuously observing thousands of features at the same time? The focus of this chapter is to describe an approach based on big data processing and computational intelligence. Big data processing capabilities [27, 47–49] allow one to process large amounts of

Attack phase	Attack methods	Attack features	Detection location	Data used for detection	Anomaly detection, pattern recognition and correlation techniques
Reconnaissance	Methods to obtain information about the structure of the network, public services and service personnel (e.g., domain names, profiles)	Port scans; automatic browsing of websites; connections from unlikely network segments	Demilitarized zone (DMZ) and network border	Firewall logs and web server logs	Use sequences; time-based traffic statistics
Gain access	Phishing; infected websites; malware; security flaws; tainted devices	Attachments; masquerading as a trustworthy entity; vulnerabilities in software	Network border, internal workstations, (web)servers	Virus scanners, firewalls; (mail)proxies; sandboxing	Command sequence analysis; file statistics; rule-based analysis
Internal recon.	Methods to gain knowledge of the attacked network (e.g., malware, rootkit, botnet)	Analysis of network traffic through system tools or through computer usage analysis	All network zones	Network traffic sensors, firewalls, host-based intrusion detection systems (HIDS) and access logs	Command sequence analysis; statistical analysis
Expand access	Methods to obtain privileges at systems; access systems in the network and access network segments (e.g., malware, exploits, rootkit)	Password sniffing or exploit software vulnerabilities to obtain higher privileges on the local system. Network activity to obtain access to other systems on the network	All network zones. Compromised workstations and servers	HIDS and network intrusion detection systems (NIDS)	Time-based analysis; content rules; statistical analysis
Gather info.	Methods to locate info. and services of interest (e.g., malware, botnet)	Network browsing and accessing locations uncommon for the used identities; incorrect program signatures	All network zones. Compromised workstations and servers	HIDS, NIDS	Use statistics; rule-based analysis
Extract info.	Malware, botnet	Network traffic to unlikely internet segments	All network zones. Compromised workstations and servers	HIDS, NIDS	Statistical analysis; rule-based analysis
Malware control	Botnet	Network traffic to and from unlikely internet segments	All network zones	Firewalls, traffic sensors, proxy servers	Statistical analysis; rule-based analysis
Erase tracks	Botnet	Overly large amount of "bad" traffic or other known means of obfuscating traffic and illegal activities	Compromised workstations and servers	Virus/malware scanners, firewalls, traffic sensors	File statistics; rule-based analysis

data at once. Computational intelligence methods allow us to employ "big data" processing capabilities to extract features that are themselves indicative of contextual anomalies (and hence indicative of intrusions and threats). The online anomaly detection strategy that is detailed in this chapter works in two domains—in *space* and *time*—as summarized below:

- *Anomaly detection and learning in space*. The objective here is to learn model parameters for normal and abnormal behaviors by exploiting the big data infrastructure that allows one to store and process large swaths of information. First, the entire data belonging to a short period of time is analyzed (*burst learning*) to decide which features and attributes need to be monitored (feature selection) and what level of data abstraction needs to be performed (processing asynchronous data). Then, *static anomaly detection* is performed and the model parameters corresponding to normal and abnormal behavior are estimated (model learning).
- *Online anomaly detection in time*. The estimated model parameters belonging to the normal and abnormal behavior during the preceding burst of learning is used to monitor the ongoing transactions using an iterated sequential probability ratio test (Page's test), which raises an alarm whenever the likelihood of anomalies in the incoming data exceeds a threshold.

Our proposed anomaly detection concept is demonstrated using a PCA based anomaly detection scheme [50] on application level features collected from video on demand systems. However, the same concept is easily extendable to employ any type of computational intelligence methods for online anomaly detection in large scale CPHS.

Figure 2 summarizes how the rest of this chapter is organized. Section 2 describes how the raw data in many different formats can be digested into some meaningful, numerical format. To demonstrate this, we introduce data from video on demand services which have many of the challenges described earlier. Section 3 describes how a small burst of data from all the devices is used for learning models for normal and abnormal behaviors. Section 4 describes how those models can be used in individual devices to quickly detect emerging anomalies. Section 5 presents numerical results and the chapter is concluded in Sect. 6.

2 Data Abstraction Methods

Figures 3 and 4 describe typical events collected from a video access device [51]. It is important to note that the most of the events occur in an asynchronous manner as a result of interactions from those who use/operate the device. Further, the frequency of each event varies significantly creating asynchronism in the data. An effective anomaly detection strategy requires (i) Accurate enough models reflecting the normal as well as abnormal nature of the device, (ii) Algorithms that are able to exploit these models for quickly detecting "emerging anomalies" buried in noise.

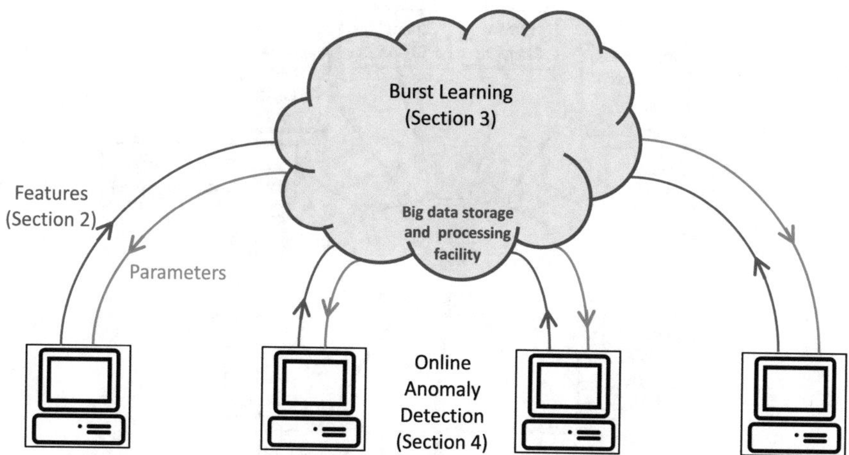

Fig. 2 Overview of the chapter. The objective is to perform anomaly detection on a large number of similar devices that are connected on the internet, such as cable modems and digital video recorders (DVR). Section 2 describes some methods to digest data from individual devices; Sect. 3 describes an approach (that is named *burst learning*) to perform anomaly detection using all the data from all the devices at once and obtain the required parameters for online anomaly detection. Section 4 describes how parameters from burst learning are used for fastest detection of potential threats as they emerge

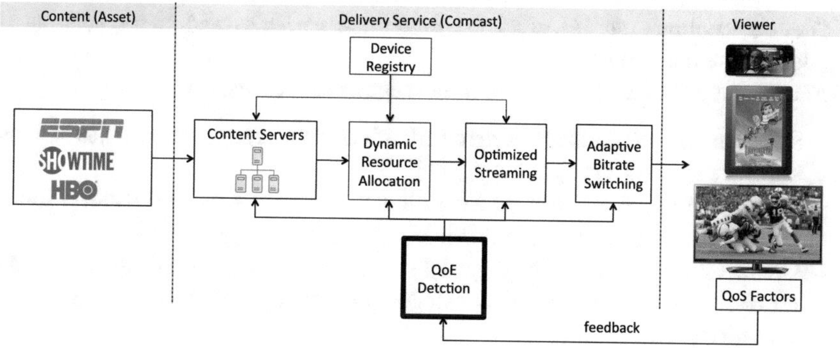

Fig. 3 Video on demand (VoD) system. A specific example of a cyber-physical-human system

The nature and volume of online transactions surpasses the ability to decide which features are needed to be stored for anomaly detection. As a result, all the possible features are streamlined for analysis: the types of data may include numbers, symbols (e.g., IP address), text, sentences (e.g., synopsis), etc. The first objective is to process these data and transform it into a matrix format. We call this the "data abstraction" process.

Let us consider the data collected from the ith device during a time interval $[t_1, t_k]$ for abstraction:

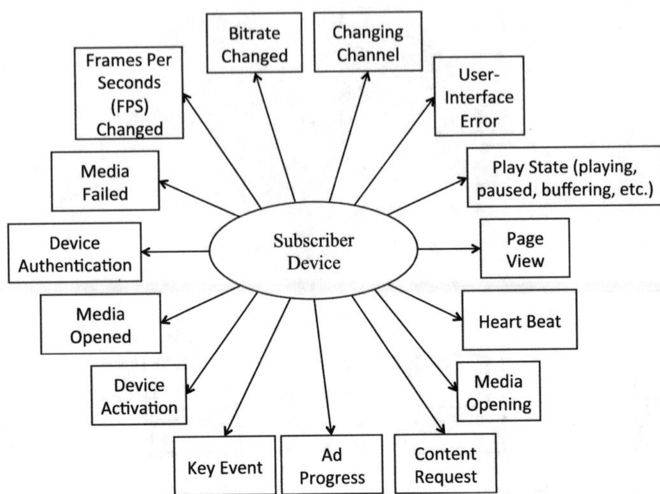

Fig. 4 Example of features of a connected device. An important characteristic of these features is the asynchronous nature of the observations

- Each device can "emit" up to n different events (see Fig. 4). Some examples of events are "communication bit-rate", "video frame rate", "device IP address", "device location", etc. Most of the events are variables and some, such as IP address, are quasi-stationary.
- Each event can take values in different ranges and formats, e.g.,
  - Communication bit-rate (say, jth event) of the device i at time t, $x_{i,j}(t)$, is in mega bits per seconds (mbps) and $x_{i,j}(t) \in [0, 7]$.
  - Location (say, $j+1$th event) of the ith device at time t, $x_{i,j+1}(t)$ denotes longitudes and latitudes both of which are in degrees, i.e., $x_{i,j}(t) \in \mathbb{R}^2$.
  - Genre of the video being played by the device (say, $j+2$th event) of the ith device at time t, $x_{i,j+2}(t)$ is categorical, e.g., $x_{i,j+2}(t) \in \{\texttt{drama}, \texttt{action}, \texttt{comedy}, \texttt{etc.}\}$.
- The event values are recorded against time $t_1, t_2, \ldots, t_k$ asynchronously, i.e., $t_k - t_{k-1} \neq t_{k-1} - t_{k-2}$.
- The asynchronous nature of the observations is indicated through "×" in grey background, which indicates that not all entries of a row are available at the same time.
- Each column in (1) denotes a particular event.
- Each row in (1) denotes the set of values for all n events observed at time t_k.

$$
\mathbf{X}_{\text{raw}}(t) = \begin{bmatrix} x_{i,1}(t_1) & \times & \cdots & x_{i,j}(t_1) & \cdots & x_{i,n-1}(t_1) & x_{i,n}(t_1) \\ x_{i,1}(t_2) & x_{i,2}(t_2) & \cdots & \times & \cdots & x_{i,n-1}(t_2) & \times \\ \times & x_{i,2}(t_3) & \cdots & \times & \cdots & x_{i,n-1}(t_3) & \times \\ \vdots & \vdots & \cdots & \vdots & \cdots & \vdots & \vdots \\ \times & x_{i,2}(t_{k-1}) & \cdots & x_{i,j}(t_{k-1}) & \cdots & \times & x_{i,n}(t_{k-1}) \\ \times & \times & \cdots & x_{i,j}(t_k) & \cdots & x_{i,n-1}(t_k) & \times \end{bmatrix} \begin{matrix} \leftarrow \\ \leftarrow \\ \leftarrow \\ \\ \leftarrow \\ \leftarrow \end{matrix} \begin{array}{|c|} \hline t_1 \\ \hline t_2 \\ \hline t_3 \\ \hline \vdots \\ \hline t_{k-1} \\ \hline t_k \\ \hline \end{array} \tag{1}
$$

Another important objective of the data abstraction process is to summarize the observations over the time window $[t_1, t_k]$ into as few parameters as possible while retaining as much information as possible. The following approaches are useful tools in this regard:

- *Mean, median and variance*: The jth event over the time window $[t_1, t_k]$ is summarized/abstracted by its mean and variance during that period.
- *Histograms and Markov chains*: These are particularly useful to abstract non-numerical outputs, such as text and categorical variables.
- *Hidden Markov models (HMM)*: This approach is well-suited to capture hidden, stochastic patterns in temporal data.

After data abstraction, (1) is compressed into the following form

$$
\mathbf{x}_i^T(t) = \left[x_{i,1}(t), \ldots, x_{i,j}(t), \ldots, x_{i,N-1}(t), x_{i,N}(t)\right] \tag{2}
$$

where t is an index representing the observation time block $[t_1, t_k]$ and $x_{i,j}(t)$ are the *abstracted* features.

Now, let us write the abstracted event data from all the devices in the following form:

$$
\mathbf{X}(t) = \left[\mathbf{x}_1(t), \mathbf{x}_2(t), \ldots, \mathbf{x}_M(t)\right]^T \tag{3}
$$

Usually $M \gg N$ in big data applications.

3 Burst Learning

Online anomaly detection strategies presented in this chapter require the knowledge of models belonging to normal and abnormal characteristics of the data. The big data framework allows us to have data belonging to a small (but the same) time duration of all the devices for learning. This concept is described in Fig. 5. Succinctly, the proposed online anomaly detection strategy works as follows:

1. *Point anomaly detection*: Use data from all the devices during time block k (shown in grey in Fig. 5) to learn model parameters θ_0 that corresponds to the normal data and θ_1 corresponding to the abnormal data.

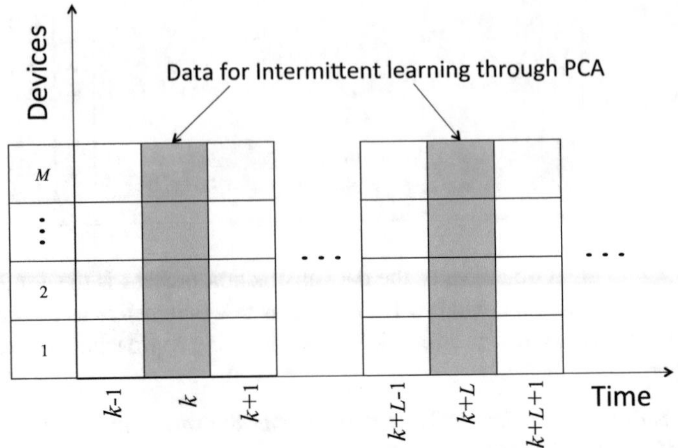

Fig. 5 Model hypothesis learning. Short bursts of data from all the devices in time block (k) is used to train models for normal and abnormal hypotheses

2. *Online anomaly detection*: Use the model parameters $\{\theta_0, \theta_1\}$ for online anomaly detection in each device (Sect. 4).
3. Repeat (1) periodically for updating the parameters.

In this section we discuss how the model parameters θ_0 and θ_1 corresponding to normal and abnormal data, respectively, can be obtained from a snapshot of data $\mathbf{X}(k)$ from all the devices during time duration k. We remove the time index k for convenience in this subsection.

First, we focus on data reduction where the number of columns N in $\mathbf{X}$ is reduced. This is achieved through principal component analysis (PCA).

As a preparation for PCA, each column of matrix $\mathbf{X}$ is scaled so that each component has zero mean and unit variance. This can be achieved by performing the following operation on each column of matrix $\mathbf{X}$

$$\underline{\mathbf{x}}_i^n \leftarrow \frac{\underline{\mathbf{x}}_i - \mu_i}{\sigma_i} \tag{4}$$

where $\underline{\mathbf{x}}_i$ is the ith column of $\mathbf{X}$, μ_i is the mean of $\underline{\mathbf{x}}_i$ and σ_i is the standard deviation of $\underline{\mathbf{x}}_i(k)$.

The $N \times N$ covariance matrix is computed as

$$\mathbf{R} = \frac{1}{M-1}\mathbf{X}^T\mathbf{X} \tag{5}$$

Performing eigendecomposition on $\mathbf{R}$ will result in

$$\mathbf{R} = \mathbf{V}\Lambda\mathbf{V}^T \tag{6}$$

The first L eigenvectors are selected to form the following *loading matrix*:

$$\mathbf{P} = [\underline{\mathbf{v}}_1, \underline{\mathbf{v}}_2, \ldots, \underline{\mathbf{v}}_L] \tag{7}$$

Now, let us define $\mathbf{T}$ as

$$\mathbf{T} = \mathbf{XP} \tag{8}$$

where the $M \times L$ matrix $\mathbf{T} = [\underline{\mathbf{t}}_1, \underline{\mathbf{t}}_2, \ldots, \underline{\mathbf{t}}_N]$ is known as the *score matrix* and each of column of $\mathbf{T}$, $\underline{\mathbf{t}}_1, \underline{\mathbf{t}}_2, \ldots, \underline{\mathbf{t}}_N$, are known as the *score vectors*.

It must be noted that each row of $\mathbf{T}$ can be written as

$$\underline{\mathbf{t}}_i^T = \underline{\mathbf{x}}_i^T \mathbf{P} \tag{9}$$

Let us post-multiply (8) by $\mathbf{P}^T$

$$\begin{aligned} \mathbf{TP}^T &= \mathbf{XPP}^T \\ &= \hat{\mathbf{X}}\left(\approx \mathbf{X} \right) \end{aligned} \tag{10}$$

Hence, the data matrix $\mathbf{X}$ can be written as

$$\begin{aligned} \mathbf{X} &= \mathbf{TP}^T + \mathbf{E} \\ &= \sum_{j=1}^{L} \underline{\mathbf{t}}_j \underline{\mathbf{p}}_j^T + \mathbf{E} \end{aligned} \tag{11}$$

where the error $\mathbf{E}$ is given by

$$\mathbf{E} = \mathbf{X} - \hat{\mathbf{X}} \tag{12}$$

3.1 Estimating the Number of PCs

It must be noted that as L increases, the error $\mathbf{E}$ will decrease. The number L is an important tuning parameter as it determines the amount of variance captured by the loading matrix $\mathbf{P}$. There are many approaches for selecting the value of L.

1. By graphically ordering the eigenvalues and selecting a cut off at the *knee* [52].
2. Based on cumulative percentage variance (*CPV*) explained by $\hat{\mathbf{X}}$ [53], which is a measure of the percentage of variance captured by the first L principal components:

$$CPV(L) = \frac{\sum_{i=1}^{L} \lambda_i}{\mathrm{trace}(\mathbf{R})} 100\% \tag{13}$$

This second approach gives a way of capturing the exact percentage of variability in the data. However, this requires the computation of trace $(\mathbf{R})$.

3.2 Anomaly Detection Using Hotelling's Statistic

Anomaly detection can be done by computing the *Hotelling's* T^2 statistic [54] for data from each device, $\mathbf{x}_i^T$, as follows:

$$T_i^2 = \mathbf{x}_i^T \mathbf{P} \Lambda_L^{-1} \mathbf{P}^T \mathbf{x}_i \tag{14}$$

The anomalies are then declared as follows:

$$\text{device } i \begin{cases} \text{normal (Hypothesis } j = 0,\ H_0) & T_i^2 \leq T_\alpha \\ \text{abnormal (Hypothesis } j = 1,\ H_1) & T_i^2 > T_\alpha \end{cases} \tag{15}$$

where T_α^2 is a threshold. Setting a threshold T_α^2 often requires expert knowledge about the nature of anomalies along with intimate familiarity with the data.

3.3 PCA in "Big Data" Using NIPALS

When the size of $\mathbf{X}$ is very large, computing $\mathbf{R}$ adds significant computational burden. The following NIPALS PCA procedure allows one to compute a small number of principal components recursively and with relatively little computational load.

$[\mathbf{P}, \mathbf{T}] =$ **NIPALS-PCA**$\left(\mathbf{X}, L\right)$

---

1. $k = 0$
2. Set $k = k + 1$
   Initialize:
   $\underline{\mathbf{t}}_k^{(i)} =$ any column of $\mathbf{X}$
   Iterate:
   Set $i = i + 1$
   $\underline{\mathbf{v}}_k^{(i)} = \mathbf{X}^T \underline{\mathbf{t}}_k^{(i-1)} / \|\mathbf{X}^T \underline{\mathbf{t}}_k^{(i-1)}\|$
   $\underline{\mathbf{t}}_k^{(i)} = \mathbf{X} \underline{\mathbf{v}}_k^{(i)}$
   until $\|\underline{\mathbf{v}}_k^{(i)} - \underline{\mathbf{v}}_k^{(i-1)}\| < \epsilon$
   $\underline{\mathbf{t}}_k \leftarrow \underline{\mathbf{t}}_k^{(i)}$
   $\underline{\mathbf{v}}_k \leftarrow \underline{\mathbf{v}}_k^{(i)}$
3. Set $\mathbf{X} \leftarrow \mathbf{X} - \underline{\mathbf{t}}_k^{(i)} \underline{\mathbf{v}}_k^{(i)^T}$
   if $k \leq L$ go to step 2

4. Obtain
   $\mathbf{T} = [\underline{\mathbf{t}}_1, \underline{\mathbf{t}}_2, \ldots, \underline{\mathbf{t}}_k]$
   $\mathbf{V} = [\underline{\mathbf{v}}_1, \underline{\mathbf{v}}_2, \ldots, \underline{\mathbf{v}}_k]$

Now, the Hoteling's statistics are obtained as follows:

$$T_i^2 \approx \underline{\mathbf{t}}_i^T \tilde{\Lambda}_L^{-1} \underline{\mathbf{t}}_i \tag{16}$$

where $\underline{\mathbf{t}}_i^T$ is the ith row of $\mathbf{T}$ and $\tilde{\Lambda}_L$ is a diagonal matrix formed by the diagonal elements of $\mathbf{T}^T\mathbf{T}$.

It must be noted that the NIPALS-PCA approach described above has its disadvantages: It is sensitive to initialization and it becomes unreliable as the number of principal components that need to be computed increase. There are other efficient ways to compute PCA [55] and some others are reported in [56].

3.4 Model Parameter Learning

For fast, online anomaly detection, we need models that represent normal as well as abnormal behaviors. This is an ongoing and challenging research problem. As an initial step, we investigated multivariate Gaussian models for this purpose [57]. A more general approach is to employ with a Gaussian mixture model: one each for normal as well as abnormal data. This requires one to employ the expectation maximization algorithm in a distributed form. Alternatively, a simpler approach is to use clustering to divide the data into groups and then employ multivariate Gaussian modeling for each group.

First, we form a group of data $\mathbf{X}_j$ (which are the outputs of a clustering algorithm)

$$\mathbf{T}_j = \mathbf{X}_j P \tag{17}$$

where each row of $\mathbf{X}_j$ is an observation corresponding to the jth cluster.

Now, the jth cluster can be represented by $\mathcal{N}\left(\boldsymbol{\mu}_j, \Sigma_j\right)$ where

$$\boldsymbol{\mu}_j = \frac{\mathbf{1}^T\mathbf{T}_j}{M_j} \tag{18}$$

$$\Sigma_j = \mathbf{T}_j \tilde{\Lambda}_j^{-1} \mathbf{T}_j^T \tag{19}$$

and $\tilde{\Lambda}_j$ is a diagonal matrix formed by the diagonal elements of $\mathbf{T}_j^T\mathbf{T}_j$.

4 Online Anomaly Detection

In this section, we describe the online anomaly detection approach assuming the models learned in Sect. 3. For more details, the reader is referred to [57].

4.1 Batch Detection

Given a series of vector outputs $\mathbf{x}_i(k),\quad k = 1, 2, \ldots, K$ from the ith device, it is desired to (quickly) detect if the device is turning into an abnormal one. This can be posed as the following hypothesis testing problem:

$$\begin{aligned} H_0 &: \mathbf{x}(k) \sim P_{\theta_0} & 1 \le k \le K \\ H_1 &: \mathbf{x}(k) \sim P_{\theta_0} & 1 \le k \le k^* - 1 \\ & \quad \mathbf{x}(k) \sim P_{\theta_1} & k^* \le k \le K \end{aligned} \tag{20}$$

where $\mathbf{x}_i(k)$ is written without the subscript i for convenience in this section.

Assuming that the change occurred in time k^*, the likelihood ratio between the hypotheses H_0 and H_1 is written as

$$\Lambda_{k^*}^K = \frac{\prod_{k=1}^{k^*-1} P_{\theta_0}(\mathbf{x}(k)) \prod_{k=k^*}^{K} P_{\theta_1}(\mathbf{x}(k))}{\prod_{k=1}^{K} P_{\theta_0}(\mathbf{x}(k))} \tag{21}$$

Now, (21) is written in the form of log-likelihood ratio as

$$S_{k^*}^K = \sum_{k=k^*}^{K} \ln \frac{P_{\theta_1}(\mathbf{x}(k))}{P_{\theta_0}(\mathbf{x}(k))} \tag{22}$$

Hence, based on a batch of K observations, the exact time of change is detected as follows

$$\hat{k}^* = \arg \max_{1 \le k^* \le K} S_{k^*}^K \tag{23}$$

Batch detection is not suitable for anomaly detection, because every time new data arrives, the batch-computation of $S_{k^*}^K$ has to be repeated. Hence, a recursive way of anomaly detection is desired. Next, we summarize *Page's test*, which is an approximate way of performing the anomaly detection recursively.

4.2 Page's Test

Consider the following log-likelihood ratio

$$S_{k^*} = \sum_{k=1}^{k^*} \ln \frac{P_{\theta_1}(\mathbf{x}(k))}{P_{\theta_0}(\mathbf{x}(k))} \tag{24}$$

where S_{k^*} can be incrementally updated as new data arrives, and an anomaly is declared when

$$S_{k^*} - m_{k^*} > h \tag{25}$$

where

$$m_{k^*} = \min_{1 \leq k \leq k^*} S_k \tag{26}$$

and h is a predefined threshold value.

Formally, the above anomaly detection time is written as

$$\hat{k}^* = \arg\min_{k^*} \left\{ k^* : S_{k^*} - m_{k^*} > h \right\} \tag{27}$$

Based on the key idea from Page [58], (27) can be recursively computed as follows

$$\text{CUSUM}_k = \max \left\{ 0, \text{CUSUM}_{k-1} + T_k \right\} \tag{28}$$

where

$$T_k = \ln \frac{P_{\theta_1}(\mathbf{x}(k))}{P_{\theta_0}(\mathbf{x}(k))} \tag{29}$$

and anomaly is declared when CUSUM_k exceeds the threshold h.

It must be noted that, for the model in Sect. 3.4-(19)

$$T_k = \frac{1}{2} \left(\mathbf{x}_k - \boldsymbol{\mu}_0 \right)^T \mathbf{R}_0^{-1} \left(\mathbf{x}_k - \boldsymbol{\mu}_0 \right) - \frac{1}{2} \left(\mathbf{x}_k - \boldsymbol{\mu}_1 \right)^T \mathbf{R}_1^{-1} \left(\mathbf{x}_k - \boldsymbol{\mu}_1 \right) + \log \left(\frac{|\mathbf{R}_0|}{|\mathbf{R}_1|} \right) \tag{30}$$

4.3 Shiryaev's Test

Shiryaev [59] proposed a Bayesian approach for change detection, which, we summarize in this subsection.

First, the *prior probability* of change towards H_1 is assumed to be geometric, i.e.,

$$\pi_0^1 = \rho(1-\rho)^{k-1} \quad \text{for } k > 0 \tag{31}$$

The probability of hypotheses at time k given the hypothesis at time $k-1$ is written as a Markov chain with the following transition matrix

$$P = \begin{pmatrix} p(H_0|H_0) & p(H_0|H_1) \\ p(H_1|H_0) & p(H_1|H_1) \end{pmatrix} = \begin{pmatrix} 1-\rho & 0 \\ \rho & 1 \end{pmatrix} \tag{32}$$

Based on Bayes rule, the *posterior probability* of change towards H_1 at time k can be written as follows:

$$\pi_k^1 = \frac{\pi_{k-1}^1 P_{\theta_1}(\mathbf{x}(k)) + (1-\pi_{k-1}^1)\rho P_{\theta_1}(\mathbf{x}(k))}{\pi_{k-1}^1 P_{\theta_1}(\mathbf{x}(k)) + (1-\pi_{k-1}^1)\rho P_{\theta_1}(\mathbf{x}(k)) + (1-\pi_{k-1}^1)(1-\rho)P_{\theta_0}(\mathbf{x}(k))} \tag{33}$$

The expression in (33) can be simplified as follows

$$\omega_k^1 = \frac{1}{1-\rho}(\omega_{k-1}^1 + \rho)T_k \tag{34}$$

where ω_k^1 is a monotonic function of π_k^1

$$\omega_k^1 = \frac{\pi_k^1}{1-\pi_k^1} \tag{35}$$

The recursive formula in (34) can be written in log-likelihood form as follows

$$g_k^1 = \ln(\rho + e^{g_k^1 - 1}) - \ln(1-\rho) + T_k \tag{36}$$

where

$$g_k^1 = \ln(\omega_k^1) \tag{37}$$

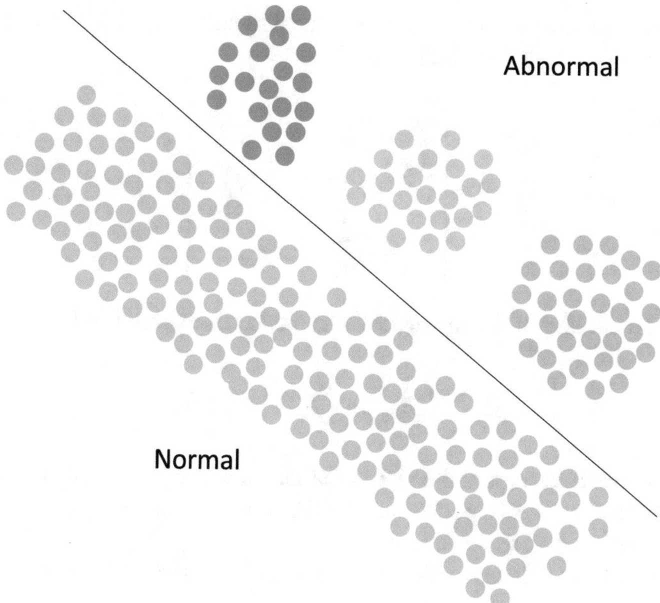

Fig. 6 Multiple anomalies. The objective of the online anomaly detection is to identify the type of anomaly

Finally, the change is declared when g_k^1 exceeds a threshold h and the change time estimate is given as

$$\hat{k}^* = \arg\min_k (g_k^1 > h) \tag{38}$$

4.4 Tagging Multiple Anomalies

Due to the large amount of data, it becomes infeasible to absorb the entire set of anomalies into one single model. As a result, the detected anomalies are clustered into several groups (see Fig. 6) and model parameters of each group are learned.

The abnormal groups are then ranked, based on expert knowledge. The objective here is to detect the type of anomaly online. This can be done by extending the Shiryaev's approach, described in Sect. 4.3.

First, let us assume that initial state of each device is denoted by the prior probabilities $\pi_0 = \left[\pi_0^0, \pi_0^1, \ldots, \pi_0^{N_a}\right]^T$, where π_0^0 correspond to the normal model and π_0^j corresponds to the jth abnormal mode.

Now, let us assume that the change in model hypothesis can be characterized by the following transition probability matrix

$$P = \begin{bmatrix} p_{0,0} & \frac{1-p_{1,1}}{N_a} & \cdots & \frac{1-p_{N_a,N_a}}{N_a} \\ \frac{1-p_{0,0}}{N_a} & p_{1,1} & \cdots & \frac{1-p_{N_a,N_a}}{N_a} \\ \vdots & \vdots & \ddots & \vdots \\ \frac{1-p_{0,0}}{N_a} & \frac{1-p_{1,1}}{N_a} & \cdots & p_{N_a,N_a} \end{bmatrix} \tag{39}$$

where the probability of staying in the same mode j from time $k-1$ to k is given by

$$p_{j,j} = 1 - \frac{1}{\tau_j}; \quad j = 0, 1, \dots, N_a \tag{40}$$

and τ_j is the expected sojourn time [60] of mode j indicating how long on average the mode j stays active. It is important to note that P above allows a particular anomaly j to be detected as well as reset.

Now, from Bayes' theorem, the mode probability of the nth hypothesis at the kth time epoch can be written as

$$\pi_k^n = \frac{\sum_{i=0}^{N_a} \pi_{k-1}^i p_{n,i} P_{\theta_n}(\mathbf{x}(k))}{\sum_{j=0}^{N_a} \sum_{i=0}^{N_a} \pi_{k-1}^i p_{n,i} P_{\theta_j}(\mathbf{x}(k))} \quad n \in \{0, 1, \dots, N_a\} \tag{41}$$

5 Simulation Results

The data for analysis comes from Comcast's Xfinity player, collected while subscriber *devices* were accessing VOD services. The model learning data is taken by setting the time duration $k = 1$ h.

The following nine features ($N = 9$) are extracted during a 1 h time frame. f_1: number of buffering events, f_2: number of media failed events, f_3: Number of User Interface(UI)-error events, f_4: Number of opening events, f_5: Number of activation events, f_6: Number of authentication events, f_7: Startup time, f_8: Number of IP addresses linked to a device, and f_9: Number of assets a device accessed. These data are collected form $N = 54306$ devices.

Figure 7 shows the extracted features $f_1 - f_9$ from all the devices.

Figure 8 shows a histogram of each feature, corresponding to Fig. 7. The histograms give a good sense of where most of the features lie.

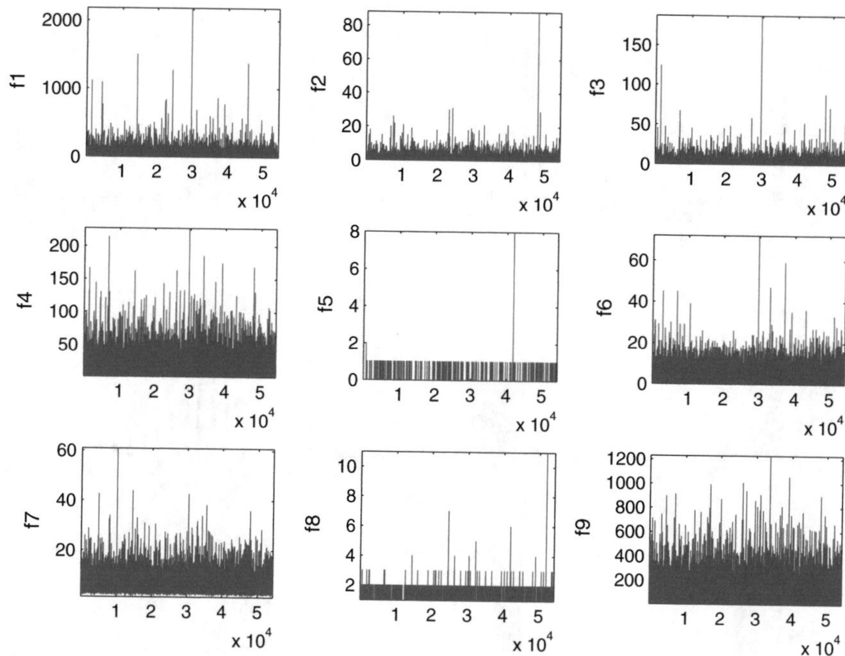

Fig. 7 **Abstracted features**. The data from each device is collected during the same time interval $[t_{k-1}, t_k]$. In each plot, the x axis ranges from 1 to 54306, denoting one of the devices, and y axis denotes the feature f_i

Figure 9 shows the result of feature reduction based on PCA. We selected the number of principal components corresponding to a cumulative percentage of variance (CPV) value of 95%, which resulted in selecting $n_{PC} = 6$ principal components.

The results of anomaly detection and model learning are shown in Fig. 10. The threshold T_α is selected as inverse of the Chi-square cumulative distribution function with n_{PC} degrees of freedom at the 99% confidence level.

Figure 11 shows the features corresponding to detected normal and abnormal features as a box plot.

Now, we demonstrate the online anomaly detection capability through simulations. We assume the number of abnormal modes to be $N_a = 1$ so that other anomaly detection approaches can be compared. The parameters corresponding $H_0(\boldsymbol{\mu}_0, \Sigma_0)$ and $H_1(\boldsymbol{\mu}_1, \Sigma_1)$ are computed based on the initial anomaly detection discussed earlier. A test data stream of length $K = 1000$ is generated as follow: The first 500 data points are generated samples of a multivariate Gaussian distribution of mean $\boldsymbol{\mu}_0$ and covariance Σ_0 and the second 500 data points are generated samples of a multivariate Gaussian distribution of mean $\boldsymbol{\mu}_1$ and covariance Σ_1. Fig. 12 shows the tracking results of all the algorithms discussed in Sect. 4.

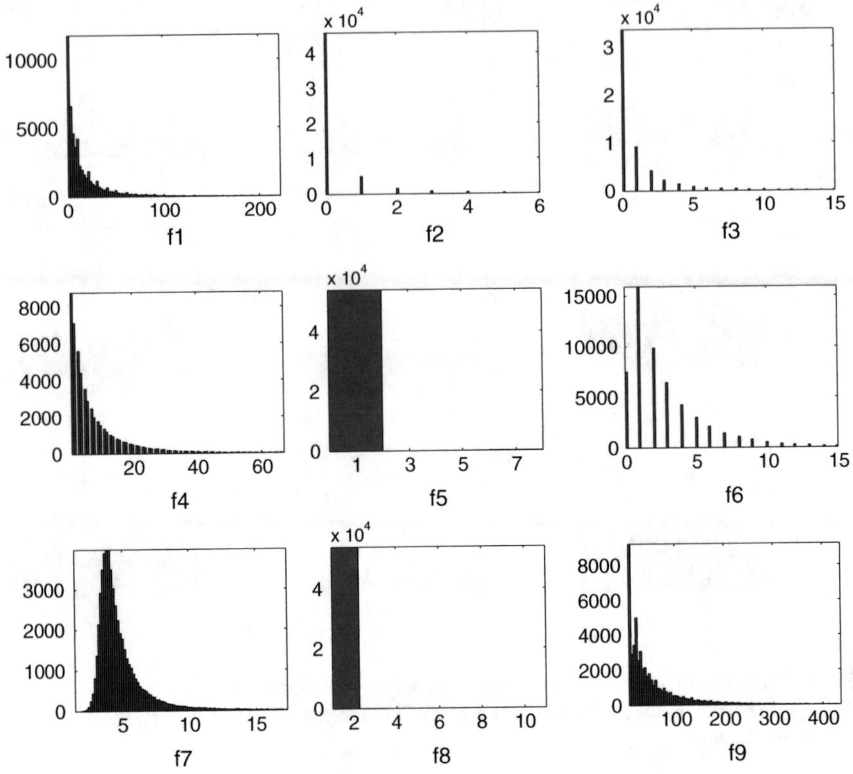

Fig. 8 PMF of the abstracted features. The PMFs correspond to the features in Fig. 7

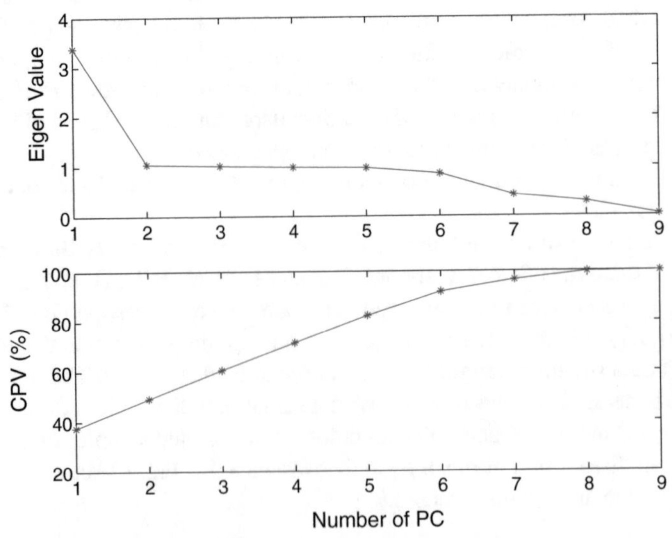

Fig. 9 Selection of the number of principal components

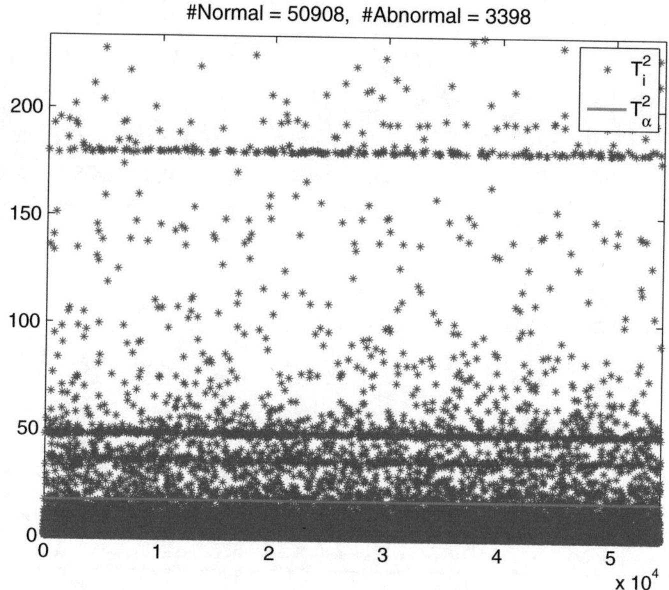

Fig. 10 Anomaly detection using Hotelling's statistics

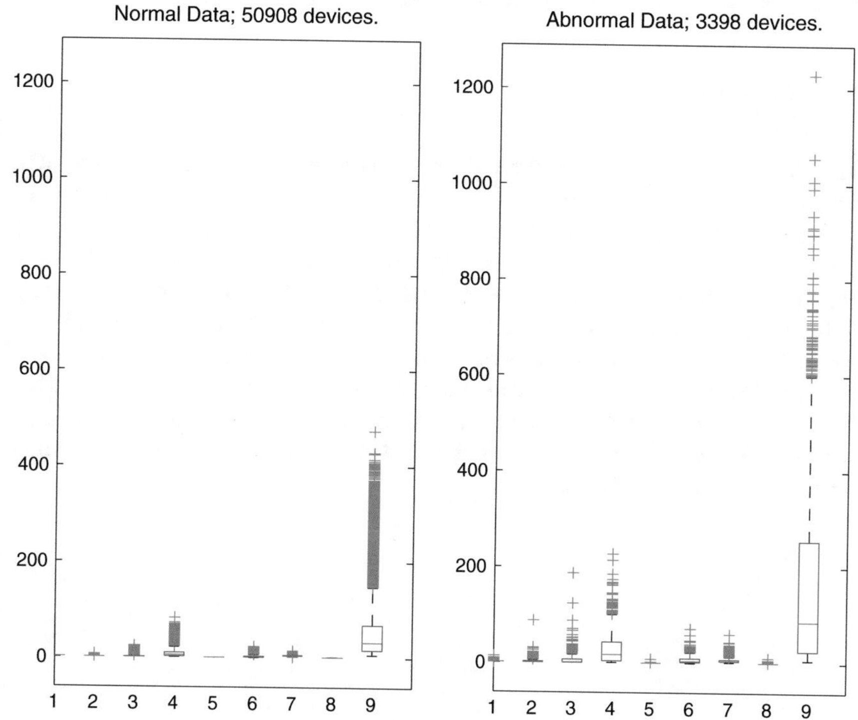

Fig. 11 Features corresponding to detected normal and abnormal devices

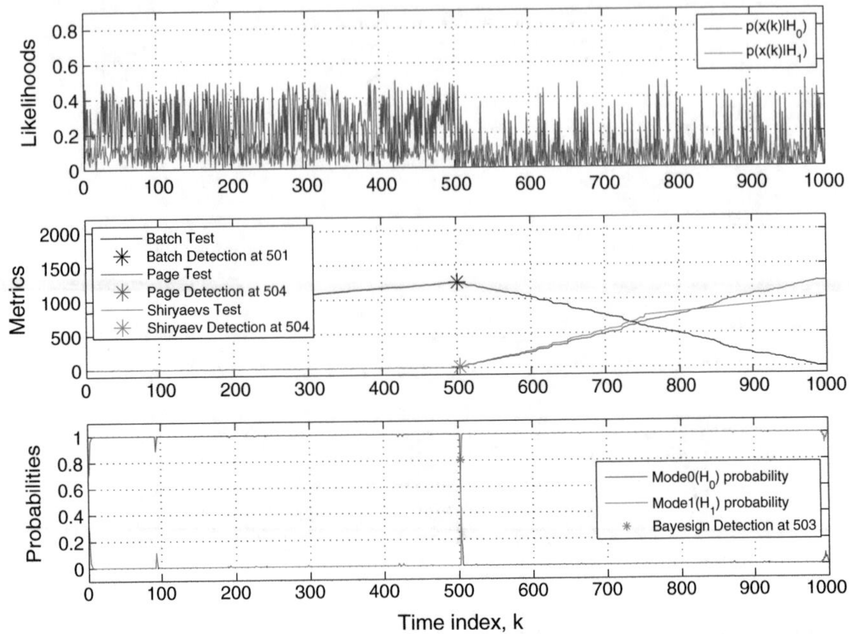

Fig. 12 Demonstration of online anomaly detection by comparing all three approaches

6 Conclusions

This chapter described an anomaly detection strategy applicable to a large number of intelligent devices (specifically, millions of cable modems and digital video recorders) connected to the Internet. The proposed anomaly detection scheme exploits the availability of distributed, big data processing for the learning of model hypotheses—from short bursts of data from all devices. Relevant parameters obtained through learning from a small interval of data (through centralized processing) are passed over to each connected device for online anomaly detection. The online anomaly detection strategy is set up as a likelihood ratio test, also known as the Page's test. Each connected device is equipped with the online anomaly detection capability, as such emerging threats can be quickly detected and the affected devices before multitudes of devices are affected.

References

1. H.-J. Liao, C.-H. R. Lin, Y.-C. Lin, and K.-Y. Tung, "Intrusion detection system: A comprehensive review," *Journal of Network and Computer Applications*, vol. 36, no. 1, pp. 16–24, 2013.
2. M. Shetty and N. Shekokar, "Data mining techniques for real time intrusion detection systems," *International Journal of Scientific & Engineering Research*, vol. 3, no. 4, 2012.
3. C. Kolias, G. Kambourakis, and M. Maragoudakis, "Swarm intelligence in intrusion detection: A survey," *computers & security*, vol. 30, no. 8, pp. 625–642, 2011.
4. S. Shin, S. Lee, H. Kim, and S. Kim, "Advanced probabilistic approach for network intrusion forecasting and detection," *Expert Systems with Applications*, vol. 40, no. 1, pp. 315–322, 2013.
5. S. X. Wu and W. Banzhaf, "The use of computational intelligence in intrusion detection systems: A review," *Applied Soft Computing*, vol. 10, no. 1, pp. 1–35, 2010.
6. L. Akoglu, H. Tong, and D. Koutra, "Graph based anomaly detection and description: a survey," *Data Mining and Knowledge Discovery*, vol. 29, no. 3, pp. 626–688, 2015.
7. G. Stringhini, C. Kruegel, and G. Vigna, "Detecting spammers on social networks," in *Proceedings of the 26th Annual Computer Security Applications Conference*, pp. 1–9, ACM, 2010.
8. D. Savage, X. Zhang, X. Yu, P. Chou, and Q. Wang, "Anomaly detection in online social networks," *Social Networks*, vol. 39, pp. 62–70, 2014.
9. W. Xu, F. Zhang, and S. Zhu, "Toward worm detection in online social networks," in *Proceedings of the 26th Annual Computer Security Applications Conference*, pp. 11–20, ACM, 2010.
10. P. Chen, L. Desmet, and C. Huygens, "A study on advanced persistent threats," in *IFIP International Conference on Communications and Multimedia Security*, pp. 63–72, Springer, 2014.
11. D. Kushner, "The real story of stuxnet," *ieee Spectrum*, vol. 3, no. 50, pp. 48–53, 2013.
12. Symantec, "Symantec internet security threat report," tech. rep., Symantec, 2011.
13. Fox-IT, "Interim report, diginotar cert authority breach," tech. rep., Fox-IT Business Unit Cybercrime, Delft, 2011.
14. U. Rivner, "Anatomy of an attack".
15. N. Villeneuve, J. T. Bennett, N. Moran, T. Haq, M. Scott, and K. Geers, *Operation" Ke3chang: Targeted Attacks Against Ministries of Foreign Affairs*. 2013.
16. D. Kindlund, X. Chen, M. Scott, and N. D. Moran, Ned anMoran, "Operation snowman: Deputydog actor compromises us veterans of foreign wars website," 2014.
17. "https://en.wikipedia.org/wiki/malware".
18. E. M. Hutchins, M. J. Cloppert, and R. M. Amin, "Intelligence-driven computer network defense informed by analysis of adversary campaigns and intrusion kill chains," *Leading Issues in Information Warfare & Security Research*, vol. 1, p. 80, 2011.
19. C. Tankard, "Advanced persistent threats and how to monitor and deter them," *Network security*, vol. 2011, no. 8, pp. 16–19, 2011.
20. L. Huang, X. Nguyen, M. Garofalakis, M. I. Jordan, A. Joseph, and N. Taft, "In-network PCA and anomaly detection," in *NIPS*, vol. 19, 2006.
21. C. C. Aggarwal, "On abnormality detection in spuriously populated data streams.," in *SDM*, SIAM, 2005.
22. D.-S. Pham, S. Venkatesh, M. Lazarescu, and S. Budhaditya, "Anomaly detection in large-scale data stream networks," *Data Mining and Knowledge Discovery*, vol. 28, no. 1, pp. 145–189, 2014.
23. X. Jiang and G. F. Cooper, "A real-time temporal bayesian architecture for event surveillance and its application to patient-specific multiple disease outbreak detection," *Data Mining and Knowledge Discovery*, vol. 20, no. 3, pp. 328–360, 2010.
24. V. Chandola, A. Banerjee, and V. Kumar, "Anomaly detection: A survey," *ACM Computing Surveys (CSUR)*, vol. 41, no. 3, p. 15, 2009.
25. V. Barnett and T. Lewis, *Outliers in statistical data*, vol. 3. Wiley New York, 1984.

26. A. Koufakou and M. Georgiopoulos, "A fast outlier detection strategy for distributed high-dimensional data sets with mixed attributes," *Data Mining and Knowledge Discovery*, vol. 20, no. 2, pp. 259–289, 2010.
27. T. White, *Hadoop: The Definitive Guide: The Definitive Guide*. O'Reilly Media, 2009.
28. D. J. Hand, "Discrimination and classification," *Wiley Series in Probability and Mathematical Statistics, Chichester: Wiley, 1981*, vol. 1, 1981.
29. K. V. Mardia, J. T. Kent, and J. M. Bibby, "Multivariate analysis (probability and mathematical statistics)," 1980.
30. T. Hastie, R. Tibshirani, J. Friedman, T. Hastie, J. Friedman, and R. Tibshirani, *The elements of statistical learning*, vol. 2. Springer, 2009.
31. S. Singh, S. Ruan, K. Choi, K. Pattipati, P. Willett, S. M. Namburu, S. Chigusa, D. V. Prokhorov, and L. Qiao, "An optimization-based method for dynamic multiple fault diagnosis problem," in *Aerospace Conference, 2007 IEEE*, pp. 1–13, IEEE, 2007.
32. M. A. Carreira-Perpinan, "A review of dimension reduction techniques," *Department of Computer Science. University of Sheffield. Tech. Rep. CS-96-09*, pp. 1–69, 1997.
33. I. K. Fodor, "A survey of dimension reduction techniques," 2002.
34. J. T. Jolliffe, *Principal Component Analysis*. New York: Springer, 2010.
35. R. Bro, "Multiway calidration. multilinear pls," *Journal of Chemometrics*, vol. 10, pp. 47–61, 1996.
36. S. Roberts and R. Everson, *Independent component analysis: principles and practice*. Cambridge University Press, 2001.
37. T.-W. Lee, *Independent component analysis*. Springer, 2010.
38. S. Kaski, "Dimensionality reduction by random mapping: Fast similarity computation for clustering," in *Neural Networks Proceedings, 1998. IEEE World Congress on Computational Intelligence. The 1998 IEEE International Joint Conference on*, vol. 1, pp. 413–418, IEEE, 1998.
39. J. B. Tenenbaum, V. De Silva, and J. C. Langford, "A global geometric framework for nonlinear dimensionality reduction," *Science*, vol. 290, no. 5500, pp. 2319–2323, 2000.
40. T. Hastie and W. Stuetzle, "Principal curves," *Journal of the American Statistical Association*, vol. 84, no. 406, pp. 502–516, 1989.
41. M. D. Ritchie, L. W. Hahn, N. Roodi, L. R. Bailey, W. D. Dupont, F. F. Parl, and J. H. Moore, "Multifactor-dimensionality reduction reveals high-order interactions among estrogen-metabolism genes in sporadic breast cancer," *The American Journal of Human Genetics*, vol. 69, no. 1, pp. 138–147, 2001.
42. M. D. Ritchie, L. W. Hahn, and J. H. Moore, "Power of multifactor dimensionality reduction for detecting gene-gene interactions in the presence of genotyping error, missing data, phenocopy, and genetic heterogeneity," *Genetic epidemiology*, vol. 24, no. 2, pp. 150–157, 2003.
43. M. Vlachos, C. Domeniconi, D. Gunopulos, G. Kollios, and N. Koudas, "Non-linear dimensionality reduction techniques for classification and visualization," in *Proceedings of the eighth ACM SIGKDD international conference on Knowledge discovery and data mining*, pp. 645–651, ACM, 2002.
44. H. Ritter and T. Kohonen, "Self-organizing semantic maps," *Biological cybernetics*, vol. 61, no. 4, pp. 241–254, 1989.
45. T. Kohonen, "The self-organizing map," *Proceedings of the IEEE*, vol. 78, no. 9, pp. 1464–1480, 1990.
46. R. H. Shumway and D. S. Stoffer, *Time series analysis and its applications: with R examples*. Springer Science & Business Media, 2010.
47. K. Singh, S. C. Guntuku, A. Thakur, and C. Hota, "Big data analytics framework for peer-to-peer Botnet detection using random forests," *Information Sciences*, vol. 278, pp. 488–497, 2014.
48. J. Camacho, G. Maciá-Fernández, J. Diaz-Verdejo, and P. Garcia-Teodoro, "Tackling the big data 4 vs for anomaly detection," in *Computer Communications Workshops (INFOCOM WKSHPS), 2014 IEEE Conference on*, pp. 500–505, IEEE, 2014.
49. M. A. Hayes and M. A. Capretz, "Contextual anomaly detection in big sensor data," in *2014 IEEE International Congress on Big Data*, pp. 64–71, IEEE, 2014.

50. B. Balasingam, M. Sankavaram, K. Choi, D. F. M. Ayala, D. Sidoti, K. Pattipati, P. Willett, C. Lintz, G. Commeau, F. Dorigo, *et al.*, "Online anomaly detection in big data," in *Information Fusion (FUSION), 2014 17th International Conference on*, pp. 1–8, IEEE, 2014.
51. D. Pasupuleti, P. Mannaru, B. Balasingam, M. Baum, K. Pattipati, P. Willett, C. Lintz, G. Commeau, F. Dorigo, and J. Fahrny, "Online playtime prediction for cognitive video streaming," in *Information Fusion (Fusion), 2015 18th International Conference on*, pp. 1886–1891, IEEE, 2015.
52. J. E. Jackson, *A user's guide to principal components*, vol. 587. John Wiley & Sons, 2005.
53. D. Zumoffen and M. Basualdo, "From large chemical plant data to fault diagnosis integrated to decentralized fault-tolerant control: pulp mill process application," *Industrial & Engineering Chemistry Research*, vol. 47, no. 4, pp. 1201–1220, 2008.
54. D. Garcıa-Alvarez, "Fault detection using principal component analysis (PCA) in a wastewater treatment plant (wwtp)," in *Proceedings of the International Student's Scientific Conference*, 2009.
55. G. H. Golub and C. F. Van Loan, *Matrix computations*, vol. 3. JHU Press, 2012.
56. Z. Meng, A. Wiesel, and A. Hero, "Distributed principal component analysis on networks via directed graphical models," in *Acoustics, Speech and Signal Processing (ICASSP), 2012 IEEE International Conference on*, pp. 2877–2880, IEEE, 2012.
57. M. Basseville, I. V. Nikiforov, *et al.*, *Detection of abrupt changes: theory and application*, vol. 104. Prentice Hall Englewood Cliffs, 1993.
58. E. Page, "Continuous inspection schemes," *Biometrika*, pp. 100–115, 1954.
59. A. N. Shiryaev, "The problem of the most rapid detection of a disturbance in a stationary process," *Soviet Math. Dokl.*, no. 2, pp. 795–799, 1961.
60. Y. Bar-Shalom, X. R. Li, and T. Kirubarajan, *Estimation with applications to tracking and navigation: theory algorithms and software*. John Wiley & Sons, 2004.

Developing Modified Classifier for Big Data Paradigm: An Approach Through Bio-Inspired Soft Computing

Youakim Badr and Soumya Banerjee

Abstract The emerging applications of big data usher different blends of applications, where classification, accuracy and precision could be identified as major concern. The contemporary issues are also being emphasized as detecting multiple autonomous sources and unstructured trends of data. Therefore, it becomes mandatory to follow suitable classification and in addition to appropriate labelling of data is required to use relevant computational intelligent techniques. This is significant, where the movement of data is random and follows linked concept of data e.g. social network and blog data, transportation data and even supporting low-carbon road transport policies. It has been agreed by the research community whether only supervised classification techniques could be useful for such diversified imbalanced classification. Subsequently, the genesis of majority and minority class detection based on supervised features following conventional data mining principle. However, the classification of majority or positive class is over-sampled by taking each minority class sample. Definitely, significant computationally intelligent methodologies have been introduced. Following the philosophy of data science and big data, the heterogeneous classification, over-sampling, mis-labelled data features cannot be standardized with hard classification. Hence, conventional algorithm can be modified to support ensemble data set for precise classification under big and random data and that can be achieved through proposed monkey algorithm dynamic classification under imbalance. The proposed algorithm is not completely supervised rather it is blended with certain number of pre-defined examples and iterations. The approach could be more specific, when more numbers of soft computing methods, if they can be hybridized with bio-inspired algorithms.

Y. Badr
CNRS, LIRIS INSA, UMR5205, INSA Lyon, Universite de-Lyon, Lyon, France
e-mail: youakim.badr@insa-lyon.fr

S. Banerjee (✉)
CNRS, LIRIS, INSA de Lyon, Lyon, France
e-mail: soumyabanerjee@bitmesra.ac.in

S. Banerjee
Birla Institute of Technology, Mesra, Ranchi, India

W. Pedrycz and S.-M. Chen (eds.), *Data Science and Big Data: An Environment of Computational Intelligence*, Studies in Big Data 24,
DOI 10.1007/978-3-319-53474-9_5

Keywords Classifier · Ensemble classifier · Over-sampling · Sequence classifiers · Pattern coverage rate · Majority class · Bio-inspired algorithm · Monkey climb algorithm · Positive and negative votes · Tweet data

1 Introduction

Big data computing needs advanced technologies or methods to solve the issues of computational time to extract valuable information without information loss. In this context, generally, Machine Learning (ML) algorithms have been considered to learn and find useful and valuable information from large value of data [1]. Combining the different textures of big data, there are methods to address the granular level of big data through granular computing and rough sets [2]. However to address the dynamic big data e.g. on-line social data [3], the use of Fuzzy sets to model raw sentiment with classification probabilities has already been developed. The artifacts and analytics based on α-cut of fuzzy sets to determine whether any given artifact expresses and impresses positive, negative, and/or neutral sentiment. The deployment of fuzzy classification and *Fuzzy Rule Based Classification* Systems found to be an emerging soft computing algorithm for big data and analytics. In addition to, as the data in big data [3] paradigm are highly skewed and hence using certain ensemble classifier could improve the machine learning and fusion process. Each base classifier is trained with a pre-processed data set. As data level approaches towards random values, the pre-processed data sets and the corresponding classifiers will be different. This will also tackle massive random under and over-sampling processes across big data interaction and movements [4]. The proposed model considers a service computing problem in big data scenario to classify the levels of different load requests on demand concerning the opinions collected from tourists in a city at real time. In multiple application domains, one obtains the predictions, over a large set of unlabeled instances, of an ensemble of different experts or classifiers with unknown reliability. Common tasks are combining these possibly conflicting predictions. However, not only the unlabeled instances with variable reliable values are not instrumental to develop the modified classifier, additionally, bio-inspired methods are found to be suitable and appropriate. Here, improved monkey algorithm is applied here to investigate dynamic classification under big data environment.

The classification is the problem of identifying to which of a set of categories a new observation belongs, on the basis of a training set of data containing observations (or instances) whose category membership is known.

The prime consideration of classifier is granularity of space decomposition, number of nearest neighbors and dimensionality. All these features could become more significant, if they are being applied to conventional MapReduce closures. Usually, the set of features is encoded as a binary vector, where each position determines if a feature is selected or not. is allows to perform feature selection with the exploration capabilities of evolutionary algorithms. However, they lack the scalability necessary to address big datasets. We find manifold computational intelligent techniques to perform the feature selection process for developing classifier.

The challenges derived from acquiring and commissioning these consolidated volume of data indicates some merits as well [5]. Categorically, the concept of Data Science evolves from the paradigm of Data Mining environment [6]. These typical categories could be addressed as core Big Data problems [7, 8]. There exist a substantial scalability of Data Mining models, the consolidation of more precise knowledge from data can be easily achieved should be accessible [9]. The conventional MapReduce model [10] has been improvised as an emerging prototype, supporting the development of accurate and scalable applications for present researchers-pertaining to data science. In order to deploy a learning procedure from a set of labeled examples (training data) it becomes mandatory to configure a prototype system, that models the problem space while utilizing their input attributes. Later, when un-known examples (test data) are being accepted by the system, an inference mechanism is evolved to determine the label of the query with respect to their appropriate instance. We investigate several big data supervised techniques that are applied on the imbalanced datasets [11, 12] such as cost-sensitive Random Forest (RF-BDCS), Random Oversampling with Random Forest (ROS + RF-BD), and the Apache Spark Support Vector Machines (SVM-BD) [12] combined with MapReduce ROS (ROS + SVM-BD).

However, the diversification of classification problem always solicits computational intelligence.

The highlight of classification problem is that the examples of one class significantly outnumber the examples of the other one [13]. Definitely, the classification problem also may represent minority class. Practically, those classes may be exceptional and significant cases. Hence, computational cost could be enhanced. In most cases, the imbalanced class problem is associated to binary classification, but the multi-class problem seldom occurs and it will be more challenging, if there exist several minority classes [14].

The suitability of standard learning algorithms normally consider a balanced training set. Therefore, majority classes are well accepted whereas the minority ones are misclassified frequently. These algorithms, which exhibits decent performance for standard classification, do not necessarily achieve the best performance for imbalanced datasets [15]. Research explores several such causes:

- The use of global performance measures for monitoring the learning paradigm, such as the standard accuracy rate, may provide an advantage to the majority class.
- Classification rules which forecast the positive classes are seldom majorly categorized and thus their coverage is very low, hence they are discarded in favor of more general rules, i.e. those that predict the negative class.
- Very small clusters of minority class examples can be identified as noise or could be error as well, and hence the chances of erroneous cancel by the classifier may persist. On the contrary, few real noisy examples can degrade the identification of the minority class, since it has limited instances to train with.

One of the solution rending computational intelligence could be resampling

Re-sampling techniques can be categorized into three groups or families:

- Under sampling methods, which create a subset of the original dataset by eliminating instances (usually majority class instances).
- Oversampling methods, which create a superset of the original dataset by replicating some instances or creating new instances from existing ones.
- Hybrids methods, which combine both sampling approaches from above. Within these families of methods, the simplest preprocessing techniques are non-heuristic methods such as random under sampling and random oversampling.

Their working procedure is straight-forward approach: they are focussed to randomly remove examples of the majority class, or replicate examples from the minority class. This process is carried out only in the training set of instances with aims at re-balancing the data distribution to the 50%. In the first case, i.e. random under sampling, the major drawback is that it can discard potentially useful data, that could be important for the learning process. In order to deal with the mentioned problems, more sophisticated methods have been proposed. Among them, the work accomplished in 2002 [16] has become one of the prominent approaches. In brief, its main idea is to create new minority class examples by interpolating several minority class instances that lie together for oversampling the training set. With this technique, the positive class is over-sampled by taking each minority class sample and introducing synthetic examples along the line segments joining any/all of the k minority class nearest neighbors. Depending upon the amount of over-sampling required, neighbors from the k nearest neighbors are randomly chosen. The standardization of computationally intelligent technique may not be possible and to create more such option this present bio-inspired soft computing is proposed.

The remaining part of the chapter could be as follows: Sect. 1 will elaborate the problem, followed by the recent similar works in Sect. 2. Section 3 introduces the elementary fundamentals of classifier and proposed approach. Section 4 presents validation on the proposed approach using soft computing. Finally, Sect. 5 gives conclusion and mentions further scope of the concept.

1.1 Motivation and Background

Emerging changes orchestrated through the rapid interaction and volume of data lead to the business analytics. The wide information generation with the help of smart devices and the rapid growth of social networks, including IoT (Internet of Things) [1, 2] contributes significant impact towards various features of business data analysis and classification. Unlikely to the conventional texture of data, big data and data grabbed through interconnected smart devices indicates different verticals e.g. text mining, opinion Mining, social network and cluster analysis, which ushers distinguished techniques to classify and analyze. The reason of such scaled differences are expected as the v3 concepts of big data which is defined as velocity, variety and volume. It is also followed by c3 concepts of cardinality, continuity and complexity. Hence, data classification has become an essential technique to support people for exploring the hidden knowledge within large-scale datasets and then using this knowledge to make informed inferences. It can be assumed there exists a sequence dataset (SD) and that L is a set of class labels. Assume further that each sequence S ∈ SD is associated with a particular class label $c_1 \in L$. The sequence classification task involves building a sequence classifier C; namely a function, which maps any sequence S, to a class label cL, i.e., C:S ∈ cL; cL ∈ L. In the second stage of the classification model, the mapping process is performed using an ensemble classifier, expressed as ESC (C1, …, Cn): S ∈ cl; cl ∈ L, where n is the number of classifiers.

The challenge of such classification is crucial in the paradigm of big data becomes crucial as multiple classes exist and linear classifier could not be adequate due to skewness of random data. Ensemble classification is a technique, in which multiple classifiers are combined in order to obtain an enhanced prediction accuracy. A good ensemble classifier with an equal voting mechanism is able to both reduce the biased prediction of r is k (from each single classifier) and to improve the forecasting accuracy of the data categories. A data set by a pairs (x_i, y_i) from (x, y) where $x \in \Re^m$ and Y is either described by a set of labels $y_i \in \sum$ or numeric values. A space described by $n \in N$ observations and $m \in N$, $m \in N$ features is here referred as a (n, m) space. The joint distribution p(x,y) and the marginals pX(x) and pY(y) all are random and unknown. Hence, the classification seems to be injected by imbalance with random big data and during the final classification. It causes several problems for standard machine learning (ML) algorithms suffers to perform with accuracy because of the unequal distribution in dependent variable. This causes the performance of existing classifiers to get biased towards majority class. The algorithms are accuracy driven i.e. they aim to minimize the overall error to which the minority class contributes proportionately less. ML algorithms assume that the data set has balanced class distributions. They also assume that errors obtained from different classes have same cost. This paper is proposing the modified classifier with respect to multi class measures considering the class imbalance attribute in the sample of big random data set. The proposed bio-inspired soft computing method addresses the problem. The embedded imbalance and strategy to tackle it, may solicit certain primary formal foundation. if there is a set of underlying stochastic regulations for a given class then any learning task will try to minimize the error in final state. Primarily, the proposed model addresses a multi-class classification problem and it also takes care-off class imbalance as well. The followings are the instances:

- One-vs.-rest (OvR) The one-vs.-rest (or one-vs.-all) strategy involves training a single classifier per class, with the samples of that class as positive samples and all other samples as negatives.
- One-vs.-one (OvO) In the one-vs.-one, one trains K(K − 1)/2 binary classifiers for a K-way multi-class problem; each receives the samples of a pair of classes from the original training set, and must learn to distinguish these two classes. At prediction time, a voting scheme is applied: all K(K − 1) / 2 classifiers are applied to an unseen sample and the class that got the highest number of "+1" predictions gets predicted by the combined classifier.

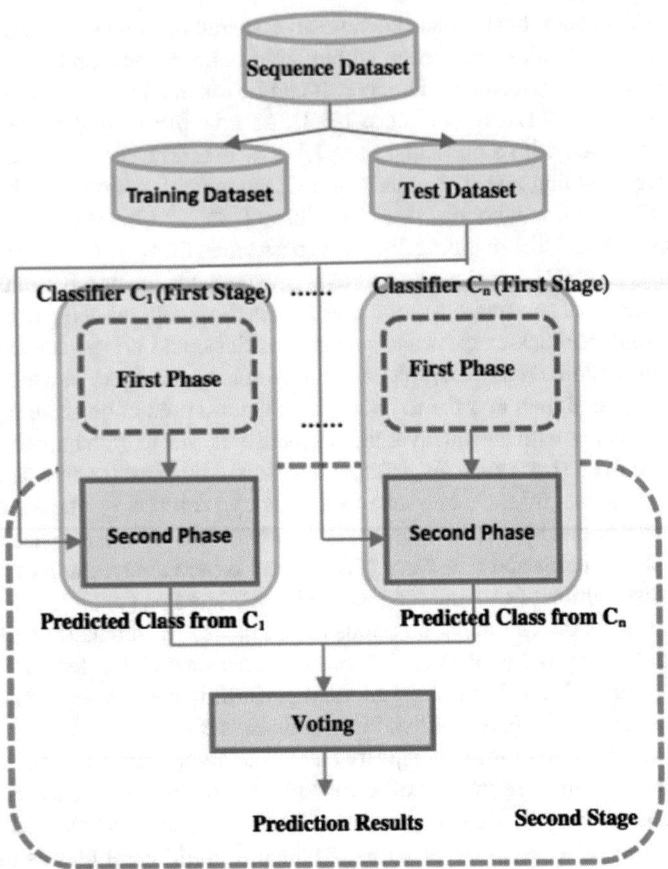

Fig. 1 Schema of learning classifier

As shown in Fig. 1, the sequence dataset is divided into a training dataset and a test dataset. The classifier consists of two stages. The first stage generates several *Sequence Classifiers* based on the *Pattern Coverage Rate* (SC-PCR) using the sequences in the training dataset.

Considering these specific ranges of classifiers and background, it has been realized that there is substantial scope to augment modified classifier schemes by incorporating bio-inspired soft computing approaches. Monkey algorithm and its vertical is well suitable for developing classifier based on bio-inspired soft computing principle to tackle randomized behavior and imbalanced classification.

2 Similar Works

There are emerging research perspectives on big data classifiers. Primarily, the conventional classification methods are unsuitable for sequence data. Hence, Ensemble Sequence Classifier (ESC) has been proposed [17]. The ESC consists of two stages. The first stage generates a Sequence Classifier based on Pattern Coverage Rate (SC-PCR) in two phases. The first phase mines sequential patterns and builds the features of each class, whereas the second phase classifies sequences based on class scores using a pattern coverage rate. Similarly, there are typical applications on Extreme learning machine (ELM) which has been developed for concerned big data and neural net based frameworks [18, 19]. Recently, classification in the presence of class imbalance [20, 21] has gained a considerable amount of attention in the last years. The principle objective is to investigate the correct identification of positive examples, without drastically deteriorating the performance on the negative class. A wide variety of solutions has been proposed to address this problem [22]. Data skewness is an often encountered challenge in real world applications and it also motivates to have better classification [23]. A sequence of open competitions for detecting irregularities in the DNA string is one of the major area of finding ensemble classification problem [24]. Data Mining Package, an open-source, MapReduce-based tool known as ICP, for the supervised classification was presented focussing large amounts of data [25].

There is a need to propose efficient classification techniques, which are suitable to handle the data stream challenges. The challenges are open data stream, concept drift, concept evolution problem and feature evolution problem [26]. The trend of feature selection in genomics experiment also solicits efficient classification solutions keeping dimensionally different gene expressions [27]. Dimension reduction is one of the biggest challenge in high-dimensional regression models ad that is well accomplished by machine learning research community [28].

While implementing the different variations of ensemble learning and classification it becomes a practice to insert the diversity into an ensemble. Sampling could be the most generic procedure. In Bagging [29], each base classifier is obtained from a random sample of the training data. In AdaBoost [30] the resampling is based on a weighted distribution, the weights are modified depending on the correctness of the prediction for the example given by the previous classifier. Bagging and AdaBoost have been modified to deal with imbalanced datasets They are referred as SMOTE-Bagging [31], SMOTEBoost [32] and RUSBoost [33]. These approaches are oriented with data level and they can manage oversampling and randomness to some extent. It is also possible to have ensembles that combine classifiers obtained with different methods [34]. The proposed algorithm is one of such combined strategical approach using monkey algorithm [35, 36] and its subsequent improvement with respect to big data environment.

3 Proposed Model

Hence, the conventional algorithm can be modified to support ensemble data set for precise classification under big and random data. The modified classifier is proposed to develop by using a novel bio-inspired mechanism. e.g. monkey algorithm. The monkey algorithm was first proposed to solve numerical optimization problems as a new swarm intelligence based algorithm stemmed from the mountain-climbing behavior of monkeys [35]. Assume that there are many mountains in a given field. At the beginning, the monkeys climb up from their respective positions to find the mountaintops (this action is called climb process). When a monkey get the top of its mountain, it will find a higher mountain within the sight and jump somewhere of the mountain from the current position (this action is called watch–jump process), then repeat the climb process. After repetitions of the climb process and the watch–jump process, each monkey will somersault to a new search domain to find a much higher mountaintop (this action is called somersault process). In this model, only climb process has been incorporated.

The improved monkey algorithm [36] will assist to formulate a scheme for dynamic classification under imbalance. The proposed model is an iterative algorithm which accumulates sequentially positive and negative votes for subsets of classes. The terminating criteria of the can be customized accordingly.

Given the initial set of classes $L = \{l_0, l_1, \ldots, l_L\}$ and an example x to classify, the algorithm will yield output at the end of the process and presents the highest mean class of positive votes as coined in ensemble learning methods. Based on *climb* and *watch jump* process of Monkey Algorithm elaborates a general scheme. At each round, a binary classifier is considered named as dichotomizer, which will compare two subsets of classes, C+ and C− ($C+ \bigcap C- = \emptyset$) and will predict if the example x belongs most likely to a class of C+ or to a class of C−. Positive and negative votes are recorded accordingly for the classes of C+ and C−. We will note in the following $f_{C+,C-}$ the classifier trained to separate the set of classes C+ from the set of classes C−: $f_{C+,C-}$ is trained such that $f_{C+,C-}(x) = 1$ for x from a class in C+ and $f_{C+,C-}(x) = -1$ for x from a class in C−.

There could be 2 levels of algorithms: firstly, the basic looping to the inserted new data for learning to be done followed by the typical module of monkey algorithm (climb process) to keep track of different iterations of sample data. However, it could be more efficient to deduce the steps of first level looping prior to enter into actual samples and iterations:

Algorithm 1 First Level of Algorithm Concerning the Initial Samples

```
D1= initial dataset with equal weights
    for i =1 to k do
            Learn new classifier Ci;
            Compute αi (importance of classifier);
            Update sample weights;
            Create new training set Di+1 (using weighted sampling)
            end for
                Construct Ensemble which uses Ci weighted by αi (i = 1, k)
```

Algorithm 2 High Level Description of Classifier with Monkey Algorithm

Require : Set S, N observations, M features

Set S of examples $(x_1,y_1) \ldots (x_n, y_n)$ /*where, $x_i \in X \subseteq \Re^n$ & $y_i \in Y = \{-1, +1\}$*/

Test : New set of S' of examples, Number of iterations

for i=1 to M do

for j= 1to N do

x[i][j]=1 to N do

x[i][j]=rand ( )

if x[i] [j] < 0.5

x[i][j]=1

end if

end for

end for

$S_N \leftarrow \{(x_i, y_i) \in S \mid y_i = -1$; /* Majority Size S_N*/

$S_P \leftarrow \{(x_i, y_i) \in S \mid y_i = +1$; /*/* Minority Size S_P*/

Climb_Monkey( )

Randomly generate 2 vectors $\Delta x' = (\Delta x', \Delta x', \ldots, \Delta x') = (\Delta x''_{i1}, \Delta x_{i2}'', \ldots, \Delta x_{in}'')$,

/*where, j = 1, 2, …, n, respectively*/

/*The parameter a(a > 0), called the step of the climb process, can be determined by specific situations. Here, set the climb step a = 1 for the 0-1 knapsack problem.*/

Set $x'_{ij} = |x_{ij} - \Delta x'_{ij}|$ and $x''_{ij} = |x_{ij} - \Delta x''_{ij}|$, /*j=1, 2, …, n, respectively, $|x|$ represents the absolute value of x.*/

Set $X_i' = (x', x', \ldots, x')$, $X'' = (x'', x'', \ldots, x'')$.

Calculate f (X_i') and f (X_i''), /*i = 1, 2, …, M, respectively*/

if f(X′) > f(X′′) and f(X′) > f(Xi),

Set $X_i = X_i'$.

end if

If f(X_i'') > f(X_i') and f(X_i'') > f(X_i)

Set $X_i = X_i''$

end if

Repeat steps until the maximum allowable number of iterations has been reached.

4 Discussion

As a justified case study, it is focussed to big data paradigm of tweets of tourists in a specific city.[1]

[1] https://github.com/DMKM1517/SmartCity/blob/master/DataScienceNotebooks/sampledata.csv.

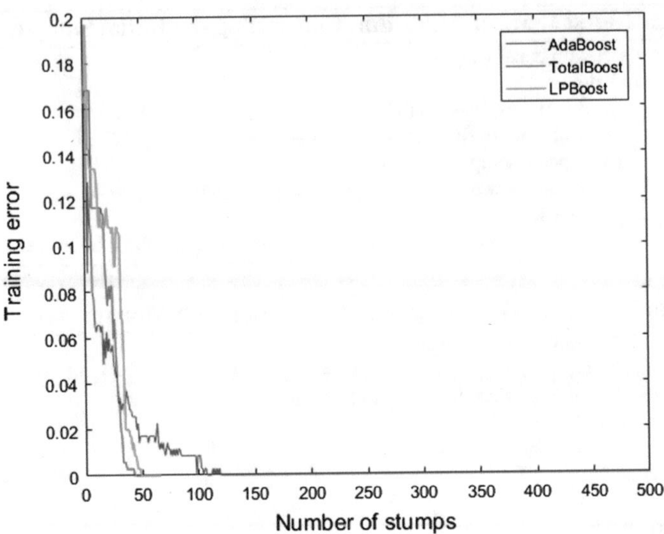

Fig. 2 Monkey climb algorithm on tweet data

However, csv format of data with 1.5M tweets having the following specifications are being considered:

- idd: Unique tweet id
- Text: Text within the tweet
- alch_scores: Score Alchemy API
- Local_score: Score
- Sentiment: Score trained classifier (sample dataset with 200 records)

Figure 2 demonstrates the training error versus number of stumps in case of monkey climb algorithm applied on the tweet data. The climb process is an incremental procedure to improve the objective function by choosing a better one between two positions that are generated around the current position. It is observed that near 100 stump the training error converges into 0.2 unit value. The experiments were performed using a desktop computer with a 3.01 GHz X4640 processor, 3 GB of RAM. For every test problem, algorithm runs 50 times individually with random initial solutions. The proposed algorithm achieve perfect prediction on the training data after a while. In brief, the following procedure could be considered:

First, we have a training dataset and pool of classifiers. Each classifier does a poor job in correctly classifying the datasets (they are usually called weak classifier). Then the pool is consulted and find the one which does the best job (minimizing classification error). The weight of the samples will be increased, which are miss-classified (so the next classifier has to work better on these samples). They are self-terminating, hence the scope of guess to include minimum numbers of members. They also produce ensembles with some very small weights, we can safely remove ensemble members.

Suppose, there are 25 base classifiers Each classifier has error rate, $\epsilon = 0.35$ Assume classifiers are independent probability that the ensemble classifier makes a wrong prediction:

$$\sum_{i=13}^{25} \binom{25}{i} \epsilon^i (1-\epsilon)^{25-i} = 0.06 \quad (1)$$

If base classifiers are $C_1, C_2, \ldots, C_N$ and error rate becomes:

$$\epsilon_i = \frac{1}{N} \sum_{j=1}^{N} w_j \delta[(C_i x_j) \neq y_j)] \quad (2)$$

This demonstrates that if any intermediate rounds yield subsequent error rate higher than 50%, then the weights w are reverted back to 1/n and the re-sampling procedure has to be invoked again.

Figure 3 shows the variation of ϵ and inversion of it, with respect to different iterations. As shown in Table 1 Record Set 4 (out of 200 records of sample) is hard to classify. Its weight is increased, therefore it is more likely to be chosen again in subsequent iterations with more climb values of monkey.

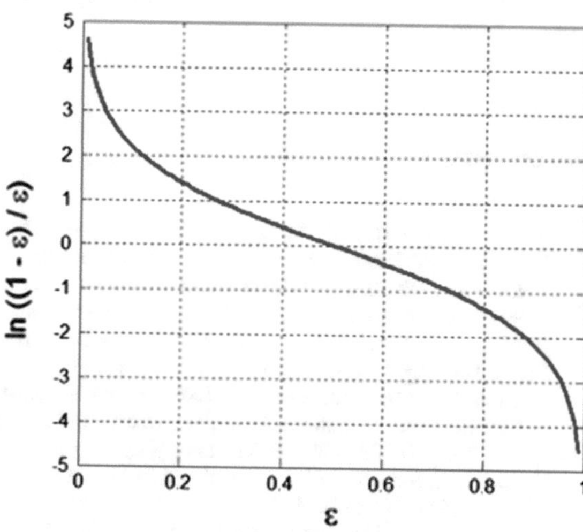

Fig. 3 Base classifier with variation of $\epsilon = 0.35$

Table 1 Snaps of different sample iterations

Original data	1	2	3	4	5	6	7	8	9	10
Iteration 1	7	3	2	8	7	9	4	10	6	3
Iteration 2	5	4	9	4	2	5	1	7	4	2
Iteration 3	**4**	**4**	8	10	**4**	5	**4**	6	3	**4**

5 Conclusion

The prime objective of existing big data oriented classifiers is not to obtain highly accurate base models, but rather to achieve base models, to quantify different kinds of errors. For example, if ensembles are used for classification, high accuracies can be accomplished, if different base models misclassify different training examples, even if the base classifier accuracy is low. Independence between two base classifiers can be evaluated in this case by measuring the degree of overlap in training examples towards misclassification. In this case, more overlap means less independence between two models. Considerable understanding of bio-inspired soft computing behavior improvise to develop the proposed model which can optimize computational effort to classify the different classes under big and skewed data paradigm. Monkey algorithm and its simple functional improved components will assist to formulate a scheme for dynamic classification under imbalance. It has been observed that the classification in ensemble mode through bio-inspired soft computing could be better accurate in-spite of the presence of noise and miss-classification. Till this report is prepared, there is not adequate research on bio-inspired and soft computing based classifier for big data environment persist. Hence, the other functional components of monkey algorithm like somersault process for all tweet records can also be incorporated towards further clarification. The search domain will be more precise and classification based on sample will be more ensembles with less data skewing and oversampling.

References

1. Shafaatunnur Hasan, Siti Mariyam Shamsuddin, Noel Lopes (2014), Machine Learning Big Data Framework and Analytics for Big Data Problems, Int. J. Advance Soft Compu. Appl, Vol. 6, No. 2, IS bSN 2074-8523; SCRG Publication.
2. Tianrui Li, Chuan Luo, Hongmei Chen, and Junbo Zhang (2015), PICKT: A Solution for Big Data Analysis, Springer International Publishing Switzerland, Ciucci et al. (Eds.): RSKT 2015, LNAI 9436, pp. 15–25. doi:10.1007/978-3-319-25754-9 2.
3. Raghava Rao Mukkamala et. al (2014), Fuzzy-Set Based Sentiment Analysis of Big Social Data IEEE 18th International Enterprise Distributed Object Computing Conference (EDOC).
4. Quan Zou, Sifa Xie Ziyu Lin Meihong Wu, Ying Ju (2016) Finding the Best Classification Threshold in Imbalanced Classification, Big Data Research, Available online 4 January.
5. Chen CP, Zhang C-Y (2014), Data-intensive applications, challenges, techniques and technologies: a survey on big data. Inf Sci 275: pp. 314–347, 2014.
6. Witten IH, Frank E, Hall MA (2011), Data mining: practical machine learning tools and techniques. Morgan Kaufmann series in data management systems. Morgan Kaufmann, Burlington, 2011.
7. Mattmann CA (2013) Computing: a vision for data science. Nature 493: pp. 473–475.
8. Provost F, Fawcett T (2013), Data science and its relationship to big data and data-driven decision making. Big Data 1(1): pp. 51–59.
9. Wu X, Zhu X, Wu G-Q, Ding W(2014), Data mining with big data. IEEE Trans Knowl Data Eng 26(1): pp. 97–107.
10. Dean J, Ghemawat S (2010), MapReduce: a flexible data processing tool. Commun ACM 53(1): pp. 72–77.

11. del Río S., López V., Benítez J. M., Herrera F (2014), On the use of MapReduce for imbalanced big data using Random Forest. Information Sciences. 284: pp. 112–137.
12. Zaharia M., Chowdhury M., Das T., et al (2012). Resilient distributed datasets: a fault-tolerant abstraction for in-memory cluster computing. Proceedings of the 9th USENIX conference on Networked Systems Design and Implementation (NSDI '12); April 2012; San Jose, Calif, USA. USENIX Association; pp. 1–14.
13. V. López, A. Fernandez, S. García, V. Palade, F. Herrera (2013). An Insight into Classification with Imbalanced Data: Empirical Results and Current Trends on Using Data Intrinsic Characteristics. Information Sciences 250, 113–141.
14. A. Fernández, V. López, M. Galar, M.J. del Jesus, F. Herrera (2013), Analysing the classification of imbalanced data-sets with multiple classes: binarization techniques and ad-hoc approaches, Knowledge-Based Systems 42. pp. 97–110.
15. A. Fernandez, S. García, J. Luengo, E. Bernadó-Mansilla, F. Herrera (2010), Genetics-based machine learning for rule induction: state of the art, taxonomy and comparative study, IEEE Transactions on Evolutionary Computation 14 (6) pp. 913–941.
16. N.V. Chawla, K.W. Bowyer, L.O. Hall, W.P. Kegelmeyer (2002) Synthetic Minority Oversampling Technique (SMOTE): Synthetic Minority Over-Sampling Technique, Journal of Artificial Intelligent Research 16, pp. 321–357.
17. I-Hui Li, I-En Liao, Jin-Han Lin, Jyun-Yao Huang (2016), An Efficient Ensemble Sequence Classifier Journal of Software, Volume 11, Number 2, pp. 133–147.
18. Jiuwen Cao and Zhiping Lin (2015), Extreme Learning Machines on High Dimensional and Large Data Applications: A Survey, Mathematical Problems in Engineering Volume 2015, Article ID 103796, pp. 1–13.
19. Extreme Learning Machines, Erik Cambria and Guang-Bin Huang (2013), IEEE Intelligent System, Published by the IEEE Computer Society.
20. V. López, A. Fernández, S. García, V. Palade, and F. Herrera (2013), An insight into classification with imbalanced data: Empirical results and current trends on using data intrinsic characteristics, Information Sciences, vol. 250, no. 0, pp. 113–141.
21. M. Galar, A. Fernandez, E. Barrenechea, H. Bustince, and F. Herrera (2012), A review on ensembles for the class imbalance problem: Bagging boosting and hybrid-based approaches, IEEE Transactions on Systems, Man, and Cybernetics, Part C: Applications and Reviews, vol. 42, no. 4, pp. 463–484.
22. S. del Río, V. López, J. Benítez, and F. Herrera (2014), On the use of Mapreduce for imbalanced big data using random forest, Information Sciences, vol. 285, pp. 112–137.
23. I. Triguero, D. Peralta, J. Bacardit, S. García, and F. Herrera (2015), MRPR: A Mapreduce solution for prototype reduction in big data classification, Neurocomputing, vol. 150, pp. 331–345.
24. Ariel Jaffe, Ethan Fetaya, Boaz Nadler, Tingting Jiang, Yuval Kluger (2016), Unsupervised Ensemble Learning with Dependent Classifiers, Appearing in Proceedings of the 19th International Conference on Artificial Intelligence and Statistics (AISTATS) 2016, Cadiz, Spain. JMLR: W&CP, Volume 51.
25. V. A. Ayma, R. S. Ferreira, P. Happ, D. Oliveira, R. Feitosa, G. Costa, A. Plaza, P. Gamba (2015), Classification algorithms for big data analysis, a Map Reduce approach, The International Archives of the Photogrammetry, Remote Sensing and Spatial Information Sciences, Volume XL-3/W2, 2015 PIA15+HRIGI15 – Joint ISPRS conference, 25–27 March Munich, Germany.
26. M. B. Chandak (2016), Role of big-data in classification and novel class detection in data streams, J Big Data 3:5, Springer-Verlag.
27. Mahmoud et al. (2014), A feature selection method for classification within functional genomics experiments based on the proportional overlapping score, BMC Bioinformatics, 15:274 http://www.biomedcentral.com/1471-2105/15/274.
28. Yengo L, Jacques J, Biernacki C (2013), Variable clustering in high dimensional linear regression models. Journal de la Societe Francaise de Statistique.
29. L. Breiman (1996), Bagging predictors, Mach. Learn. 24, pp. 123–140.

30. Y. Freund, R.E. Schapire (1996), Experiments with a new boosting algorithm, in: Machine Learning, Proceedings of the Thirteenth International Conference (ICML '96), Bari, Italy, July 3–6, pp. 148–156.
31. S. Wang, X. Yao (2009), Diversity analysis on imbalanced data sets by using ensemble models, in: IEEE Symposium Series on Computational Intelligence and Data Mining (IEEE CIDM 2009), pp. 324–331.
32. N. Chawla, A. Lazarevic, L. Hall, K. Bowyer (2003), Smoteboost: improving prediction of the minority class in boosting, in: 7th European Conference on Principles and Practice of Knowledge Discovery in Databases (PKDD 2003), pp. 107–119.
33. C. Seiffert, T. Khoshgoftaar, J. Van Hulse, A. Napolitano (2010), Rusboost: a hybrid approach to alleviating class imbalance, IEEE Trans. Syst. Man Cybern., Part A: Syst. Hum. 40 (1) pp. 185–197.
34. S.B. Kotsiantis, P.E. Pintelas (2003), Mixture of expert agents for handling imbalanced data sets, Ann. Math. Comput. Teleinform. 1 (1) pp. 46–55.
35. R.Q. Zhao, W.S. Tang (2008), Monkey algorithm for Global numerical optimization, J. Uncertain Syst. 2 (3) pp. 164–175.
36. Yongquan Zhoua, Xin Chena, Guo Zhou (2016), An improved monkey algorithm for a 0-1 knapsack problem, Applied Soft Computing, Elsevier 38, pp. 817–830.

Unified Framework for Control of Machine Learning Tasks Towards Effective and Efficient Processing of Big Data

Han Liu, Alexander Gegov and Mihaela Cocea

Abstract Big data can be generally characterised by 5 Vs—Volume, Velocity, Variety, Veracity and Variability. Many studies have been focused on using machine learning as a powerful tool of big data processing. In machine learning context, learning algorithms are typically evaluated in terms of accuracy, efficiency, interpretability and stability. These four dimensions can be strongly related to veracity, volume, variety and variability and are impacted by both the nature of learning algorithms and characteristics of data. This chapter analyses in depth how the quality of computational models can be impacted by data characteristics as well as strategies involved in learning algorithms. This chapter also introduces a unified framework for control of machine learning tasks towards appropriate employment of algorithms and efficient processing of big data. In particular, this framework is designed to achieve effective selection of data pre-processing techniques towards effective selection of relevant attributes, sampling of representative training and test data, and appropriate dealing with missing values and noise. More importantly, this framework allows the employment of suitable machine learning algorithms on the basis of the training data provided from the data pre-processing stage towards building of accurate, efficient and interpretable computational models.

Keywords Big data · Computational intelligence · Data mining · Machine learning · Data processing · Predictive modelling · Model selection · Decision tree learning · Bayesian learning · Instance based learning

---

H. Liu (✉) · A. Gegov · M. Cocea
School of Computing, University of Portsmouth, Buckingham Building, Lion Terrace, Portsmouth PO1 3HE, UK
e-mail: Han.Liu@port.ac.uk

A. Gegov
e-mail: Alexander.Gegov@port.ac.uk

M. Cocea
e-mail: Mihaela.Cocea@port.ac.uk

W. Pedrycz and S.-M. Chen (eds.), *Data Science and Big Data: An Environment of Computational Intelligence*, Studies in Big Data 24,
DOI 10.1007/978-3-319-53474-9_6

1 Introduction

Big data can generally be characterized by 5Vs—Volume, Velocity, Variety, Veracity and Variability. In particular, volume generally reflects the space required to store data. Velocity reflects the speed of data transmission and processing, i.e. how effectively and efficiently real-time data is collected and processed on the platform of cloud computing. Variety reflects the type of data, i.e. data can be structured or unstructured and can also be in different forms such as text, image, audio and video. Veracity reflects the degree to which data can be trusted. Variability reflects the dissimilarity between different instances in a data set. More details on big data can be found in [1–4].

In many studies, machine learning has been considered as a powerful tool of big data processing. As introduced in [5], the relationship between big data and machine learning is very similar to the relationship between resources and human learning. In this context, people can learn from resources to deal with new matters. Similarly, machines can learn from big data to resolve new problems. More details on big data processing by machine learning can be found in [6–12].

Machine learning is regarded as one of the main approaches of computational intelligence [13]. In general, computational intelligence encompasses a set of nature or biology inspired computational approaches such as artificial neural networks, fuzzy systems and evolutionary computation. In particular, artificial neural networks are biologically inspired to simulate the human brains in terms of learning through experience. Also, fuzzy systems involve using fuzzy logic, which enables computers to understand natural languages [14]. Moreover, evolutionary computation works based on the process of nature selection, learning theory and probabilistic methods, which helps with uncertainty handling [13]. As stated in [15], learning theories, which help understand how cognitive, emotional and environmental effects and experiences are processed in the context of psychology, can help make predictions on the basis of previous experience in the context of machine learning. From this point of view, machine learning is naturally inspired by human learning and would thus be considered as a nature inspired approach. In addition, most machine learning methods involve employing heuristics of computational intelligence, such as probabilistic measures, fuzziness and fitness, towards optimal learning. In particular, C4.5, Naïve Bayes and K nearest neighbors, which are selected for the experimental studies in Sect. 4, involve employing such heuristics.

In machine learning context, learning algorithms are typically evaluated in four dimensions, namely accuracy, efficiency, interpretability and stability, following the concepts of computational intelligence. These four dimensions can be strongly related to veracity, volume, variety and variability, respectively.

Veracity reflects the degree to which data can be trusted as mentioned above. In practice, data needs to be transformed to information or knowledge for people to use. From this point of view, the accuracy of information or knowledge discovered from data can be highly impacted by the quality of the data and thus is an effective way of evaluation against the degree of trust.

Volume reflects the size of data. In the areas of machine learning and statistics, the data size can be estimated through the product of data dimensionality and sample size [16]. Increase of data dimensionality or sample size can usually increase the computational costs of machine learning tasks. Therefore, evaluation of the volume for particular data is highly related to estimation of memory usage for data processing by machine learning methods.

Variety reflects the format of data, i.e. data types and representation. Typical data types include integer, real, Boolean, string, nominal and ordinal [17]. In machine learning and statistics, data types can be simply divided into two categories: discrete and continuous. On the other hand, data can be represented in different forms, e.g. text, graph and tables. All the differences mentioned above in terms of data format can impact on the interpretability of models learned from data.

Variability reflects the dissimilarity between different instances in a data set. In machine learning, the performance of learning algorithms can appear to be highly unstable due to change of data samples, especially when the data instances are highly dissimilar to each other. Therefore, the stability of a learning algorithm can be highly impacted by data variability.

The above four aspects (accuracy, efficiency, interpretability and stability) are also impacted greatly by the selection of different machine learning algorithms. For example, data usually needs to be pre-processed by particular algorithms prior to the training stage, which leads to a particular level of impact on data modelling. Also, inappropriate sampling of training and test data can also lead to building a poor model and biased estimation of accuracy, respectively. Further, different learning algorithms can usually lead to different quality of models learned from the same training data. In addition, in the context of online learning, velocity, which is related to the learning speed of an algorithm, is an important impact factor for data streams to be processed effectively and efficiently. However, this chapter focuses on offline learning, which analyses in depth how the nature of learning algorithms is related to the nature of static data.

This chapter is organized as follows: Sect. 2 introduces fundamental concepts of machine learning and how computational intelligence contributes to the design of learning algorithms. Section 3 presents a framework proposed in a nature inspired way for control of machine learning tasks towards appropriate employment of learning algorithms and efficient processing of big data. Section 4 reports experimental studies on employment of learning algorithms and efficient processing of big data and discusses the obtained results in both quantitative and qualitative terms. Section 5 highlights the contributions of this chapter and suggests further directions towards advancing this research area by using computational intelligence approaches.

2 Fundamentals of Machine Learning

Machine learning is a branch of artificial intelligence and involves two stages: training and testing [18]. The first stage aims to learn something from known properties by using learning algorithms and the second stage aims to make predictions on unknown properties by using the knowledge learned in the first stage. From this point of view, training and testing are also referred to as learning and prediction, respectively. In practice, a machine learning task is aimed at building a model, which is further used to make predictions, through the use of learning algorithms. Therefore, this task is usually referred to as predictive modelling.

Machine learning could be divided into two special types: supervised learning and unsupervised learning [19], in terms of the form of learning. Supervised learning means learning with a teacher, because all instances from a training set are labelled, which makes the learning outcomes very explicit. In other words, supervised learning is naturally inspired by student learning with the supervision of teachers. In practice, the aim of this type of learning is to build a model by learning from labelled data and then to make predictions on other unlabeled instances with regard to the value of a predicted attribute. The predicted value of an attribute could be either discrete or continuous. Therefore, supervised learning could be involved in both classification and regression tasks for categorical prediction and numerical prediction, respectively. In contrast, unsupervised learning means learning without a teacher. This is because all instances from a training set are unlabeled and thus the learning outcomes are not explicit. In other words, unsupervised learning is naturally inspired by student learning without being supervised. In practice, the aim of this type of learning is to discover previously unknown patterns from data sets. It includes association and clustering. The former aims to identify correlations between attributes whereas the latter aims to group objects on the basis of their similarity to each other.

According to [18], machine learning algorithms can be put into several categories: decision tree learning, rule learning, instance based learning, Bayesian learning, perceptron learning and ensemble learning. All of these learning algorithms show the characteristic of nature inspiration.

Both decision tree learning and rule learning aim to learn a set of rules on an inductive basis. However, the difference between the two types of learning is that the former generates rules in the form of a decision tree and the latter generates if-then rules directly from training instances [1, 20, 21]. The above difference is mainly due to the fact that the former follows the divide and conquer approach [22] and the latter follows the separate and conquer approach [23]. In particular, the divide and conquer approach is naturally similar to the top-down approach of student learning, such as dividing a textbook into several levels: parts, chapters, sections and subsections. The separate and conquer approach is naturally similar to the iterative approach of student learning, which means by reading through an entire material in the first iteration and then focusing on more important parts of the material for deeper understanding in the subsequent iterations.

Instance based learning generally involves predicting test instances on the basis of their similarity to the training instances, such as K nearest neighbor [24]. This type of learning is also referred to as lazy learning, due to the fact that it does not aim to learn in depth to gain some pattern from data but just to make as many correct predictions as possible [16]. In other words, this type of learning is naturally similar to the exam centered approach of student learning, which means that students mainly aim to answer correctly the exam questions without deep understanding of knowledge.

Bayesian learning essentially employs the Bayes theorem [25]. In particular, this type of learning is based on the assumption that all the input attributes are totally independent of each other. In this context, each attribute-value pair would be independently correlated to each of the possible classes, which means that a posterior probability is provided between the attribute-value pair and the class. A popular method of Bayesian learning is Naive Bayes [26]. This type of learning is naturally similar to the prior-knowledge based approach of human reasoning, which means that people make decisions, reasoning and judgments based on the knowledge they obtained before, towards having the most confident choice.

Perceptron learning aims to build a neural network topology that consists of a number of layers and that has a number of nodes, each of which represents a perceptron. Some popular algorithms include backpropagation [17] and probabilistic neural networks [18]. This type of learning is biology inspired as stated in Sect. 1. Ensemble learning generally aims to combine different learning algorithms in the training stage or computational models in the testing stage towards improvement of overall accuracy of predictions. Some popular approaches of ensemble learning include bagging [27] and boosting [28]. This type of learning is naturally similar to the approach of group learning for students to collaborate on a group assignment.

In terms of evaluation of a machine learning task, there are generally two main approaches: cross-validation and split of data into a training set and a test set. Cross-validation generally means to split a data set into n disjoint subsets. In this context, there would be n iterations in total for the evaluation, while at each iteration a subset is used for testing and the other $n - 1$ subsets are used for training. In other words, each of the n subsets is in turn used as the test set at one of the n iterations, while the rest of the subsets are used together as the training set. In laboratory research, ten-fold cross-validation is used more popularly, i.e. the original data set is split into 10 subsets. Cross-validation is generally more expensive in terms of computational cost. Therefore, researchers sometimes instead choose to take the approach of splitting a data set into a training set and a test set in a specific ratio, e.g. 70% of the data is used as the training set and the rest of the data is used as the test set. This data split can be done randomly or in a fixed way. However, due to the presence of uncertainty in data, the random split of data is more popular for researchers in machine learning or similar areas.

In this chapter, new perspectives of the two approaches of evaluating machine learning tasks are used in Sect. 4. In particular, cross-validation is used towards measuring effectively the learnability of an algorithm, i.e. the extent to which the algorithm is suitable to build a confident model on the provided training data. This is in order to help employ appropriately the suitable learning algorithms for building predictive models on the basis of existing data. The other approach for splitting a data set into a training set and a test set is adopted towards learning a model that covers highly complete patterns from the training data and evaluating the model accuracy using highly similar but different instances from the test data. This is in order to ensure the model accuracy evaluated by using the test data is trustworthy. Details on the use of the new perspectives are presented in Sect. 4.

3 Framework for Control of Machine Learning Tasks

This section presents a framework for control of machine learning tasks towards appropriate employment of learning algorithms and effective processing of big data. In particular, the key features of the proposed framework are described in detail. Also, the motivation of developing this framework is justified by analyzing the impact of big data on machine learning, i.e. this is to argue the relevance for effective control of machine learning tasks in a big data environment.

3.1 Key Features

A unified framework for control of machine learning tasks is proposed in a nature inspired way in [1] as a further direction. The purpose is to effectively control the pre-processing of data and to naturally employ learning algorithms and the generated predictive models. As mentioned in [1], it is relevant to deal with issues on both the algorithms side and the data side for improvement of classification performance. In fact, a database is daily updated in real applications, which could result in the gradual increase of data size and in changes to patterns that exist in the database. In order to avoid lowering computational efficiency, the size of a sample needs to be determined in an optimal way. In addition, it is also required to avoid the loss of accuracy. From this point of view, the sampling is critical not only in terms of the size of a sample but also in the representativeness of the sample.

Feature selection/extraction is another critical task for pre-processing of data. As mentioned in [1], high dimensional data usually results in high computational costs. In addition, it is also very likely to contain irrelevant attributes which result in noise and coincidental patterns. In some cases, it is also necessary to effectively detect noise if the noise is introduced naturally or artificially. In other words, noise may be introduced in a dataset due to mistakes in typing or illegal modifications from hackers. A potential way of noise handling is using association rules to detect that the value of an attribute is incorrect on the basis of the other attribute-value pairs in the same data instance. Also, appropriate employment of learning algorithms and predictive models are highly required, due to the fact that there are many existing machine learning algorithms, but no effective ways to determine which of them are suitable for a particular data set. Traditionally, the decision is made by experts based on their knowledge and experience. However, it is fairly difficult to judge the correctness of the decision prior to empirical validation. In real applications, it is not realistic to frequently change decisions after it has been confirmed that the chosen algorithms are not suitable.

The arguments above outline the necessity to develop the framework for control of machine learning tasks in a nature inspired way. In other words, this framework aims to adopt computational intelligence techniques to control machine learning tasks. In this framework, the actual employment of both learning algorithms and predictive models follows computational intelligent approaches. The suitability of a learning algorithm and the reliability of a model are measured by statistical analysis on the basis of historical records. In particular, each algorithm in the algorithms base, as illustrated in Fig. 1, is assigned a weight which is based on its performance in previous machine learning tasks. The weight of an algorithm is naturally similar to the impact factor of a journal which is based on its overall citation rate. Following the employment of suitable learning algorithms, each model generated is then also assigned a weight which is based on its performance on the latest version of validation data in a database. Following the employment of high quality models, a knowledge base is finalized and deployed for real applications as illustrated in Fig. 1.

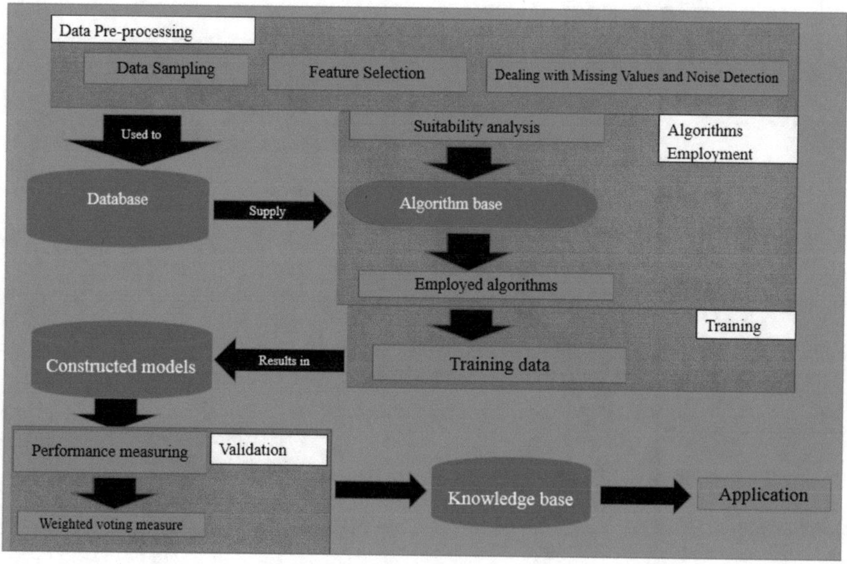

Fig. 1 Unified framework for control of machine learning tasks [1]

3.2 Justification

As mentioned in Sect. 1, machine learning algorithms are usually evaluated against accuracy, efficiency, interpretability and stability. The presence of big data has deeply affected machine learning tasks in the four aspects mentioned above.

In terms of accuracy, overfitting of training data can be significantly reduced in general as the size of data is greatly increased. There is evidence reported in [29] that learning from a large training set can significantly improve the performance in predictive modelling. The evidence is illustrated in Fig. 2, which was provided with an illustration by Banko and Brill in 2001 [30] that the complex problem of learning on automated word disambiguation would keep improving after the size of training data is beyond billions of words. In particular, each of the four learning algorithms shows an increase of at least 10% in terms of test accuracy, while the number of words is increased from 0.3 million to 1 billion. For example, the memory-based algorithm gets the test accuracy increased from 83 to 93%, and the winnow algorithm achieves to increase the test accuracy from 75 to 97%. The improvement in learning performance is due to the fact that the increase in data size can usually improve the completeness of the pattern covered. In other words, small data may cover only a small part of a pattern in a hypothesis space. Therefore, overfitting of training data is likely to result in the case that a learning algorithm may build a

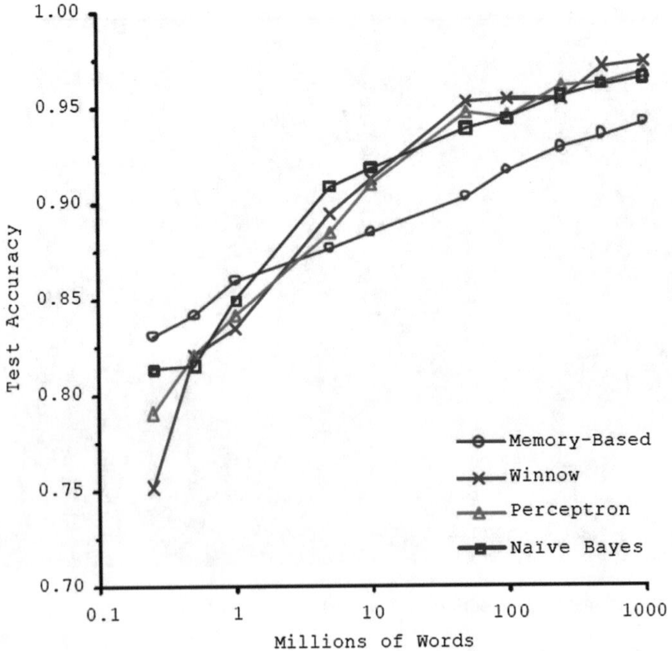

Fig. 2 Improvement of words disambiguation by learning from big data [30]

model that performs greatly on training data but poorly on test data. This case occurs especially when the training data covers a pattern that is highly different from the one in the test data. When the size of data is increased, the training data is likely to cover a pattern that is more similar to the one present in the test data.

On the other hand, the increase in the size of data may also increase the chance to have noise and coincidental patterns present in the data. This is due to the fact that the biased improvement in the quantity of data may result in the loss of quality. Also, large training data is likely to cover some patterns which occur in fairly low frequencies. This could mean that the patterns covered by the training data are purely coincidental rather than scientifically confident.

The above issues regarding accuracy can be solved through scaling up algorithms or scaling down data. As specified in [1], the former way is to reduce the bias on the algorithms side. In particular, the algorithms can be designed to be more robust against noise and thus avoid being confused by coincidental patterns. In the context of rule learning, the reduction of bias can be achieved through direct advancement of rule generation methods or employment of rule simplification algorithms; similar ways also apply to other types of learning algorithms. The latter way is to reduce the variance on the data side. In particular, data can be pre-processed through removal of irrelevant attributes by feature selection techniques or through the merger of redundant attributes by feature extraction techniques. In addition, data can also be resampled by selecting only those instances that are more representative.

In terms of efficiency, the increase in the size of data is likely to increase the computational costs in both training and testing stages. In the training stage, it may slow down the process of building a predictive model by learning from big data. In the testing stage, the predictive model is likely to have a high level of complexity, which significantly increases the computational complexity in predicting on unseen instances. In particular to rule learning algorithms, the presence of big data may result in the generation of a large number of complex rules.

As stressed in [7, 29, 31], processing of big data needs decomposition, parallelism, modularity and recurrence. In this case, these machine learning algorithms, which are inflexible and work in black box manners, would fail in dealing with big data. This case would immediately happen to those algorithms that are quadratically complex ($O\ (n^2)$), when encountering data with millions of points (instances).

The above issues regarding efficiency can also be resolved through scaling up algorithms or scaling down data. In the former way, the algorithms can be designed to have a low level of computational complexity in the training stage and thus be less affected by the increase in the size of training data. In the context of rule learning, the improvement of efficiency can be achieved through the employment of pruning algorithms, as some of such algorithms can stop the process of rule learning earlier. In the latter way, the size of data can be reduced through dimensionality reduction and data sampling. This not only reduces the computational costs in the training stage, but also results in the generation of simpler models and thus speeds up the process of predicting on unseen instances in the testing stage.

In terms of interpretability, the increase in the size of data usually decreases the interpretability. As analyzed in [1, 16], interpretability can be affected by the size of training data in terms of model complexity. In the context of rule learning, big data may result in the generation of a large number of complex rules, which would make it difficult for people to read and understand.

The above issues regarding interpretability can also be solved through scaling up algorithms or scaling down data. In the former way, the algorithms can be designed to be robust against noise and irrelevant or redundant attributes. In other words, the presence of noise and irrelevant/redundant attributes would not result in irrelevant patterns being learned by the algorithms. In the context of rule learning, algorithms for rule generation may decide to skip some attributes or attribute-value pairs for generation of decision trees or if-then rules due to the irrelevance of these attributes or attribute-value pairs. In addition, the employment of rule simplification methods also helps improve the interpretability since such employment usually results in the generation of a smaller number of simpler rules. In the latter way, the data size is reduced through dimensionality reduction and data sampling as mentioned above. In particular, as discussed in [1, 16], the reduction of data dimensionality decreases the maximum length (the maximum number of rule terms) of each single rule. The data sampling also reduces the maximum number of rules. In this approach, the interpretability can be improved if the dimensionality reduction and data sampling are effectively undertaken.

In terms of stability, the increase in the size of data usually leads to the increase in dissimilarity between different instances, and thus results in high variance in terms of the performance of learning algorithms when an experiment is repeated independently on the same data. In other words, big data could generally have high variability, which results in low stability of performance being shown from repeated experiments on the same data, especially when unstable algorithms are used. Some unstable algorithms include neural network learning and rule learning [17, 32].

The above stability issues can also be resolved through scaling up algorithms or scaling down data. As introduced in [33, 34], prediction accuracy in machine learning tasks can be affected by bias and variance. Bias generally means errors originated from use of statistical heuristics and can be reduced through scaling up algorithms. Variance generally means errors originated from random processing of data and can be reduced by scaling down data. From this point of view, heuristic based algorithms usually show high bias and low variance on fixed training and test data. In contrast, random algorithms usually show low bias and high variance on fixed training and test data. However, if both the training data and the test data are randomly sampled from the original data, heuristic based algorithms usually appear to be more sensitive to the change of sample and thus perform less stably. In the same situation, random algorithms, in contrast, usually appear to be less sensitive and perform more stably when an experiment is repeated independently on the basis of random sampling of training and test data.

On the basis of the above descriptions, it is highly relevant to develop the unified framework illustrated in Fig. 1 towards effective control of machine learning tasks in a big data environment. In particular, learning algorithms need to be employed appropriately by measuring their accuracy, efficiency, interpretability and stability on the basis of particularly selected data. It is also important to have effective data pre-processing so that employment of algorithms can be done on the basis of high quality data provided following the data pre-processing.

4 Experimental Studies

This section presents two case studies on big data. The first case study addresses the veracity aspect, and is designed to confirm that cross-validation can be used to measure the learnability of algorithms on particular training data towards effective employment of learning algorithms for predictive modelling. The second case study addresses variability, and is designed to show how to measure the data variability through checking the variance of the performance of a particular algorithm, while independently repeated experiments are undertaken on the same data. The other two aspects of big data on volume and variety respectively have been studied in [1, 16] through theoretical analysis and empirical investigations in terms of efficiency and interpretability of computational models.

4.1 Measure of Learnability

This case study is done using 10 data sets retrieved from the biomedical repository [35]. The characteristics of these data sets are described in Table 1.

In particular, all these selected data are highly dimensional and have additional test sets supplied. This selection is in order to support the experimental setup, which employs cross-validation [36] to measure the learnability of particular algorithms on the training data and then employ suitable algorithms to build models that are

Table 1 Medical data sets

Name	Attribute types	#Attributes	#Instances	#Classes
ALL-AML	Continuous	7130	72	2
DLBCL-NIH	Continuous	7400	160	2
LungCancer	Continuous	12534	32	2
MLL_Leukemia	Continuous	12583	72	3
BCR-ABL	Continuous	12559	327	2
E2A-PBX1	Continuous	12559	327	2
Hyperdip50	Continuous	12559	327	2
MLL	Continuous	12559	327	2
T-ALL	Continuous	12559	327	2
TEL-AML1	Continuous	12559	327	2

evaluated by using test instances. In other words, for each of the selected data sets, the whole training set is provided in order to measure the extent to which a particular algorithm is suitable to build a model on the training set, and the test set is used to evaluate the performance of the model learned by the algorithm. In this setup, the results would show the extent to which the learnability of an algorithm measured by using cross validation on training data can provide a good basis for judging whether the algorithm can build a confident model that performs well on additional test data. In this case study, C4.5, Naïve Bayes and K nearest neighbor are chosen as learning algorithms for testing due to their popularity in real applications. In addition, these three algorithms can also be seen as nature inspired approaches as justified in Sect. 2. The results of this experimental study can be seen in Table 2.

Table 2 shows that in almost all cases the learnability of an algorithm measured by cross validation is effective for judging the suitability of an algorithm to a particular training set, which leads to expected performance on the corresponding test set. In other words, the results show that if an algorithm is judged to be suitable for a particular training set through measuring its learnability, then the model learned by the algorithm from the training set usually performs well on the additionally supplied test set.

On the other hand, when an algorithm is judged to be unsuitable for a particular training set through cross-validation, the results generally indicate the phenomenon that the model learned by the algorithm from the training set performs a low level of classification accuracy on the additionally supplied test set. In particular, it can be seen on the DLBCL-NIH data that all these three algorithms are judged to be less suitable for the training set and the models learned by these algorithms from the training set fail to perform well on the corresponding test set. Another similar case can be seen on the MLL-Leukemia data that Naïve Bayes is judged to be unsuitable

Table 2 Learnability on training data and prediction accuracy on test data

Dataset	C4.5 I (%)	C4.5 II (%)	NB I (%)	NB II (%)	KNN I (%)	KNN II (%)
ALL-AML	93	100	70	71	88	97
DLBCL-NIH	44	58	55	63	56	63
LungCancer	94	89	25	90	88	97
MLL_Leukemia	79	100	22	53	89	100
BCR-ABL	91	95	96	95	97	96
E2A-PBX1	96	87	92	95	98	88
Hyperdip50	91	88	81	80	94	98
MLL	94	97	94	95	97	100
T-All	91	100	87	87	55	99
TEL-AML1	95	95	76	76	98	98

NB: C4.5 I means testing the learnability of the algorithm by cross validation on the basis of training data and C4.5 II means testing the performance of the predictive model using the additionally supplied test data. The same also applies to NB and KNN for the training set and the model learned by the algorithm fails to perform well on the corresponding test set.

In addition, there are two exceptional cases on the lung-cancer and T-All data. In the first case, Naïve Bayes is judged to be very unsuitable for the training set but the performance on the test set by the model learned by the algorithm from the training set is very good. In the second case, K nearest neighbor is judged to be less suitable for the training set but the actual performance on the test set by the model learned by the algorithm from the training set is extremely good. For both cases, it could be because the training set essentially covers the complete information and the split of the training set for the purpose of cross validation could result in incompleteness to which both Naïve Bayes and K nearest neighbor are quite sensitive. However, when the algorithm learns from the whole training set, the resulted model covers the complete information from the training set and thus performs well on the test set.

4.2 Measure of Data Variability

This case study is conducted using 20 data sets retrieved from the UCI [37] and the biomedical repositories. The characteristics of these chosen data sets are described in Table 3.

Table 3 Data sets from UCI and biomedical repositories

Name	Attribute types	#Attributes	#Instances	#Classes
Anneal	Discrete, continuous	38	798	6
Balance-scale	Discrete	4	625	3
Car	Discrete	6	1728	4
Credit-a	Discrete, continuous	15	690	2
Credit-g	Discrete, continuous	20	1000	2
Diabetes	Discrete, continuous	20	768	2
Heart-statlog	Continuous	13	270	2
Hepatitis	Discrete, continuous	20	155	2
Ionosphere	Continuous	34	351	2
Iris	Continuous	4	150	3
Lymph	Discrete, continuous	19	148	4
Wine	Continuous	13	178	3
Zoo	Discrete, continuous	18	101	7
Sonar	Continuous	61	208	2
Segment	Continuous	19	2310	7
ColonTumor	Continuous	2001	62	2
DLBCLOutcome	Continuous	7130	58	2
DLBCLTumor	Continuous	7130	77	2
DLBCL-Stanford	Continuous	4027	47	2
Lung-Michigan	Continuous	7130	96	2

The data sets selected from the UCI repository are all considered as small data as they are of lower dimensionality and sample size, except for the segment data which is considered to be big data due to its larger sample size. On the other hand, the last five data sets selected from the biomedical repository are all considered as big data due to the fact that they are of high dimensionality. This selection is in order to put the case study in the context of data science by means of processing data with different scalability. In addition, all these chosen data sets are not supplied additional test sets. The selection of the data sets was also made so that both discrete and continuous attributes are present, which is in order to investigate how the different types of attributes could impact on the data variability.

On the basis of the chosen data, the experiment on each data set is undertaken by independently repeating the training-testing process 100 times and checking the variance of the performance over the 100 repetitions, on the basis of random sampling of training and test data in the ratio of 70:30. This experimental setup is in order to measure the extent to which the data is variable leading to variance in terms of performance in machine learning tasks. In this context, C4.5, Naïve Bayes and K nearest neighbor are chosen as learning algorithms for testing the variance due to the fact that these algorithms are not stable, i.e. they are sensitive to the changes in data sample. The results are presented in Table 4.

It can be seen from Table 4 that on each data set, while different algorithms are used, the standard deviation of the classification accuracy over 100 independently

Table 4 Data variability measured by standard deviation of classification accuracy

Dataset	C4.5	NB	KNN
Anneal	0.007	0.017	0.023
Balance-scale	0.028	0.022	0.020
Car	0.011	0.019	0.028
Credit-a	0.026	0.021	0.030
Credit-g	0.027	0.023	0.022
Diabetes	0.027	0.028	0.027
Heart-statlog	0.044	0.039	0.045
Hepatitis	0.046	0.042	0.073
Ionosphere	0.031	0.043	0.035
Iris	0.030	0.033	0.027
Lymph	0.057	0.057	0.055
Wine	0.048	0.027	0.054
Zoo	0.045	0.068	0.063
Sonar	0.057	0.059	0.052
Segment	0.010	0.015	0.010
ColonTumor	0.094	0.105	0.089
DLBCLOutcome	0.122	0.104	0.109
DLBCLTumor	0.074	0.067	0.072
DLBCL-Stanford	0.133	0.060	0.096
Lung-Michigan	0.040	0.041	0.028

repeated experiments appears to be in a very similar level, except for the DLBCL-Standford data set on which Naïve Bayes displays a much lower level of standard deviation.

On the other hand, while looking at different data sets, the standard deviation for them appears to be very different no matter which one of the three algorithms is adopted. In particular, for the 15 UCI data sets, the standard deviation is lower than 5% in most cases or a bit higher than 5% in several cases (e.g. on the lymph and sonar data sets). In contrast, for the last five data sets selected from the biomedical repository, the standard deviation is usually higher than 5% and is even close to or higher than 10% in some cases (e.g. on the colonTumer and DLBCLOutcome data sets). An exceptional case happens from the lung-Michigan data set, which appears to have the standard deviation lower than 5%, no matter which one of the three algorithms is used.

In addition, it can also be seen from Table 4 that the data sets that contain only continuous attributes appear to have the standard deviation higher than the data sets that contain discrete attributes. Some data sets that contain both discrete and continuous attributes also appear to have the standard deviation higher than the data sets that contain only discrete attributes. In fact, the presence of continuous attributes generally increases the attribute complexity, and thus makes the data more complex, which leads to the potential increase of the data variability.

The results shown in Table 4 generally indicate that attribute complexity, data dimensionality and sample size impact on the size of data and that data with a larger size is likely to be of higher variability, leading to higher variance in terms of performance in machine learning tasks, especially when the training and test data are sampled on a purely random basis.

5 Conclusion

This chapter has proposed a unified framework in a nature inspired way for control of machine learning tasks in Sect. 3.1, and the necessity of the proposal has been justified in Sect. 3.2 through analyzing the impact of big data on machine learning. Two case studies have been conducted experimentally following computational intelligence methodologies in Sect. 4. The results from the case studies also indicate the necessity of proposing the unified framework through using computational intelligence concepts.

The results from the first case study indicate that cross-validation is an effective way to measure the extent to which an algorithm is suitable to build a predictive model on the basis of the existing data. In fact, a test set is not actually available in reality and instead a set of unseen instances are given for the model to predict the values of unknown attributes of each particular instance. From this point of view, the framework proposed in Sect. 3 is highly relevant in order to achieve appropriate employment of learning algorithms on the basis of the existing data. However, it is difficult to guarantee in reality that the existing data can cover the full population.

Therefore, the framework proposed in Sect. 3 can be modified further to work towards achieving natural selection of learning algorithms. In other words, the learnability of an algorithm measured through cross-validation can be used as the chance of being employed for predictive modelling, towards predicting unseen instances through natural selection of the predefined classes. Similar ideas have been applied to voting based classification in [38].

On the other hand, the results from the second case study indicate that data can be of high variability, which could lead to high variance in terms of performance in machine learning tasks while training and test data are sampled on a purely random basis. In fact, as described in Sect. 3.2, while training and test data are sampled randomly, the algorithms based on statistical heuristics generally display higher variance than those algorithms with high randomness. However, these heuristics-based algorithms, such as C4.5, Naïve Bayes and K nearest neighbor, are highly popular in practical applications. This indicates the necessity to have effective and efficient pre-processing of data prior to the training stage in order to avoid any high variance due to random sampling. In particular, effective sampling of training and test data can be achieved through data clustering in order to ensure that the training instances are of high similarity to the test instances. This is naturally inspired by the principle of student examination that the exam questions should all cover what the students actually learned from learning materials rather than anything else outside of these sources. In other words, representative sampling of training and test data would make the model learned from the training data cover more complete patterns and the model accuracy evaluated by using the test data more trustworthy. On the basis of the above descriptions, clustering-based sampling of training and test data is strongly recommended as a further direction.

References

1. H. Liu, A. Gegov and M. Cocea, Rule Based Systems for Big Data: A Machine Learning Approach, 1 ed., vol. 13, Switzerland: Springer, 2016.
2. "What is Big Data," SAS Institute Inc, [Online]. Available: http://www.sas.com/big-data/. [Accessed 17 May 2015].
3. "Master Data Management for Big Data," IBM, [Online]. Available: http://www.01.ibm.com/software/data/infosphere/mdm-big-data/. [Accessed 17 May 2015].
4. W. Pedrycz and S. M. Chen, Eds., Information Granularity, Big Data, and Computational Intelligence, vol. 8, Switzerland: Springer, 2015.
5. P. Levine, "Machine Learning + Big Data," WorldPress, [Online]. Available: http://a16z.com/2015/01/22/machine-learning-big-data/. [Accessed 15 May 2015].
6. T. Condie, P. Mineiro, N. Polyzotis and M. Weimer, "Machine learning for big data," in *ACM SIGMOD/PODS Conference*, San Francisco, USA, 2013.
7. L. Wang and C. A. Alexander, "Machine Learning in Big Data," *International Journal of Mathematical, Engineering and Management Sciences,* vol. 1, no. 2, pp. 52–61, 2016.
8. X. Wu, X. Zhu, G. Q. Wu and W. Ding, "Data Mining with Big Data," *IEEE Transactions on Knowledge and Data Engineering,* vol. 26, no. 1, pp. 97–107, 2014.

9. S. Suthaharan, “Big data classification: problems and challenges in network intrusion prediction with machine learning,” *ACM SIGMETRICS Performance Evaluation Review,* vol. 41, no. 4, pp. 70–73, 2014.
10. O. Y. Al-Jarrah, P. D. Yoo, S. Muhaidat and G. K. Karagiannidis, “Efficient Machine Learning for Big Data: A Review,” *Big Data Research,* vol. 2, no. 3, pp. 87–93, 2015.
11. D. E. O’Leary, “Artificial Intelligence and Big Data,” *IEEE Intelligent Systems,* vol. 28, no. 2, pp. 96–99, 2013.
12. C. Ma, H. H. Zhang and X. Wang, “Machine learning for Big Data Analytics in Plants,” *Trends in Plant Science,* vol. 19, no. 12, pp. 798–808, 2014.
13. H. Adeli and N. Siddique, Computational Intelligence: Synergies of Fuzzy Logic, Neural Networks and Evolutionary Computing, New Jersey: John Wiley & Sons, 2013.
14. L. Rutkowski, Computational Intelligence: Methods and Techniques, Heidelberg: Springer, 2008.
15. J. Worrell, “Computational Learning Theory: 2014-2015,” University of Oxford, 2014. [Online]. Available: https://www.cs.ox.ac.uk/teaching/courses/2014-2015/clt/. [Accessed 20 9 2016].
16. H. Liu, M. Cocea and A. Gegov, “Interpretability of Computational Models for Sentiment Analysis,” in *Sentiment Analysis and Ontology Engineering: An Environment of Computational Intelligence*, vol. 639, W. Pedrycz and S. M. Chen, Eds., Switzerland, Springer, 2016, pp. 199–220.
17. P.-N. Tan, M. Steinbach and V. Kumar, Introduction to Data Mining, New Jersey: Pearson Education, 2006.
18. T. Mitchell, Machine Learning, New York: McGraw Hill, 1997.
19. D. Barber, Bayesian Reasoning and Machine Learning, Cambridge: Cambridge University Press, 2012.
20. H. Liu, A. Gegov and F. Stahl, “Categorization and Construction of Rule Based Systems,” in *15th International Conference on Engineering Applications of Neural Networks*, Sofia, Bulgaria, 2014.
21. H. Liu, A. Gegov and M. Cocea, “Network Based Rule Representation for Knowledge Discovery and Predictive Modelling,” in *IEEE International Conference on Fuzzy Systems*, Istanbul, 2015.
22. R. Quinlan, “Induction of Decision Trees,” *Machine Learning,* vol. 1, pp. 81–106, 1986.
23. J. Furnkranz, “Separate-and-Conquer rule learning,” *Artificial Intelligence Review,* vol. 13, pp. 3–54, 1999.
24. J. Zhang, “Selecting typical instances in instance-based learning,” in *The 9th International Conference on Machine Learning*, Aberdeen, Scotland, 1992.
25. H. e. Michiel, “Bayes formula,” in *Encyclopedia of Mathematics*, Springer, 2001.
26. I. Rish, “An Empirical Study of the Naïve Bayes Classifier,” *IJCAI 2001 workshop on empirical methods in artificial intelligence,* vol. 3, no. 22, pp. 41–46, 2001.
27. L. Breiman, “Bagging predictors,” *Machine Learning,* vol. 24, no. 2, pp. 123–140, 1996.
28. Y. Freund and R. E. Schapire, “Experiments with a New Boosting Algorithm,” in *Machine Learning: Proceedings of the Thirteenth International Conference (ICML ‘96)*, 1996.
29. “Machine Learning on Big Data,” EBTIC, 19 August 2014. [Online]. Available: http://www.ebtic.org/pages/ebtic-view/ebtic-view-details/machine-learning-on-big-data-d/687. [Accessed 15 May 2015].
30. M. Banko and E. Brill, “Scaling to very very large corpora for natural language disambiguation,” in *Proceedings of the 39th Annual Meeting on Association for Computational Linguistics*, 2001.
31. K. M. Tarwani, S. Saudagar and H. D. Misalkar, “Machine Learning in Big Data Analytics: An Overview,” *International Journal of Advanced Research in Computer Science and Software Engineering,* vol. 5, no. 4, pp. 270–274, 2015.
32. I. Kononenko and M. Kukar, Machine Learning and Data Mining: Introduction to Principles and Algorithms, Chichester, West Sussex: Horwood Publishing Limmited, 2007.

33. H. Liu, A. Gegov and M. Cocea, "Collaborative Rule Generation: An Ensemble Learning Approach," *Journal of Intelligent and Fuzzy Systems,* vol. 30, no. 4, pp. 2277–2287, 2016.
34. H. Liu, A. Gegov and M. Cocea, "Hybrid Ensemble Learning Approach for Generation of Classification Rules," in *International Conference on Machine Learning and Cybernetics*, Guangzhou, 2015.
35. J. Li and H. Liu, "Kent Ridge Bio-medical Dataset," I2R Data Mining Department, 2003. [Online]. Available: http://datam.i2r.a-star.edu.sg/datasets/krbd/. [Accessed 18 May 2015].
36. S. Geisser, Predictive Inference, New York: Chapman and Hall, 1993.
37. M. Lichman, "UCI Machine Learning Repository," University of California, School of Information and Computer Science, 2013. [Online]. Available: http://archive.ics.uci.edu/ml. [Accessed 12 May 2015].
38. H. Liu, A. Gegov and M. Cocea, "Nature and Biology Inspried Approach of Classification towards Reduction of Bias in Machine Learning," in *International Conference on Machine Learning and Cybernetics*, Jeju Island, South Korea, 2016.

An Efficient Approach for Mining High Utility Itemsets Over Data Streams

Show-Jane Yen and Yue-Shi Lee

Abstract *Mining frequent itemsets* only considers the number of the occurrences of the itemsets in the transaction database. *Mining high utility itemsets* considers the purchased quantities and the profits of the itemsets in the transactions, which the profitable products can be found. In addition, the transactions will continuously increase over time, such that the size of the database becomes larger and larger. Furthermore, the older transactions which cannot represent the current user behaviors also need to be removed. The environment to continuously add and remove transactions over time is called a data stream. When the transactions are added or deleted, the original high utility itemsets will be changed. The previous proposed algorithms for mining high utility itemsets over data streams need to rescan the original database and generate a large number of candidate high utility itemsets without using the previously discovered high utility itemsets. Therefore, this chapter proposes an approach for efficiently mining high utility itemsets over data streams. When the transactions are added into or removed from the transaction database, our algorithm does not need to scan the original transaction database and search from a large number of candidate itemsets. Experimental results also show that our algorithm outperforms the previous approaches.

Keywords Data mining · Knowledge discovery · High utility itemset · Frequent itemset · Closed itemset · Data stream · Large databases · Utility threshold · Mining algorithm · Information maintenance

---

S.-J. Yen (✉) · Y.-S. Lee
Department of Computer Science and Information Engineering,
Ming Chuan University, Taoyuan County, Taiwan
e-mail: sjyen@mail.mcu.edu.tw

W. Pedrycz and S.-M. Chen (eds.), *Data Science and Big Data: An Environment of Computational Intelligence*, Studies in Big Data 24,
DOI 10.1007/978-3-319-53474-9_7

1 Introduction

In this section, we first introduce some preliminaries for mining high utility itemsets [7]. Let I = {i1, i2,..., im} be the set of all the items. An itemset X is a subset of I and the length of X is the number of items contained in X. An itemset with length k is called a k-itemset. A transaction database D = {T1, T2,..., Tn} contains a set of transactions and each transaction has a unique transaction identifier (TID). Each transaction contains the items purchased in this transaction and their purchased quantities. The purchased quantity of item ip in a transaction Tq is denoted as o(ip, Tq). The utility of item ip in Tq is u(ip, Tq) = o(ip, Tq) × s(ip), in which s(ip) is the profit of item ip. The utility of an itemset X in Tq is the sum of the utilities of the items contained in X ⊆ Tq, which is shown in expression (1). If X ⊄ Tq, u(X, Tq) = 0. The utility of an itemset X in D is the sum of the utilities of X in all the transactions containing X, which is shown in expression (2).

The transaction utility (tu) of a transaction Tq is the sum of the utilities of the items in Tq, which is shown in expression (3). The total utility of the whole transaction database D is the sum of the transaction utilities of all the transactions in D. A utility threshold is a user specified percentage and a minimum utility (MU) can be obtained by multiplying total utility of D and the user-specified utility threshold. An itemset X is a high utility itemset if the utility of X in D is no less than the minimum utility.

$$u(X, T_q) = \sum_{i_p \in X \subseteq T_q} u(i_p, T_q) \quad (1)$$

$$u(X) = \sum_{X \subseteq T_q \in D} u(X, T_q) \quad (2)$$

$$tu(T_q) = \sum_{i_p \in T_q} u(i_p, T_q) \quad (3)$$

For example, Table 1 is a transaction database, in which each integer number represents the purchased quantity for an item in a transaction. Table 2 is a Profit Table which records the profit for each item in Table 1. Suppose the user-specified utility threshold is 60%. Because the total utility of Table 1 is 224, the minimum utility is 226 * 60% = 134.4. The utility of itemset {D} is u ({D}) = 3×6 = 18 ≦

Table 1 A transaction database

Item / TID	A	B	C	D	E
T_1	0	0	16	0	1
T_2	0	6	0	1	1
T_3	2	0	1	0	0
T_4	0	10	0	1	1
T_5	1	0	0	1	1

Table 2 Profit table

Item	Profit ($) (Per Unit)
A	3
B	10
C	1
D	6
E	5

135, which is not a high utility itemset. The utility of itemset {BD} is u({BD}) = $(6 \times 10 + 1 \times 6) + (10 \times 10 + 1 \times 6) = 172 \geq 135$. Therefore, itemset {BD} is a high utility itemset.

For mining frequent itemset [1], all the subsets of a frequent itemset are frequent, that is, there is a downward closure property for frequent itemsets. However, the property is not available for high utility itemsets, since a subset of a high utility itemset may not be a high utility itemset. For the above example, itemset {BD} is a high utility itemset in Table 1, but its subset {D} is not a high utility itemset. Therefore, Liu et al. [7] proposed a *Two-Phase algorithm* for mining high utility itemsets. They defined the *transaction weighted utility* (twu) for an itemset X, which is shown in expression (4).

$$\mathrm{twu}(X) = \sum_{X \subseteq T_q \in D} \mathrm{tu}(T_q) \tag{4}$$

If the twu of an itemset is no less than MU, then the itemset is a high transaction weighted utility (HTWU) itemset. According to expression (4), the twu for an itemset X must be greater than or equal to the utility of X in D. Therefore, if X is a high utility itemset, then X must be a HTWU itemset. All the subsets of a HTWU itemset are also HTWU itemsets. Therefore, there is a downward closure property for HTWU itemsets. The first phase for the Two-Phase algorithm [7] is to find all the HTWU itemsets which are called candidate high utility itemsets by applying Apriori algorithm [1]. Two-Phase algorithm scans the database again to compute the utilities for all the candidate high utility itemsets and find high utility itemsets in the second phase.

Although some approaches [2, 7, 9, 15, 16, 18] have been proposed for mining high utility itemsets in a static transaction database, these approaches cannot efficiently discover high utility itmesets in a data stream environment, since they need to rescan the original database and re-discover all the high utility itemsets when some transactions are added into or removed from the database. In a data stream environment, the transactions are generated or removed in an extremely fast way. We need to immediately identify which itemsets can be turn out to be high utility itemsets, and vice versa. Besides, in this environment, we need to keep the information for all the itemsets, otherwise some high utility itemsets may be lost. However, the memory space is limited. It is very difficult to retain the utilities of all the itemsets in a large database.

Recently, some approaches [3, 6, 10, 14] have been proposed to find high utility itemsets in a data stream, which can be divided into Apriori-like [6, 14] and Tree-based approaches [3, 10]. However, these approaches just tried to find candidate high utility itemsets, that is HTWU itemsets. They still need to take a lot of time to rescan the original database and search for high utility itemsets from the large number of candidate itemsets without using the previous found information. Therefore, in this chapter, we propose an efficient algorithm HUIStream(mining High Utility Itemset in data Stream) for mining high utility itemsets in a data stream. When the transactions are added or deleted, our algorithms can just update HTWU itemsets according to the added or deleted transactions and directly calculate the utilities of HTWU itemsets without rescan the original database and search for high utility itemsets from the HTWU itemsets.

2 Related Work

The early approaches for mining frequent itemsets [1, 4, 12] are based on Apriori-like approach, which iteratively generate candidate (k + 1)-itemsets from the frequent k-itemsets (k $\geq$ 1) and check if these candidate itemsets are frequent. However, in the cases of extremely large input sets or low minimum support threshold, the Apriori-like algorithms may suffer from two main problems of repeatedly scanning the database and searching from a large number of candidate itemsets.

Since Apriori-like algorithms require multiple database scans to calculate the number of occurrences of each itemset and record a large number of candidate itemsets, Tree-based algorithms [5, 11, 21] improve these disadvantages, which transform the original transaction database into an FP-tree and generate the frequent itemsets by recursively constructing the sub-trees according to the FP-Tree. Because all the transactions are recorded in a tree, Tree-based algorithms do not need multiple database scans and do not need to generate a large number of candidate itemsets.

Although Tree-based algorithms have been able to efficiently identify frequent itemsets from the transaction database, because of the number of the frequent itemsets may be very large, the execution time and memory usage would increase significantly. Therefore, some researchers have proposed the concept of closed itemsets [17, 20]. The number of the closed frequent itemsets is far less than the number of the frequent itemsets in a transaction database, and all the frequent itemsets can be derived from the frequent closed itemsets, so either memory usage or execution time for mining frequent closed itemsets is much less than that of mining frequent itemsets.

Liu et al. [7] proposed Two-Phase algorithm for mining high utility itemsets. Since the subset of a high utility itemsets may not be a high utility itemsets, that is, there is no downward closure property for high utility itemset, Liu et al. proposed the transaction weighted utility (twu) of an itemset to find out high utility itemsets.

In the first stage, Two-Phase algorithm applied Apriori algorithm [1] to find all the HTWU itemsets as the candidate itemsets, and then scans the transaction database to calculate the utility for each candidate itemset in order to identify which candidate itemsets are high utility itemsets. Although Two-Phase algorithm can find all the high utility itemsets from a transaction database, a large number of HTWU itemsets would be generated in the first phase, such that much time would be taken to search for high utility itemsets from these candidates in the second phase, since the twu of an itemset is much greater than the utility for the itemset.

Tseng et al. [14] proposed an algorithm THUI-Mine for mining high utility itemsets in a data stream, which only stores length two HTWU itemsets and applies Two-Phase algorithm to find all the HTWU itemsets. When a set of transactions is added, if there are new items in the added transactions, THUI-Mine will only determine whether the new items satisfy the utility threshold in the added transactions. If the items in the added transactions already exist in the original database, THUI-Mine will judge if the items are still HTWU items. Because THUI-Mine uses Two-Phase algorithm to re-mine the high utility itemsets, it still needs to take a lot of time to scan the database many times. HUP-HUI-DEL algorithm [8] also applies Two-Phase algorithm and only considers the transaction deletion. It still needs to generate a large number of candidate high utility itemsets and repeatedly scans the database to find high utility itemsets.

Li et al. [6] proposes MHUI algorithm, which discovers high utility itemsets in a specific sliding window. MHUI takes use of BITvector or TIDlist to store the transaction IDs in which each item is contained to avoid repeatedly scanning the database. MHUI stores length 2 HTWU itemsets in the structure LexTree-2HTU (Lexicographical Tree with 2-HTU itemset). When the transactions are added or deleted, MHUI generates all the length 2 itemsets from the added or deleted transactions and updates the structure LexTree-2HTU. MHUI uses level-wise method to generate all the HTWU itemsets from the length 2 HTWU itemsets, and re-sacan the database to find high utility itmesets.

Ahmed et al. [3] proposes a tree structure IHUP to stores the utility for each transaction and divides IHUP into three types according the order of the items which appear in the tree nodes: $IHUP_L$, $IHUP_{TF}$ and $IHUP_{TWU}$-. When a transaction is added, $IHUP_L$ stores each item of the transaction in the tree node according to the alphabetic order, but $IHUP_{TF}$ and $IHUP_{TWU}$ need to adjust the tree nodes to make sure that the items of each transaction are ordered by support and twu, respectively. IHUP needs to spend a large amount of memory space to store the whole database in a tree structure and applies FP-Growth algorithm [5] to repeatedly generate subtree structure. Finally, IHUP still needs to rescan the whole database to calculate the utility for each HTWU itmesets and generate high utility itemsets.

Yun and Ryang proposes HUPID-Growth algorithm [19] and SHU-Grow algorithm [13], respectively. HUPID-Growth scans the database once to construct HUPID-Tree and TIList and adjust the order of the items in the tree nodes to reduce the over-estimated value of the utility for each node in a path, that is to reduce the over-estimated utility for each itemsets. SHU-Grow uses the tree structure IHUP and stores the accumulated utility for each node when a set of transactions are added. SHU-Grow applies the strategies of UP-Growth algorithm [16] to reduce the over-estimated utility and the number of the candidate high utility itemsets. UPID-Growth and SHU-Grow still apply FP-Growth algorithm to find HTWU itemsets and search for high utility itemsets from a large number of candidate high utility itemsets.

3 Mining High Utility Itemsets in a Data Stream

In this section, we first introduce the storage structure for our algorithm HUIStream. When a transaction is added into or deleted from the transaction database, HUI-Stream updates HTWU itemsets related to the added or deleted transaction, respectively. In the following, we propose two algorithms HUIStream+ and HUIStream− for maintaining the HTWU itemsets and generates all the high utility itemsets from the HTWU itemsets when a transaction is added and deleted, respectively.

In order to avoid rescanning the original database and searching from the candidate high utility itemsets when the transactions are added or deleted, we have the following definitions. An itmeset X is a closed twu itemset if there is no superset of X, which has the same twu as X. An itemset is a closed HTWU itemset if X is a closed twu itemset and the twu of X is no less than user-specified minimum utility. For any two itemsets X and Y ($X \subseteq Y$), if the twu of Y is the same as the twu of X and Y is not contained in any other itemset with the same twu as Y, then Y is the closure of X. For a closed HTWU itemset X, the proper subset of X, which has the same twu as X, is called the Equal TWU itemset of X, and X is the closure of the Equal TWU itemset.

In order to efficiently find the high utility itemsets in a data stream without information loss, HUIStream first determines which itemsets in the transaction are closed twu itemsets when a transaction is added or deleted. All the closed twu itemsets are recorded in a *Closed Table*, since the number of the closed itemsets is much less than the number of the itemsets and all the itemsets can be generated by the closed itemsets in a transaction database. There are three fields included in the Closed Table: *Cid* records the identification of each closed twu itemset; *CItemset* records the closed twu itemset with utility of each item in the closed twu itemset; *twu* records the twu of the closed twu itemset.

Table 3 shows the content of the Closed Table after the previous four transactions in Table 1 are added. There are five closed twu itemsets, in which the utility and twu of the closed twu itemset {E} with Cid 3 are 15 and 205, respectively. For each item, we use the table *Cid List* to record the Cids of the closed twu itemsets which contain the item. Table 4 shows the content of the Cid List after the previous four transactions in Table 1 are added, in which the field CidSet for item C is {1, 4, 5}, since item C is contained in the three closed twu itemsets with Cids 1, 4 and 5.

For example, the total utility of the previous four transactions in Table 1 is 210. If the utility threshold is 60%, that is the minimum utility is 210 * 60% = 126, then

Table 3 The closed table after processing the previous four transactions in Table 1

Cid	CItemset	twu
0	0	0
1	C:16, E:5	21
2	B:160, D:12, E:10	182
3	E:15	205
4	A:6, C:1	7
5	C:19	30

Table 4 The Cid list after processing the previous four transactions in Table 1

Item	CidSet
A	4
B	2
C	1, 4, 5
D	2
E	1, 2, 3

Table 5 Closed HTWU itemsets and their equal TWU itemsets

Closed HTWU itemset	Equal TWU itemset
{BDE}	{B}, {D}, {BD}, {BE}, {DE}, {BDE}
{E}	{E}

the closed HTWU itemsets are {BDE} and {E}. The Equal TWU itemsets for the two closed twu itemsets are shown in Table 5. The closed HTWU itemsets and their Equal TWU itemsets form the candidate high utility itemsets. HUIStream only needs to update the closed HTWU itemsets, that is, update the content of Closed Table and Cid List, and then the twu values of all the Equal TWU itemsets can be computed without rescanning the database.

3.1 *The Algorithm HUIStream*$^{+}$

In this subsection, we describe how HUIStream finds the closed twu itemsets which need to be updated and all the HTWU itemsets after adding a transaction. When a transaction T_{ADD} is added, the twu value will be increased just for T_{ADD} and the subsets of T_{ADD}. Therefore, HUIStream only considers whether T_{ADD} and the subsets of T_{ADD} are closed twu itemsets or not. If $X \subseteq T_{ADD}$ is a closed twu itemset before adding the transaction T_{ADD}, it must be a closed twu itemset after adding the transaction, because the twu value for the supersets ($\not\subset T_{ADD}$) of X would not be changed [20]. Therefore, all the closed twu itemsets which need to be updated can be obtained by performing the intersections on T_{ADD} and all the closed twu itemsets in the Closed Table. However, the intersections of T_{ADD} and most of the closed twu itemsets would be empty. It will waste a lot of unnecessary time to intersect T_{ADD} with all the closed twu itemsets. In order to avoid that the intersection is empty, HUIStream identifies which closed twu itemsets contain some items in $T_{ADD} = \{i_1, i_2, \ldots, i_m\}$ from Cid List according to expression (5).

$$\mathrm{SET}(\{\mathrm{TADD}\}) = \mathrm{CidSet}(i_1) \cup \mathrm{CidSet}(i_2) \cup \ldots \cup \mathrm{CidSet}(i_m) \quad (5)$$

The closed twu itemset X obtained by the intersection of each closed twu itemset Y with Cid in SET($\{T_{ADD}\}$) and T_{ADD} need to be updated when a transaction T_{ADD} is added. The closed twu itemsets which need to be updated after adding transaction T_{ADD} are recorded in the table $Temp_{ADD}$, which includes the two fields: UItemset records the closed twu itemset X which needs to be updated; Closure_Id records the Cid of the closure of X before adding transaction T_{ADD}, that is, the Cid of itemset Y. If there is the same closed twu itemset X generated by the intersections of different closed twu itemsets and T_{ADD}, then the closure of X is the closed twu itemset with the largest twu value among the different closed twu itemsets. Because an added transaction T_{ADD} must be a closed twu itemset [20], T_{ADD} is recorded in $Temp_{ADD}$ and the corresponding Closure_Id is set to be 0, which represents that we cannot know the closure of T_{ADD} before adding the transaction so far. HUIStream can update the content of Closed Table according to the $Temp_{ADD}$.

For each record in $Temp_{ADD}$, HUIStream compares the itemset X in UItemset and the itemset Y with Cid in Closure_Id. If X and Y are the same, that is, X is a closed twu itemset before adding the transaction T_{ADD}, then the utility of each item in X is increased by adding the utility of the item in T_{ADD}, and the twu of X is increased by adding the tu of T_{ADD}. If X and Y are different, that is, X is not a closed twu itemset before adding the transaction T_{ADD} and turns out to be a closed twu itemset after adding the transaction, then HUIStream assigns X a unique Cid and adds it into The Closed Table and Cid List, in which the twu of X is the twu of Y plus the tu of T_{ADD} and the utility of each item is the utility of the item in Y in Closed Table plus the utility of the item in T_{ADD}, since the twu of X is equal to the twu of Y before adding the transaction T_{ADD}.

If itemset X is not a closed HTWU itemset before adding transaction T_{ADD}, but is closed HTWU itemsets after adding the transaction, then the Equal TWU itemsets of the Closure Y of X becomes the Equal TWU itemsets of X. If the Closure of X is not a closed HTWU Itemset, then the Equal TWU itmesets of X are the subsets of X, which have the same twu as X. HUIStream uses the following method to determine if the subset Z of X is an Equal TWU itemset of X: If the twu of X is the largest twu among all the itemsets with Cids in SET(Z) according to expression (5), then Z is an Equal TWU itemset of X, that is, X is the Closure of Z. If X is a closed twu itemset before adding the transaction, then HUIStream only needs to justify if X is a closed HTWU itemset after adding the transaction.

For example, suppose the utility threshold is 60% for Table 1. The Close Table and Cid List are shown in Tables 3 and 4 after adding the previous four transactions in Table 1. When the transaction T_5 is added, HUIStream records the itemset T_5 = {ADE} in the field UItemset of $Temp_{ADD}$ and set 0 to Closured_Id. Because SET(T_5) = CidSet(A) ∪ CidSet(D) ∪ CidSet(E) = {1, 2, 3, 4} according to expression (1), HUIStream performs the intersections of T_5 and the closed twu itemsets with Cids 1, 2, 3 and 4 from Closed Table, respectively. Firstly, because Cid 1 is itemset {CE} and {ADE} ∩ {CE} = {E}, HUIStream adds UItemset {E} and Closure_Id 1 in the $Temp_{ADD}$. Secondly, because Cid 2 is itemset {BDE} and {ADE} ∩ {BDE} = {DE}, HUIStream adds UItemset {DE} and Closure_Id 2 in the TempADD. Thirdly, Cid 3 is itemset {E} and {ADE} ∩ {E} = {E} has existed in TempADD. Because the twu of the closed twu itemset with Cid 3 is greater than the twu of the closed twu itemset with Cid 1, the corresponding Closure_Id of UItemset {E} is replaced with Cid 1. Finally, because Cid 4 is itemset {AC} and {ADE} ∩ {AC} = {A}, HUIStream adds UItemset {A} and Closure_Id 4 in the TempADD. After adding the transaction T_5, the $Temp_{ADD}$ is shown in Table 6.

HUIStream updates Closed Table and Cid List according to $Temp_{ADD}$. For Table 6, because the first UItemset {ADE} of $Temp_{ADD}$ is not in Closed Table and the corresponding Closure_Id is 0, which means that itemset {ADE} is not a closed twu itemset before adding transaction T_5, but turns out to be a closed twu itemset after adding transaction T_5, HUIStream adds the CItemset {A:3, D:6, E:5} with Cid 6 and twu = 3+6 + 5 = 14 into the Closed Table. HUIStream also inserts Cid 6 into Cid List for items A, D and E. Because the minimum utility is 135 after adding transaction T_5, itemset {ADE} is not a closed HTWU itemset.

For the second UItemset {E} and the corresponding Closure_Id 3 in Table 6, because Cid 3 in Closed Table (Table 3) is also {E}, which means that itemset {E} is a closed twu itemset before adding transaction T_5, the twu of {E} after adding the transaction T_5 is the twu of {E} in the Closed Table plus the tu of T5, that is 205 + 14 = 219. Because the utility of {E} is 5 in T_5, HUIStream updates the CItemset {E:15} in Table 3 as {E:20 (=15 + 5)}.

For the third UItemset {DE} and the corresponding Closure_Id 2 in Table 6, because Cid 2 in Closed Table (Table 3) is {BDE}, which means that itemset {DE} is not a closed twu itemset and the Closure of {DE} is {BDE} before adding transaction T_5, the twu of {DE} after adding transaction T5 is the twu of {BDE} in the Closed Table plus the tu of T_5, that is, 182 + 14 = 196. HUIStream adds the

Table 6 The TempADD after adding transaction T_5

UItemset	Closure_Id
{ADE}	0
{E}	3
{DE}	2
{A}	4

Table 7 The closed table after adding transaction T_5 in Table 1

Cid	CItemset	twu
0	0	0
1	C:16, E:5	21
2	B:160, D:12, E:10	182
3	E:20	219
4	A:6, C:1	7
5	C:17	28
6	A:3, D:6, E:5	14
7	D:18, E:15	196
8	A:9	21

Table 8 The Cid list after adding transaction T_5 in Table 1

Item	CidSet
A	4, 6, 8
B	2
C	1, 4, 5
D	2, 6, 7
E	1, 2, 3, 6, 7

Table 9 The closed HTWU itemsets after adding transaction T_5 in Table 1

Closed HTWU itemset	Equal TWU itemset
{BDE}	{B}, {BD}, {BE}, {BDE}
{E}	{E}
{DE}	{D}, {DE}

CItemset {D:18 (=12 + 6), E:15 (=10 + 5)} in the Closed Table and assigns the new closed twu itemset {DE} a Cid 7, which is added to Cid List for items D and E. Because the two itemsets {D} and {DE} are the Equal TWU itemsets of {BDE} before adding transaction T_5, and {D} and {DE} both are contained in {DE}, the two itemsets become the Equal TWU itemset of {DE} after adding transaction T_5.

For the fourth record in $Temp_{ADD}$, the UItemset is {A} and the corresponding Closure_Id is 4. Because Cid 4 in the Closed Table (Table 3) is {AC}, which means that itemset {DE} is not a closed twu itemset and the Closure of {A} is {AC} before adding transaction T_5. The twu of {A} after adding transaction T_5 is the twu of {AC} in the Closed Table plus the tu of T5, that is 7 + 14 = 21, which is not a closed HTWU itemset. HUIStream adds the CItemset {A:9 (=6 + 3)} in the Closed Table and assigns the new closed twu itemset {A} a Cid 8, which is added to Cid List for item A. After adding transaction T_5, the Closed Table and Cid List are shown in Table 7 and Table 8, and the closed HTWU itemsets and the corresponding Equal TWU itemsets are shown in Table 9.

3.2 The Algorithm HUIStream⁻

In this subsection, we describe how HUIStream finds the closed twu itemsets which need to be updated and all the HTWU itemsets after deleting a transaction. The itemsets need to be updated after deleting a transaction are the subsets of T_{DEL}, since the twu of the subsets of T_{DEL} would be decreased after deleting the transaction. Therefore, the closed twu itemsets which need to be updated can be obtained by performing the intersections on T_{DEL} and all the closed twu itemsets before deleting transaction T_{DEL}. In order to avoid that the intersection is empty, HUI-Stream only performs the intersections on T_{DEL} and the closed twu itemsets with Cids in SET({T_{DEL}}) according to expression (5) after deleting transaction TDEL. The closed twu itemsets which need to be updated after deleting a transaction are recorded in a table $Temp_{DEL}$, which includes the two fields: DItemset records the closed twu itemset X which needs to be updated; C1 records the Cid of X before deleting the transaction; C2 records the information which can be used to determine if X is still a closed twu itemset after deleting the transaction. Because a deleted transaction T_{DEL} is a closed twu itemset before the deletion [20], HUIStream firstly puts T_{DEL} in the first record of $Temp_{DEL}$, and sets the corresponding C1 and C2 to be 0.

Because the intersection of T_{DEL} and different closed twu itemsets S may obtain the same itemset X, the field C1 in $Temp_{DEL}$ records the Cid p of the closed twu itemset with the largest twu among all the closed twu itemsets in S. Cid p is the Cid of itemset X, since itemset X is a closed itemset before deleting the transaction T_{DEL}. Because itemset X may not be a closed twu itemset after deleting the transaction, the field C2 in $Temp_{DEL}$ records the Cid q of the closed twu itemset with the largest twu among all the closed twu itemsets in S except the closed twu itemset with Cid p. If the twu of Cid q is equal to the twu of Cid p minus the tu of T_{DEL}, which means that the itemset with Cid q has the same twu as X, then X is not a closed twu itemset any more after deleting the transaction. HUIStream updates the content of the Closed Table according to the Table $Temp_{DEL}$.

For each record with values X, p and q for DItemset, C1 and C2 in $Temp_{DEL}$, respectively, the twu values of the closed twu itemsets with Cid p and Cid q can be obtained from Closed Table. If the twu of X minus the tu of T_{DEL} is equal to 0, which means that X is not contained in any transaction after deleting T_{DEL}, then itemset X with Cid p is removed from the Closed Table. If itemset Y with Cid q is not itemset X and the twu of X minus the tu of T_{DEL} is equal to the twu of Y, then X is not a closed twu itemset after deleting transaction T_{DEL}, since Y is a superset of X and they have the same twu values.

If X is still a closed twu itemset after the deletion, then HUIStream updates the twu of X and the utility of each item in X in the Closed Table as follows: the updated twu of X is the twu of X minus the tu of T_{DEL} and the updated utility of each item in X is the utility of the item in X minus the utility of the item in T_{DEL}. If X is not a closed HTWU itemset before deleting the transaction but is a closed

Table 10 The Temp_{DEL} after deleting the transaction T_1

DItemset	C1	C2
{CE}	1	0
{E}	3	7
{C}	5	4

HTWU itemset after the deletion, then HUIStream finds all the subsets of X, which have the same twu as X, that is all the Equal TWU itemsets of X.

For example, in Table 1, when the transaction T_1 = {CE} is deleted, HUIStream firstly puts {CE} in the field DItemset in Temp_{DEL} and the corresponding C1 and C2 are set to be 0. Because SET(T1) = CidSet(C) ∪ CidSet(E) = {1, 2, 3, 4, 5, 6, 7} according to expression (1) and Cid List in Table 8, HUIStream performs the intersections on T_1 and the closed twu itemsets with Cids 1, 2, 3, 4, 5, 6 and 7 from Closed Table in Table 7, respectively. Firstly, Cid 1 is itemset {CE} and {CE} ∩ {CE} = {CE} which exists in Temp_{DEL} and the corresponding C1 and C2 are 0 s. Because from Table 7, we can see that the twu of Cid 1 is greater than the twu of Cid 0, the Cid in C1 is changed to 1 and the Cid in C2 remains 0 for DItemset {CE} in the table Temp_{DEL}.

Secondly, because Cid 2 is itemset {BDE} and {CE} ∩ {BDE} = {E} which is not in Temp_{DEL}, the itemset {E} is added to Temp_{DEL}, and the corresponding C1 and C2 are set to 2 and 0, respectively. Thirdly, Cid 3 is itemset {E} and {CE} ∩ {E} = {E} has existed in Temp_{DEL}. Because the twu of Cid 3 is greater than the twu of Cid 2 and the twu of Cid 2 is greater than the twu of Cid 0, the Cid in C1 is replaced with 3 and the Cid in C2 is replaced with 2 for DItemset {E} in the table Temp_{DEL}. HUIStream continuously performs the intersections on T_1 and the closed twu itemsets with Cids 4, 5, 6, and 7, respectively, and updates the content of Temp_{DEL}. After deleting the transaction T_1, the Temp_{DEL} is shown in Table 10.

HUIStream updates the content of Closed Table and Cid List according to TempDEL. For example, in Table 10, the first record in DItemset is {CE} and C1 is Cid 1. Because the twu of {CE} with Cid 1 in Table 7 minus the tu of T_1 is equal to 0, itemset {CE} is not a closed twu itemset after deleting transaction T_1. Therefore, HUIStream removes the information about {CE} from Closed Table and Cid List. The second record in DItemset is {E} and C1 is Cid 3. Because the twu of {E} with Cid 3 in Table 7 minus the tu of T_1 is equal to the twu of {DE} with Cid 7 in C2, itemset {E} is not a closed twu itemset after deleting transaction T_1, since there exists a superset {DE} of {E} and they have the same twu values. HUIStream removes the information about {E} from the Closed Table and Cid List, and moves all the Equal TWU itemsets of {E} to the Equal TWU itemsets of {DE} with Cid 7. There is the same situation with the second record for the third record in TempDEL. All the information about itemset {C} is removed from the Closed Table and Cid List. After deleting transaction T1 from Table 1, the Closed Table and Cid List are shown in Table 11 and Table 12, respectively, and the closed HTWU itemsets and their Equal TWU itemsets are shown in Table 13.

Table 11 The closed table after deleting the transaction T_1 from Table 1

Cid	CItemset	twu
0	0	0
2	B:160, D:12, E:10	182
4	A:6, C:1	7
6	A:3, D:6, E:5	14
7	D:18, E:15	196
8	A:9	21

Table 12 The Cid list after deleting the transaction T_1 from Table 1

Item	CidSet
A	4, 6, 8
B	2
C	4
D	2, 6, 7
E	2, 6, 7

Table 13 The closed HTWU itemsets and their equal TWU itemsetsafter deleting T_1

Closed HTWU itemset	Equal TWU itemset
{BDE}	{B}, {BD}, {BE}, {BDE}
{DE}	{D}, {E}, {DE}

Table 14 The candidate high utility itemsets and their utilities

Closed HTWU Itemset	Equal TWU itemsets
{B:160, D:12, E:10}	u({B}) = 160 u({BD}) = 172 u({BE}) = 170 u({BDE}) = 182
{D:18, E:15}	u({D}) = 18 u({E}) = 15 u({DE}) = 33

3.3 High Utility Itemset Generation

After processing the added and deleted transactions, all the Equal TWU itemsets of the closed HTWU itemsets are the candidate high utility itemsets. The utility of each Equal TWU itemset X for each closed HTWU itemset Y can be obtained by accumulating the utility of each item of X in Y from the Closed Table, and then all the high utility itemsets can be obtained without scanning the database.

For example, from Table 13, we can see that the itemsets {BDE} and {DE} are closed HTWU itemsets. For itemset {BDE} with Cid 2, the utility of {BDE} is 182(=B:160 + D:12 + E:10), which can be obtained from Closed Table in Table 11. The utility of the Equal TWU itemset {BD} of {BDE} is 172 (=B:160 + D:12), The utility of the Equal TWU itemset {BE} of {BDE} is 170 (=B:160 + E:10). All the candidate high utility itemsets and their utilities after deleting transaction T_1 are shown in Table 14, in which itemsets {B},{BD},{BE} and {BDE} are high utility itemsets.

4 Experimental Results

In this section, we evaluate the performance of our HUIStream algorithm and compare it with IHUP algorithm [3]. Our experiments are performed on Intel(R) Core(TM) 2 Quad CPU Q9400 @ 2.66 GHz with 4 GB RAM and running on Windows XP. The two algorithms are implemented in JAVA language.

We first generate two synthetic datasets T5I2D100 K and T5I4D100 K by using IBM Synthetic Data Generator [22], in which T is the average length of the transactions, I is the average size of maximal potentially frequent itemsets and D is the total number of the transactions. The number of distinct items is set to 1000. For the profit of each item, we use the log Normal Distribution [2, 7] and set the range of the profits between 0.01 and 10, which is shown in Fig. 1. The purchased quantity for an item in a transaction is randomly set to the number between 1 and 10.

Figure 2 and Fig. 3 show the execution time of IHUP and HUIStream, which the utility threshold is set to be 0.1%, the number of transactions is increased from 10 K to 100 K, and the size of sliding window is set to be 1 K and 10 K, respectively. From the experiments, we can see that HUIStream outperforms IHUP, and the performance gap increases as the number of transactions increases and the times of window size movements increases, since HUIStream only updates the closed twu

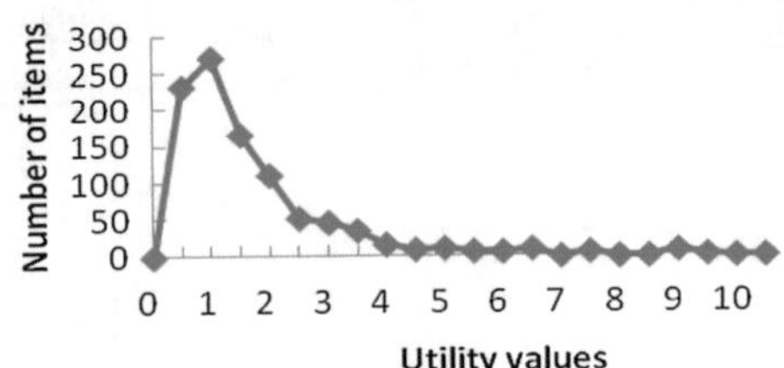

Fig. 1 Utility value distribution in utility table

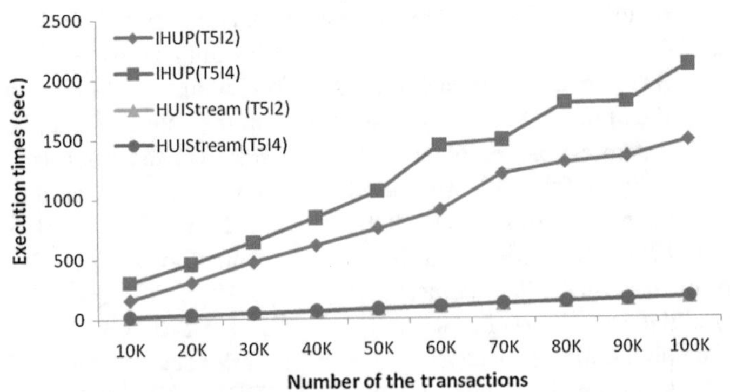

Fig. 2 The execution time for the two algorithms on window size = 1 K

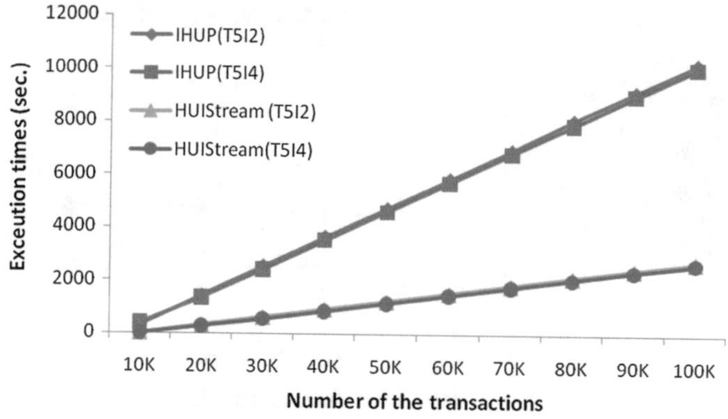

Fig. 3 The execution time for the two algorithms on window size = 10 K

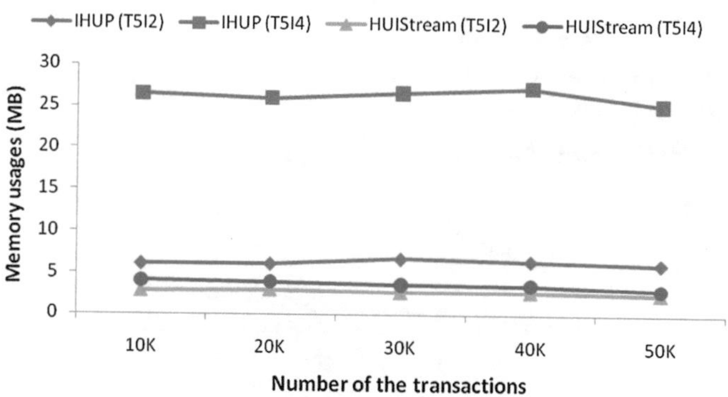

Fig. 4 The memory usages for the two algorithms on window size = 1 K

itemsets related to the added or deleted transactions. However, IHUP needs to re-mine HTWU itemsets by using FP-Growth algorithm [5] and rescan the database to find high utility itemsets from the HTWU itemsets.

Figures 4 and 5 show the memory usages for the two algorithms IHUP and HUIStream when the number of transactions is increased from 10 K to 50 K and the size of sliding window is set to be 1 K and 10 K, respectively, from which we can see that the memory usage for IHUP is significantly larger than that of HUI-Stream. This is because IHUP needs to recursively construct the subtrees for re-mining HTWU itemsets when a transactions are added or deleted, but HUI-Stream only needs to store and update the Closed Table and Cid List.

In the following experiments, we generate the two datasets T10I4D100K and T10I6D100K. The number of distinct items is set to 2000, and the utility threshold is set to be 0.1%. s Figures 6 and 7 show the execution time of IHUP and HUIStream, which the number of transactions is increased from 10 K to 50 K, and the size of sliding window is set to be 1 K and 10 K, respectively. From Fig. 6,

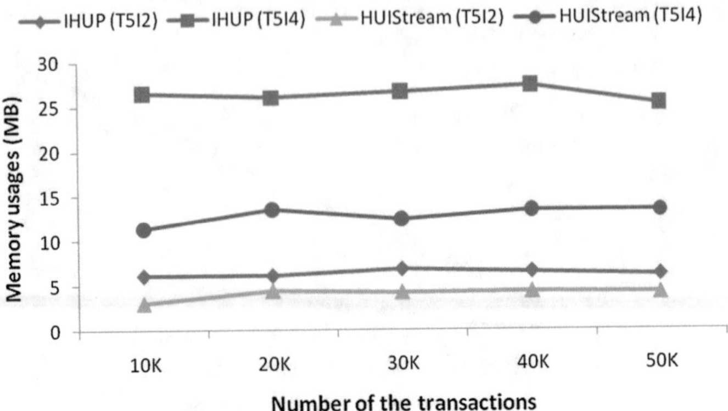

Fig. 5 The memory usages for the two algorithms on window size = 10 K

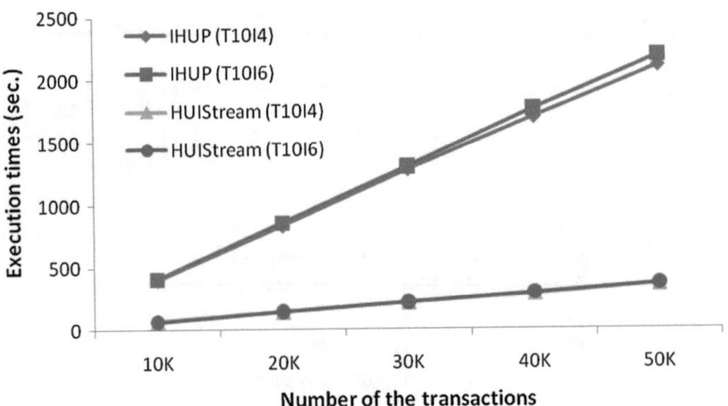

Fig. 6 The execution time for the two algorithms on window size = 1 K

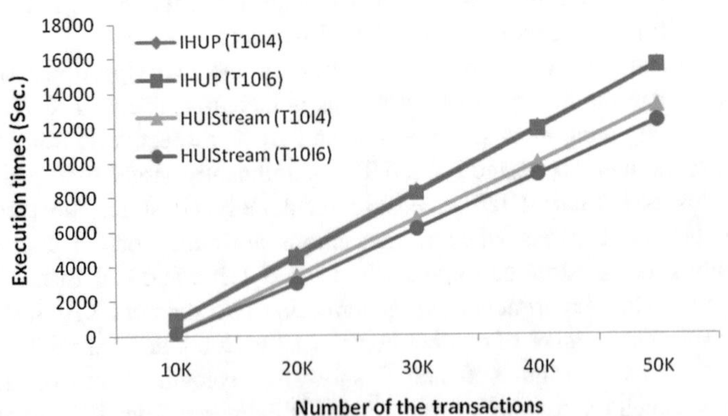

Fig. 7 The execution time for the two algorithms on window size = 10 K

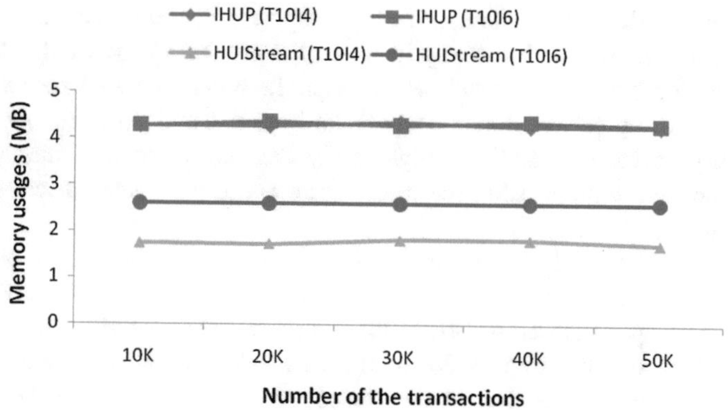

Fig. 8 The memory usages for the two algorithms on window size = 1 K

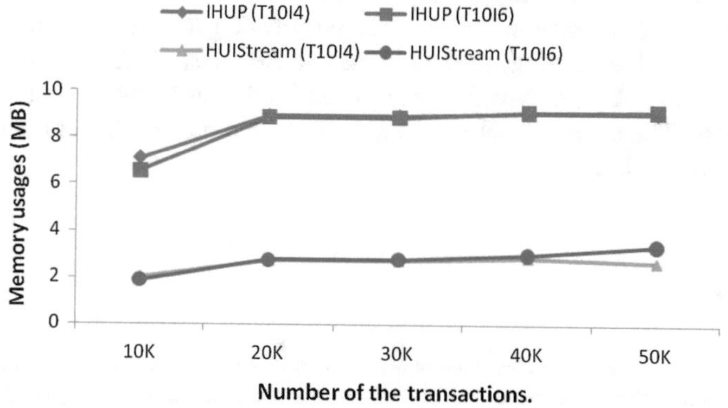

Fig. 9 The memory usages for the two algorithms on window size = 10 K

we can see that HUIStream outperforms IHUP and the performance gap increases as the number of transactions increases. Although the performance gaps are similar when the number of transactions increases from 10 K to 50 K, HUIStream still outperforms IHUP in Fig. 7. The memory usages for IHUP and HUIStream in this experiment are shown in Figs. 8 and 9, from which we can see that the memory usage for HUIStream still less than the memory usage for IHUP on the datasets with longer transaction size.

5 Conclusion

There are many previous approaches for mining high utility itemsets in a data stream. However, they all first need to generate a large number of candidate high utility itemsets and then scan the whole database to caculate the utility for each high utility itemset. Although some approaches propose some strategies to reduce the number of the candidate high utility itemsets, the number of the candidates is still large when the size of the database is large. In order to avoid rescanning the database, some approaches store the whole database in a tree structure, but they also need to re-generate the candidate high utility itemsets when the transactions are added or deleted without using the information about previously discovered high utility itemsets.

In order to improve the performance of the previous approaches, we propose an algorithm HUIStream for mining high utility itemsets over a data stream. We take use of the concept of closed itemsets [20] and propose the definition of closed twu itemsets which can be used to derive the twu values of all the itemsets in the database. Because the number of the closed twu itemsets is much less than the number of the itemsets in the database, HUIStream only keeps all the closed twu itemsets, such that the twu values of all the itemsets in the database can be reserved. Therefore, our approach only needs to update the closed twu itemsets about the added or deleted transaction without any information loss when a transaction is added or deleted. According to the closed twu itemsets, HUIStream can directly obtain the high utility itemsets from the closed HTWU itemsets without rescanning the database. The experimental results also show that our HUIStream outperforms the other approaches.

References

1. Agrawal, R., Srikant, R.: Fast Algorithms for Mining Association Rules. In: Proceedings of 20th International Conference on Very Large Databases, Santiago, Chile, 487–499 (1994)
2. Ahmed, C. F., Tanbeer, S. K., Jeong B.S., Lee. Y.K.: An Efficient Candidate Pruning Technique for High Utility Pattern Mining. In: Proceedings of the 13th Pacific-Asia Conference on Knowledge Discovery and Data Mining, 749–756 (2009)
3. Ahmed, C. F., Tanbeer, S. K., Jeong, B. S., Lee, Y. K.: Efficient tree structures for high utility pattern mining in incremental databases. IEEE Transactions on Knowledge and Data Engineering, **21**(12), 1708–1721 (2009)
4. Brin, S., Motwani, R., Ullman, J., Tsur, S.: Dynamic itemset counting and implication rules for market basket data. In: Proceedings ACM SIGMOD International Conference on management of Data, 255–264 (1997)
5. Han, J., Pei, J., Yin, Y., Mao, R.: Mining frequent patterns without candidate generation: a frequent-pattern tree approach. Data Mining and Knowledge Discovery, **8**(1), 53–87 (2004)
6. Li, H.F., Huang, H.Y., Chen, Y.C., Liu, Y.J., Lee, S.Y.: Fast and memory efficient mining of high utility itemsets in data streams. In: Proceedings of the 8th IEEE International Conference on Data Mining, 881–886 (2008)
7. Liu, Y., Liao, W. K., Choudhary, A.: A Fast High Utility Itemsets Mining Algorithm. In: Proceedings of the International. Workshop on Utility-Based Data Mining, 90–99 (2005)
8. Lin, C. W., Lan, G. C., Hong, T. P.: Mining high utility itemsets for transaction deletion in a dynamic database. Intelligent Data Analysis **19**(1), 43–55 (2015)
9. Li, Y.C., Yeh, J.S., Chang, C.C.: Isolated Items Discarding Strategy for Discovering High Utility Itemsets. Data and Knowledge Engineering, **64**(1), 198–217 (2008)
10. Morteza, Z., Aijun, A.: Mining top-k high utility patterns over data streams. Information Sciences, **285**(1), 138–161 (2014)

11. Mohammad, E., Osmar, R. Z.: COFI approach for mining frequent itemsets revisited. In: Proceedings of the 9th ACM SIGMOD workshop on Research issues in data mining and knowledge discovery, 70–75 (2004)
12. Park, J. S., Chen, M. S., Yu, P. S.: An Effective Hash-Based Algorithm for Mining Association Rules. ACM SIGMOD **24**(2), 175–186 (1995)
13. Ryang, H., Yun, U.: High utility pattern mining over data streams with sliding window technique. Expert Systems with Applications, **57**, 214–231(2016)
14. Tseng, S.M., Chu, C. J., Liang, T.: Efficient mining of temporal high utility itemsets from data streams. In: Proceedings of the ACM International Conference on Utility-Based Data Mining Workshop, 18–27 (2006)
15. Tseng, S.M., Shie, B.E., Philip Yu, S.: Efficient algorithms for mining high utility itemsets from transactional databases. IEEE Transactions on Knowledge and Data Engineering, **25**(8), 1772–1786 (2013)
16. Tseng, S.M., Wu, C.W., Shie, B.E., Philip Yu, S: UP-Growth: an efficient algorithm for high utility itemset mining. In: ACM SIGKDD, 253–262 (2010)
17. Wang, J., Han, J., Pei, J.: CLOSET+: Searching for the Best Strategies for Mining Frequent Closed Itemsets. In: Proceedings of the 9th ACM SIGKDD International Conference on Knowledge Discovery and Data Mining, 236–245 (2003)
18. Yen, S.J., Chen, C.C., Lee, Y.S.: A fast algorithm for mining high utility Itemsets. In: Proceedings of International Workshop on Behavior Informatics, joint with the 15th Pacific-Asia Conference on Knowledge Discovery and Data Mining (PAKDD), 171–182 (2011)
19. Yun, U., Ryang, H.: Incremental high utility pattern mining with static and dynamic databases. Applied Intelligence, **42**(2), 323–352(2015)
20. Yen, S.J., Wu, C.W., Lee, Y.S., Vincent Tseng, S.: A Fast Algorithm for Mining Frequent Closed Itemsets over Stream Sliding Window. In: Proceedings of IEEE International Conference on Fuzzy Systems (FUZZ-IEEE), 996–1002 (2011)
21. Yen, S. J., Wang, C. K., Ouyang, L. Y.: A search space reduced algorithm for mining frequent patterns. Journal of Information Science and Engineering, **28** (1), 177–191 (2012)
22. IBM Synthetic Data Generator http://www.almaden.ibm.com/software/quest/Resorces/index.shtml

Event Detection in Location-Based Social Networks

Joan Capdevila, Jesús Cerquides and Jordi Torres

Abstract With the advent of social networks and the rise of mobile technologies, users have become ubiquitous sensors capable of monitoring various real-world events in a crowd-sourced manner. Location-based social networks have proven to be faster than traditional media channels in reporting and geo-locating breaking news, i.e. Osama Bin Laden's death was first confirmed on Twitter even before the announcement from the communication department at the White House. However, the deluge of user-generated data on these networks requires intelligent systems capable of identifying and characterizing such events in a comprehensive manner. The data mining community coined the term, *event detection*, to refer to the task of uncovering emerging patterns in data streams. Nonetheless, most data mining techniques do not reproduce the underlying data generation process, hampering to self-adapt in fast-changing scenarios. Because of this, we propose a probabilistic machine learning approach to event detection which explicitly models the data generation process and enables reasoning about the discovered events. With the aim to set forth the differences between both approaches, we present two techniques for the problem of event detection in Twitter: a data mining technique called Tweet-SCAN and a machine learning technique called WARBLE. We assess and compare both techniques in a dataset of tweets geo-located in the city of Barcelona during its annual festivities. Last but not least, we present the algorithmic changes and data processing frameworks to scale up the proposed techniques to big data workloads.

J. Capdevila (✉) · J. Torres
Universitat Politècnica de Catalunya (UPC), Barcelona Supercomputing Center (BSC), Barcelona, Spain
e-mail: jc@ac.upc.edu; capdevila.pujol.joan@gmail.com

J. Torres
e-mail: torres@ac.upc.edu

J. Cerquides
Artificial Intelligence Research Institute (IIIA), Spanish National Research Council (CSIC), Madrid, Spain
e-mail: cerquide@iiia.csic.es

W. Pedrycz and S.-M. Chen (eds.), *Data Science and Big Data: An Environment of Computational Intelligence*, Studies in Big Data 24,
DOI 10.1007/978-3-319-53474-9_8

Keywords Event detection · Social networks · Geolocation · Twitter · Anomaly detection · DBSCAN · Topic models · Probabilistic modeling · Variational inference · Apache spark

1 Introduction

Sensor networks are systems composed of several tenths of spatially-distributed autonomous devices capable of monitoring their surroundings and communicating with their neighbors [2]. Detecting abnormal behaviors in these networks have attracted the interest of different communities ranging from communications [36] to data mining [14]. In particular, the task of detecting and characterizing anomalous subgroups of measurements that emerge in time has been coined as *event detection* and it has found many applications in surveillance systems, environmental monitoring, urban mobility, among many others [43].

In contrast, social networks came about to interconnect users mainly for communication purposes. However, the rise of mobile technologies and positioning systems have turned users into ubiquitous sensors capable of monitoring and reporting real-world events (i.e. music concert, earthquakes, political demonstration). Most of these events are very challenging to detect through sensor networks, but location-based social networks, which incorporate geo-tagging services, have shown to report them even faster than traditional media [48]. For example, Mumbai terrorist attacks were instantly described on Twitter by several eyewitness in the crime area [40] and Osama Bin Laden's death was first revealed on the same platform before the communication department at the White House had even confirmed his death [33].

Therefore, there has recently been a growing interest to build intelligent systems which are able to automatically detect and summarize interesting events from online social content [34]. In particular, Twitter has attracted most of the attention in both research and industry because of its popularity[1] and its accessibility[2] [3]. Tweet messages respond to the *What's happening?* question through a 140-character-long text message, and tweet meta-data might also contain details about the when, where and who [45]. Social networks in general, and Twitter in particular, are classic big data scenarios in which large volumes of heterogeneous data are generated in streaming by millions of uncoordinated users. Applications such as event detection have to consider these challenges in order to generate veracious knowledge from this data. In other words, event detection in Twitter has to deal with the 5 Vs defined in big data: *volume*, *velocity*, *variety*, *veracity* and *variability*.

In this chapter, we present two techniques for retrospective event detection, that is to say that both techniques seek to discover events from historical data, not from a stream. As a result, *velocity* is disregarded for this retrospective study, but left for future work in online or prospective setups. Both techniques deal with tweet *variety*

[1] https://blog.twitter.com/2013/new-tweets-per-second-record-and-how.

[2] https://dev.twitter.com/rest/public.

by modeling the spatial, temporal and textual dimensions of a tweet independently. They could also be extended to take into account other forms of data (image, video, etc.).

The first technique, called Tweet-SCAN [9], is based on the Density-based Spatial Clustering of Applications with Noise (DBSCAN) [18]. This is a well-known algorithm for bottom-up event detection from the data mining community [43]. This algorithm identifies as events groups of densely packed tweets which are about similar themes, location and time period. However, such techniques do not consider uncertain measurements (i.e. GPS errors) or partial information (i.e. tweets without location), compromising the *veracity* of the results. Moreover, these detection techniques lack of knowledge about the data generation process hampering them to adapt in *varying* scenarios. Nonetheless, parallel and distributed versions of DBSCAN [21] are enabling to scale up event detection in large datasets [12].

On the other hand, computational intelligent approaches like probabilistic models and learning theory can help to mitigate some of these issues by accounting for the uncertainty in a very principled way [5]. WARBLE [10], the second technique presented here, follows this approach and tackles the event detection problem through heterogeneous mixture models [4]. These are probabilistic models that represent subpopulations within an overall population, and each sub-population might be generated by a different statistical distribution form. Last but not least, recent advances in approximate inference have mitigated the high computational cost in learning probabilistic models in scenarios with large *volumes* of data [24].

The rest of this chapter is structured as follows. In Sect. 2, we define the problem of event detection in location-based social networks. We then provide the necessary background regarding DBSCAN and mixture models in Sect. 3. Section 4 contains detailed explanation about the two event detection techniques and their scaling in the presence of large data volumes. Tweet-SCAN is described in Sect. 4.1 and WARBLE, in Sect. 4.2. The experimental setup and results is in Sect. 5. We first introduce "La Mercé" dataset for local event detection in Sect. 5.1, we then present the metrics to evaluate the detection performance in Sect. 5.2 and we ultimately evaluate both techniques in Sect. 5.3. Finally, Sect. 6 presents some conclusions out of this chapter and points out to several future steps.

2 Problem Definition

Event detection in social networks lacks of a formal definition for an event, hampering the progress of this field. Broadly speaking, [30] defined an event as "a *significant* thing that happens at some specific time and place". However, this definition does not specify what *significant* means in the context of social networks. Lately, [34] built on top of this definition to provide the following one:

Definition 1 Event (e): In the context of Online Social Networks (OSN), (significant) event (e) is something that cause (a large number of) actions in the OSN.

Note first that it does not constrain an event to happen at some specific time and place, in contrast to [30]. This enables to have a more general definition from which we can then distinguish several event types (Global, Local or Entity-related) depending on the constraints. However, this definition still lacks of some sort of formalization regarding the significant number of actions in the social network (e.g. post new content or accept a friend request). With the aim to unify and formalize this, we add to Definition 1 the idea that events are caused by abnormal occurrences:

Definition 2 Event (e): In the context of Online Social Networks (OSN), (significant) event (e) is something that cause an *abnormal* number of actions in the OSN.

This definition resembles that of event detection in sensor networks [43], in which events are anomalous occurrences that affect a subgroup of the data. Note also that this captures more complex events than Definition 1. For example, an abnormal decrease of actions in the social network, as it might initially happen during a shooting in a crowded area, should be also considered a significant event.

Moreover, location-based social networks have enabled to narrow down the scope of events to geo-located events [48], enabling the identification of many real-world occurrences such as music concerts, earthquakes or political demonstrations. Moreover, by restricting the geographical dimension of such events, we are able to identify local events taking place in urban environments, which will be the application of the techniques presented in this chapter.

Therefore, the task of event detection in a social network consists of identifying and characterizing a set of events that are anomalous with respect to a baseline. This task can be performed either retrospectively or prospectively. While the former aims to retrieve events from historical data in a batch mode, the latter seeks to identify them in streaming data in an online fashion. In the following sections, we will present two different approaches to retrospectively uncover these anomalous patterns:

1. Tweet-SCAN: A data mining approach based on DBSCAN [18] in which events are groups of posts (i.e. tweets) that are more densely packed than the baseline.
2. WARBLE: A probabilistic approach based on heterogeneous mixture models [4] in which events are groups of posts (i.e. tweets) generated by a statistical distribution different from that of non-event tweets.

Both techniques follows the anomaly-based approach to event detection by assuming that events are groups of similar tweets (in space, time and textual meaning) and they are masked by tones of non-event tweets such as *memes*, user conversations or re-post activities. While Tweet-SCAN considers distance as the metric for similarity, WARBLE uses probability to assess the pertinence to event or non-event.

3 Background

In this section, we present digested background regarding DBSCAN and mixture models, methods that are used by the later proposed techniques. Both methods have been used for clustering in applications with noise. We instead propose them for retrospective event detection, given that the noise component can be used to for modeling the baseline or expected behavior.

3.1 DBSCAN

DBSCAN [18] was initially proposed to uncover clusters with arbitrary shapes whose points configure a dense or packed group. This means that for each point in a cluster its neighborhood at a ϵ distance must contain at least a minimum number of points, *MinPts*. Formally, this implies the definition of two predicates:

1. $NPred(o, o') \equiv N_\epsilon(o, o') = |o - o'| \leq \epsilon$.
2. $MinWeight(o) \equiv |\{o' \in D \mid |o - o'| \leq \epsilon\}| \geq MinPts$.

The fulfillment of both predicates allows to define the notion of a point p being directly density-reachable from another point q, see (left) Fig. 1, where ϵ is given by the circle radius and *MinPts* is set to 2. In this scenario, q is a *core point* because it satisfies both predicates and p is a *border point* since it breaks the second predicate. The notion of being direct reachable is extended to density-reachable points when p and q are far apart, but there is a chain of points in which each pair of consecutive points are directly density-reachable, as it is the case in (middle) Fig. 1. Finally, it might happen that p and q are not density-reachable, but there is a point o from which they are both density-reachable, that is when p and q are said to be density-connected, for example in (right) Fig. 1. Note that both points, p and q, are here *border points*, while o is a *core point*.

Consequently, a cluster in DBSCAN is defined to be a set of density-connected points that contains all possible density-reachable points. Furthermore, *noise points* can now be defined as those points which do not belong to any cluster since they are not density-connected to any.

GDBSCAN [39] generalizes DBSCAN by redefining the above-mentioned predicates to cope with spatially extended objects. For example, the neighborhood of a set of polygons is defined by the intersect predicate instead of a distance function. It is also the case for a set of points with financial income attributes within a region whose *MinWeight* predicate is a weighted sum of incomes instead of mere point cardinality, so that clusters become regions with similar income. Therefore, both predicates can be generalized as follows:

1. $NPred(o, o')$ is binary, reflexive and symmetric.
2. $MinWeight(o) \equiv wCard(\{o' \in D \mid NPred(o, o')\}) \geq MinCard$, where *wCard* is a function that $2^D \rightarrow \mathbb{R}^{\geq 0}$.

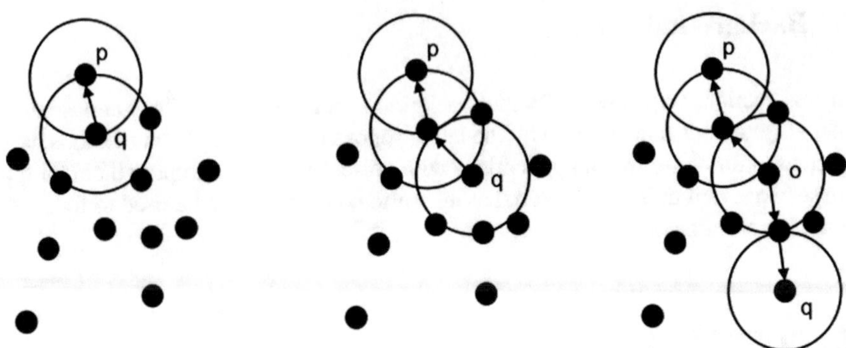

Fig. 1 Directly density-reachable (*left*), density-reachable (*middle*) and density-connected (*right*) points

These new predicates enable to extend the concept of density-connected points to objects and thus generalize density-based clustering to spatially extended objects, like geo-located tweets. Moreover, we note that DBSCAN-like techniques have been considered for event detection in sensor networks as a *bottom-up* approach [43].

3.2 Mixture Models

Mixture models are probabilistic models for representing the presence of subpopulations within an overall population and they have been very popular for clustering and unsupervised learning. Mixture of Gaussians or Gaussian Mixture Models (GMM) are the most widely used mixture model [31]. In this model, each mixture component is a multivariate Gaussian with mean μ_k and covariance Σ_k parameters. This means that given the component assignment c_n, the generative process for the n-th observations is,

$$x_n \sim N\left(\mu_{c_n}, \Sigma_{c_n}\right). \tag{1}$$

[4] proposed a more general model in which not all mixture components share the same distributional form. In particular, observations from one of the mixture components came from a Poisson process associated with noise. Therefore, the generative process can be rewritten as,

$$x_n \sim \begin{cases} N\left(\mu_{c_n}, \Sigma_{c_n}\right) & c_n < K \\ U(x_{min}, x_{max}) & c_n = K \end{cases} \tag{2}$$

where $U(x_{min}, x_{max})$ corresponds to a multivariate uniform distribution with x_{min}, the most south-western point and x_{max}, the most north-eastern point. This model has shown to perform reasonably well in cluster recovery from noisy data in both synthetic and real datasets [20].

Heterogeneous mixture models enable to propose generative models for event detection, in which event-related observations, i.e. tweets, are drawn from a distributional ($c_n < K$) form that entails locality in the temporal, spatial and textual dimensions, while non-event data points are generated from the background distribution ($c_n = K$).

4 Event Detection Techniques

4.1 *Tweet-SCAN: A Data Mining Approach*

Tweet-SCAN [9] is defined by specifying the proper neighborhood and MinWeight predicates introduced in Sect. 3.1 for GDBSCAN in order to associate density-connected sets of tweets to real-world events. Next, we introduce both predicates and the text model for the textual component of a tweet.

4.1.1 Neighborhood Predicate

Most event-related tweets are generated throughout the course of the event within the area where it takes place. Consequently, we need to find sets of tweets density-connected in space and time, as well as in meaning.

We also note that closeness in space is not comparable to time, nor to meaning. Because of this, Tweet-SCAN is defined to use separate positive-valued ϵ_1, ϵ_2, ϵ_3 parameters for space, time and text, respectively. Moreover, specific metrics will be chosen for each dimension given that each feature contains different type of data.

The neighborhood predicate for a tweet o in Tweet-SCAN can be expressed as follows,

$$NPred(o, o') \equiv |o_1 - o'_1| \leq \epsilon_1,\ |o_2 - o'_2| \leq \epsilon_2,\ |o_3 - o'_3| \leq \epsilon_3 \qquad (3)$$

where $|o_i - o'_i|$ are distance functions defined for each dimension, namely space, time and text. The predicate symmetry and reflexivity are guaranteed as long as $|o_i - o'_i|$ are proper distances. Particularly, we propose to use the Euclidean distance for the spatial and temporal dimensions given that latitude and longitude coordinates as well as timestamps are real-valued features and the straight line distance seems a reasonable approximation in this scenario. The metric for the textual component will be defined later once we present the text model for Tweet-SCAN.

4.1.2 MinWeight Predicate

Tweet-SCAN seeks to group closely related tweets generated by a diverse set of users instead of a reduced set of them. User diversity is imposed to avoid that a single user continuously posting tweets from nearby locations could trigger a false event in Tweet-SCAN. Forcing a certain level of user diversity within a cluster can be achieved through two conditions in the *MinWeight* predicate that must be satisfied at the same time,

$$MinWeight(o) \equiv |N_{NPred}(o)| \geq MinPts, \;\; UDiv(N_{NPred}(o)) \geq \mu \tag{4}$$

where $N_{NPred}(o)$ is the set of neighboring tweets of o such that $\{o' \in D \mid NPred(o, o')\}$ w.r.t. the previously defined Tweet-SCAN neighborhood predicate. The first condition from the MinWeight predicate establishes that neighboring tweets must have a minimum cardinality *MinPts* as in DBSCAN. While in the second condition, the user diversity *UDiv*() ratio, which is defined as the proportion of unique users within the set $N_{NPred}(o)$, must be higher than a given level μ of user diversity.

4.1.3 Text Model

The text message in a tweet is a 140-character-long field in which users type freely their thoughts, experiences or conversations. The fact that users tweet in different languages, argots and styles dramatically increases the size of the vocabulary, making the use of simple Bag of Words (BoW) models [38] not viable. Therefore, we propose to use probabilistic topic models, which are common dimensionality reduction tools in text corpus [6]. In this approach, a tweet message is encoded into a K-dimensional vector which corresponds to the Categorical probability distribution over the K topics. K is often much smaller than the vocabulary size and the resulting topics are represented by semantically similar words.

Nonparametric Bayesian models like Hierarchical Dirichlet Process (HDP) [41] can automatically infer the number of topics K, overcoming the limitation of their parametric counterparts like Latent Dirichlet Allocation (LDA) [7]. The HDP topic model basically consists of two nested Dirichlet Process: G_o, with base distribution H and concentration parameter γ, and G_i, with base distribution G_o and concentration parameter α_o. Although the number of topics is automatically inferred, the hyperparameters γ and α_o might strongly influence the number of components. Because of this, vague informative gamma priors such as, $\gamma \sim Gamma(1, 0.1)$ and $\alpha_o \sim Gamma(1, 1)$ are usually considered [17, 41].

The straightforward use of HDP models on raw tweets does not provide meaningful topic distributions [25] due to the lack of word co-occurrence in short texts like tweets. Because of this, we propose the scheme from Fig. 2 which aims to alleviate these shortcomings. First, raw tweets, modeled as Bag of Words, are preprocessed and **cleaned** through classical data cleaning techniques from Natural Lan-

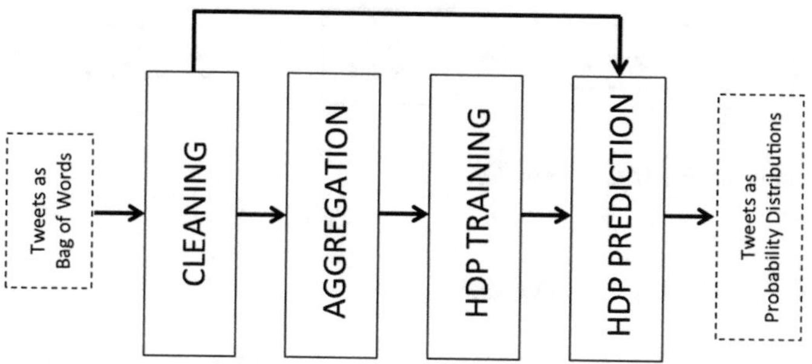

Fig. 2 Text model scheme. Stages are highlighted in *bold* in the text

guage Processing (NLP): lowering case, removing numbers and special characters, and stripping white-spaces. Then, processed tweets are **aggregated** to build longer training documents from a group of concatenated tweets. These aggregated documents are used to **train** the HDP model. Finally, the trained HDP model is employed to **predict** the topic distributions per each single tweet in order to obtain the Categorical probability distributions over the K topics that summarize each tweet message.

In the aggregation stage, we consider the aggregation scheme by top key terms proposed in [25]. This consists in first identifying a set of top key terms through the TF-IDF statistic [37], and then aggregating tweets that contains each of these top keywords. Thus, there will be as many training documents as top key terms and very few tweets will be unassigned as long as we choose a reasonable number of top keywords.

Finally, we propose to use the Jensen-Shannon (JS) distance for the textual component in Tweet-SCAN neighborhood predicate. JS is a proper distance metric for probability distributions [16]. It is defined as,

$$JS(p, q) = \sqrt{\frac{1}{2}D_{KL}(p||m) + \frac{1}{2}D_{KL}(q||m)} \tag{5}$$

where p, q and m are probability distributions and $D_{KL}(p||m)$ is the Kullback-Leibler divergence between probability distribution p and m written as,

$$D_{KL}(p||m) = \sum_i p(i)log_2\frac{p(i)}{m(i)} \quad m = \frac{1}{2}(p + q) \tag{6}$$

where m is the average of both distributions.

In Tweet-SCAN, p and q from Eq. (5) are two Categorical probability distributions over topics which are associated to two tweet messages. Given that Jensen-Shannon distance is defined through base 2 logarithms, JS distance will output a real value within the [0, 1]. Documents with the similar topic distribution will have a Jensen-Shannon distance close to 0 and those topic distributions which are very far apart, distance will tend to 1.

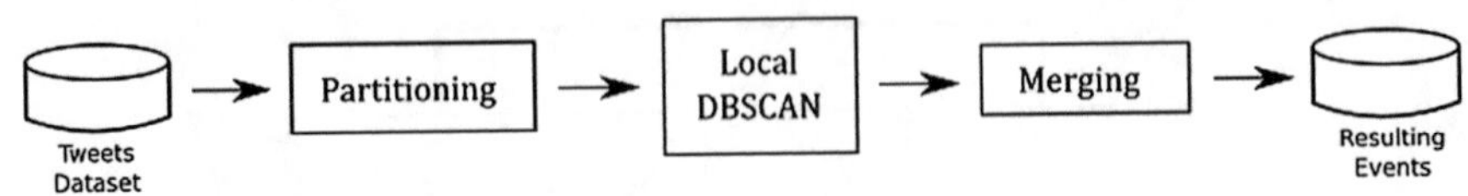

Fig. 3 Simplified MR-DBSCAN workflow

4.1.4 Scaling up to Large Datasets

To scale up Tweet-SCAN to large datasets, we propose to build on current parallel versions of DBSCAN such as MR-DBSCAN [22] which parallelizes all the critical sub-procedures of DBSCAN. The MR-DBSCAN workflow, shown in Fig. 3, first partitions the full dataset, then performs local DBSCAN clustering in each partition, and finally merges the local clusters into global ones, which correspond to events in our case.

An implementation of MR-DBSCAN in Apache Spark named RDD-DBSCAN was proposed by [15]. Apache Spark [46] is a computing framework in which distributed data collections, called Resilient Distributed Datasets (RDD), can be cached into memory for fast map-reduce operations.

The extension of DBSCAN algorithm for large scale event detection based on RDD-DBSCAN was developed by [12] and preliminary results show that by increasing parallelism we can reduce computation time.

4.2 WARBLE*: A Machine Learning Approach*

Next, we introduce WARBLE [10] a probabilistic model and learning scheme to uncover events from tweets through heterogeneous mixture models introduced in Sect. 3.2.

4.2.1 Probabilistic Model

[29] proposed a probabilistic model for event detection based on homogeneous mixture models in which each mixture component shares the same distributional form. Formally, they assume that the n-th tweet $\mathbb{T}_n$ is generated according to,

$$\mathbb{T}_n \sim f\left(\beta_{e_n}\right) \tag{7}$$

where f is the probability distribution function (pdf), common for all mixture components and β_k are the distribution parameters corresponding to the k-th mixture component.

As argued in the introduction, a vast majority of tweets is not event-related. Therefore, we would like to address rarity of event data by introducing a new mixture com-

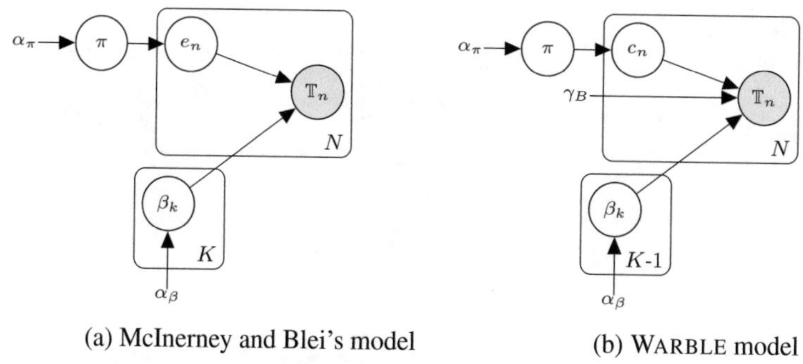

(a) McInerney and Blei's model (b) WARBLE model

Fig. 4 Simplified probabilistic graphical models (PGMs)

ponent, to which we will refer as **background**, which contains those tweets which are not part of any event. In probabilistic terms, it seems clear that the distribution of tweets inside the background component should be widely different from that inside events.

Accordingly, the WARBLE model generalizes McInerney and Blei's model to handle heterogeneous components as introduced in Sect. 3. To do that, for each component k, we enable a different base function f_k as

$$\mathbb{T}_n \sim f_{c_n}\left(\beta_{c_n}\right) \tag{8}$$

where the latent variables are now symbolized as c_n to denote that a tweet might be generated by event component ($c_n < K$) or by background ($c_n = K$).

Figure 4 shows simplified probabilistic graphical models (PGMs) [27] for McInerney and Blei's and our proposals. The proposed WARBLE model uses a different distributional form γ_B for the K-th mixture component.

Moreover, geo-located tweets tends to be unevenly distributed through space and time. For example, it is known that users are more likely to tweet during late evening and from highly populated regions [28]. Consequently, the background component ($c_n = K$) needs to cope with density **varying spatio-temporal distributions**.

In particular, we propose to model the distributional form γ_B for the background component through two independent histogram distributions for time and space with parameters T_B and L_B, respectively. The temporal histogram distribution is represented through a piecewise-continuous function which takes constant values (T_{B_1}, T_{B_2}, ... $T_{B_{I_T}}$) over the I_T contiguous intervals of length b. Similarly, the spatial background is modeled through a 2d-histogram distribution over the geographical space, which is represented in a Cartesian coordinate system. The 2d-piecewise-continuous function is expressed through I_L constant values ($L_{B_1}, L_{B_2}, \ldots L_{B_{I_L}}$) in a grid of squares with size $b \times b$ each.

Figure 5 shows the complete probabilistic graphical model for the WARBLE model, where tweets $\mathbb{T}_n$ are represented by their temporal t_n, spatial l_n and textual w_n features.

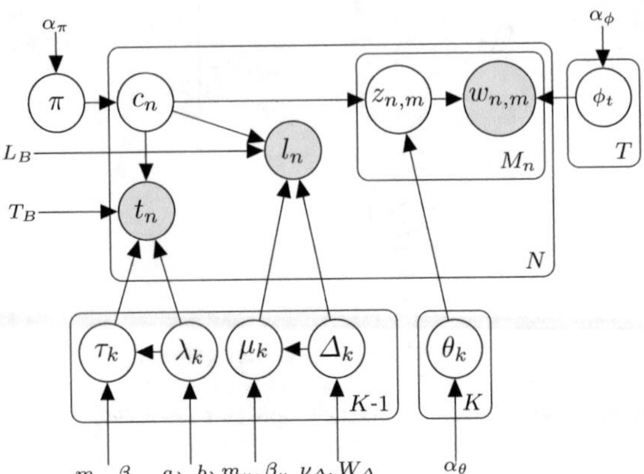

Fig. 5 The complete WARBLE model

The event-related components ($k < K$) generate the temporal, spatial and textual features from a Gaussian distribution with mean τ_k and precision λ_k, a Gaussian distribution with mean μ_k and precision matrix Δ_k and a Categorical distribution with proportions θ_k, respectively. Moreover, priors over these distributions are assumed with hyperparameters m_τ, β_τ, a_λ, b_λ, m_μ, β_μ, ν_Δ, W_Δ and α_θ.

The background component ($k = K$) accounts for the spatio-temporal features of non-event tweets, which are drawn from the histogram distributions with parameters (L_B) and (T_B) introduced earlier. However, textual features of the K-th component are not constrained by any textual background, but drawn from a Categorical distribution with proportions θ_K and hyperparameter α_θ.

Finally, we consider T topic distributions over words $\phi = \{\phi_1, \dots, \phi_T\}$ generated from a Dirichlet distribution with hyperparameter α_ϕ. The topic distributions ϕ are learned simultaneously with component assignments c_n which has lately been found very promising in modeling short and sparse text [35] and we refer here as **simultaneous topic-event learning**. In contrast to traditional topic modeling, where distributions over topics are document-specific [7], the WARBLE model assumes that topics $z_{n,m}$ are drawn from component-specific distributions θ_k. This enables to directly obtain topics that are event-related or background-related, providing also an interesting approach for automatic event summarization.

4.2.2 Learning from Tweets

Next, we describe how we can learn the WARBLE model from tweets to identify a set of events in a region during a period of interest. We first show how to learn the background model and later explain the assignment of tweets to events or background components.

Learning the background model. To learn the spatio-temporal background, we propose to collect geo-located tweets previous to the period of interest in order to add a sense of 'normality' to the model.

From the collected tweets, the temporal background is built by first computing the daily histogram with I_T bins. Then, the daily histogram is smoothed by means of a low pass Fourier filter in order to remove high frequency components. The cut-off frequency f_c determines the smoothness of the resulting signal. The normalized and smoothed histogram provides the parameters for the temporal background $T_{B_1}, T_{B_2}, \ldots, T_{B_{I_T}}$.

The spatial background is built following the same procedure. However, geographical location has to be first projected into a Cartesian coordinate system in order to consider locations in a 2-d Euclidean space. The spatial range limits can be determined from the most southwestern and northeastern points. We consider now a two dimensional Gaussian filter with a given variance σ. The resulting 2d-histogram provides the parameter for the spatial background $L_{B_1}, L_{B_2}, \ldots, L_{B_{I_L}}$.

We suggest to set the number of bins for the temporal and spatial histograms as well as the cut-off frequency and variance empirically. Future work will examine how to automatically adjust these parameters.

Assigning tweets to mixture components. To assign tweets to mixture components, we need to find the most probable assignment of tweets to mixture components, given the data at hand. That is finding c^*,

$$c^* = \operatorname*{argmax}_c p(c|l, t, w; \Gamma) \tag{9}$$

where Γ stands for the model hyperparameters $L_B, T_B, \alpha_\pi, \alpha_\theta, \alpha_\phi, m_\tau, \beta_\tau, a_\lambda, b_\lambda, m_\mu, \beta_\mu, \nu_\Delta$ and W_Δ. Exactly assessing c^* is computationally intractable for the WARBLE model.

Therefore, we propose to first use mean-field variational Bayesian inference [19, 26] to approximate $p(X|D; \Gamma)$ (where X stands for the set of random variables containing $c, z, \pi, \tau, \lambda, \mu, \Delta, \theta$ and ϕ, and D stands for our data, namely l, t, and w) by a distribution $q(X; \eta)$ (where η stands for the variational parameters). Then, assess c^* from the approximation, that is

$$c^* = \operatorname*{argmax}_c q(c; \eta) = \operatorname*{argmax}_c \int_{X-c} q(X; \eta). \tag{10}$$

The functional forms for the mean-field approximation $q(X; \eta)$ and the updates for the variational parameters can be found in a separate technical report [11]. Variational parameters are updated in an iterative fashion one at a time as in coordinate descent.

4.2.3 Scaling up to Large Datasets

Recent advances in approximate inference are enabling to scale up inference of probabilistic models to large high-dimensional datasets [24]. In particular, the application of stochastic optimization techniques to variational inference has enabled to process datasets in an online fashion [23], avoiding to have the whole dataset cached in memory or even in a local machine.

The stochastic variational inference paradigm [24] sets a variational objective function which also uses the factorized mean-field distribution $q(X; \eta)$. However, the variational updates are now computed from noisy estimates of the objective function instead of the true gradient. As a result, the computation of noisy gradients does not require the local variational parameters for the whole dataset, but only those associated with the randomly sampled data point.

Although stochastic algorithm are sequential in nature, their parallelization have been actively researched in order to preserve the statistical correctness while speeding up the run time of the algorithm in multicore machines [1]. The straightforward application of such techniques on distributed systems with commodity hardware is not obvious due to the high latency introduced by the network. Recently, some have distributed the inference of specific probabilistic models such as Latent Dirichlet Allocation (LDA) [32], but their parallel scheme is tailored to this model.

System for Parallelizing Learning Algorithm with Stochastic Methods (Splash) has been introduced as general framework for parallelizing stochastic algorithms on distributed systems [47]. It is build on top of Apache Spark [46] and it benefits from the abstraction of this data processing engine. Splash consist of a programming interface in which the user defines the sequential stochastic algorithm and a execution engine in which it averages and reweights local updates to build the global update.

Our approach to scale up WARBLE is to use the general Splash framework built on top of Apache Spark.

5 Experimental Setup and Results

5.1 "La Mercé": A Dataset for Local Event Detection

We have collected data through the Twitter streaming API[3] via Hermes [13]. In particular, we have established a long standing connection to Twitter public stream which filters all tweets geo-located within the bounding box of Barcelona city. This long standing connection was established during the local festivities of "La Mercè", that took place during few days in September 2014 and 2015.[4]

[3] http://dev.twitter.com/streaming/overview.

[4] Dataset published in https://github.com/jcapde87/Twitter-DS.

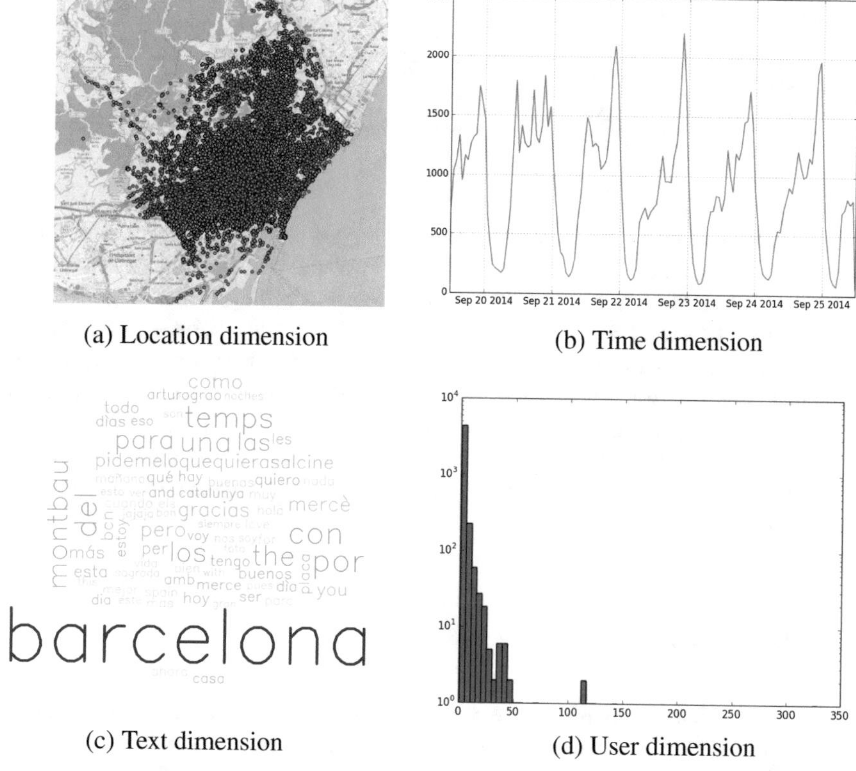

(a) Location dimension

(b) Time dimension

(c) Text dimension

(d) User dimension

Fig. 6 Tweets dimensions from "La Mercè" 2014

"La Mercè" festivities bring with several social, cultural and political events that happen in different locations within a considerably short period of time. This scenario is a suitable test bed for evaluating the accuracy of Tweet-SCAN on discovering these local events from tweets. Moreover, the abundance of events during these days causes that some of them overlap in time and space, making text more relevant to distinguish them. However, these events are apparently not distinguishable by analyzing tweet dimensions separately as shown in Fig. 6, where event patterns are not visible. Figure 6a shows the spatial distribution of tweets within the borders of Barcelona city, where different tweet density levels can be appreciated in the map. Figure 6b represents the time series of tweets from the 19th to the 25th of September and daily cycles are recognizable. Figure 6c is a wordcloud in which more frequent words are drawn with larger font size, such as "Barcelona". The multilingualism at Twitter is also reflected at this wordcloud although this work does not considered translating between different languages. Last, Fig. 6d is a histogram of the number of tweets per user, which shows that most of the users tweet very few times, while there are a few, although non-negligible number of users, who tweet very often. All four dimensions play a key role in Tweet-SCAN to uncover events.

Table 1 "La Mercè" local festivities data sets

	Tweets	Tagged tweets	Tagged events
"La Mercè" 2014	43.572	511	14
"La Mercè" 2015	12.159	476	15

As shown in Table 1, we have also manually tagged several tweets with the corresponding events as per the agenda in "La Mercè" website[5] and our own expert knowledge as citizens. With this tagged subset of tweets, we will experimentally evaluate the goodness of Tweet-SCAN. We also note that the number of tweets collected in 2015 is much less than in 2014. This is because Twitter released new smart-phone apps in April 2015 for Android and IOS that enable to attach a location to a tweet (such as a city or place of interest) apart from the precise coordinates.[6] Since tweets generated during "La Mercè" 2014 data set did not contain this functionality, we only consider tweets whose location is specified through precise coordinates for "La Mercè" 2015 data set (12.159 tweets).

5.2 Detection Performance Metrics

Clustering evaluation metrics have been applied in retrospective event detection given that this problem is defined to look for groups of tweets which are clustered together. The task of evaluating clustering against a tagged data set or *gold standard* is known as extrinsic cluster evaluation, in contrast to intrinsic evaluation, which is based on the closeness/farness of objects from the same/different clusters. Among extrinsic measures, we find out that purity, inverse purity and, specially, the combined F-measure have been extensively used for event discovery [44].

Purity is the weighted average of the maximum proportion of tweets from cluster C_i labeled as L_j over all clusters C_i, and it is expressed as follows,

$$Purity = \sum_i \frac{|C_i|}{N} max_j \frac{|C_i \cap L_j|}{|C_i|} \tag{11}$$

where higher purity means that more tweets clustered as C_i are from the same labeled event, and lower purity represents that they are from more different labels. Given that the number of clusters is not fixed, we note that purity is trivially maximum when each object is set to a different cluster, but it is minimum when all objects are set to the same cluster.

[5]http://lameva.barcelona.cat/merce/en/.

[6]https://support.twitter.com/articles/78525.

To compensate the trivial solution of purity, inverse purity is introduced. Inverse purity is the weighted average of the maximum proportion of tweets labeled as event L_i that belongs to cluster C_j over all labels L_i, and it is defined as follows,

$$Inv.\ Purity = \sum_i \frac{|L_i|}{N} max_j \frac{|C_j \cap L_i|}{|L_i|} \tag{12}$$

where higher inverse purity means that more tweets labeled as event L_i are from the same cluster, and lower inverse purity represents that they are from more different clusters. Hence, Inverse Purity is trivially maximum when grouping all tweets into a unique cluster, but it is minimum if each tweet belongs to a different cluster.

[42] combined both measures through the harmonic mean into the Van Rijsbergen's F-measure to mitigate the undesired trivial solutions from purity and inverse purity.

The F-measure score is defined as,

$$F = \sum_i \frac{|L_i|}{N} max_j\, 2 \cdot \frac{Rec(C_j, L_i) \cdot Prec(C_j, L_i)}{Rec(C_j, L_i) + Prec(C_j, L_i)} \tag{13}$$

where L_i is the set of tweets labeled as event i and C_j is the set of tweets clustered as j and N is the total number of tweets. Recall and precision are defined over these sets as the proportions $Rec(C_j, L_i) = \frac{|C_j \cap L_i|}{|L_i|}$ and $Prec(C_j, L_i) = \frac{|C_j \cap L_i|}{|C_j|}$.

5.3 *Assessment*

This section assesses Tweet-SCAN and WARBLE techniques in "La Mercé" data sets presented in Sect. 5.1 through the detection metrics introduced earlier in Sect. 5.2. In particular, we will first show how to determine Tweet-SCAN and WARBLE parameters and we will then evaluate both tuned up techniques during the main day of "La Mercé" 2014.

5.3.1 Determining Tweet-SCAN Density Thresholds

We aim to determine the best performing neighborhood sizes for Tweet-SCAN in terms of its spatio-temporal, textual and user diversity parameters.

First, we assess Tweet-SCAN in terms of F-measure scores when varying ϵ_1, ϵ_2 and ϵ_3. Figure 7 shows four possible ϵ_1, ϵ_2 configurations as function of ϵ_3 for "La Mercé" 2014 and 2015 data sets. Note that, we consider a value of *MinPts* equal to 10, which implies that an event will have at least 10 tweets.[7]

[7] Although we have tested several different *MinPts* values, $MinPts = 10$ outperforms all others given that labeled events had at least 10 tweets.

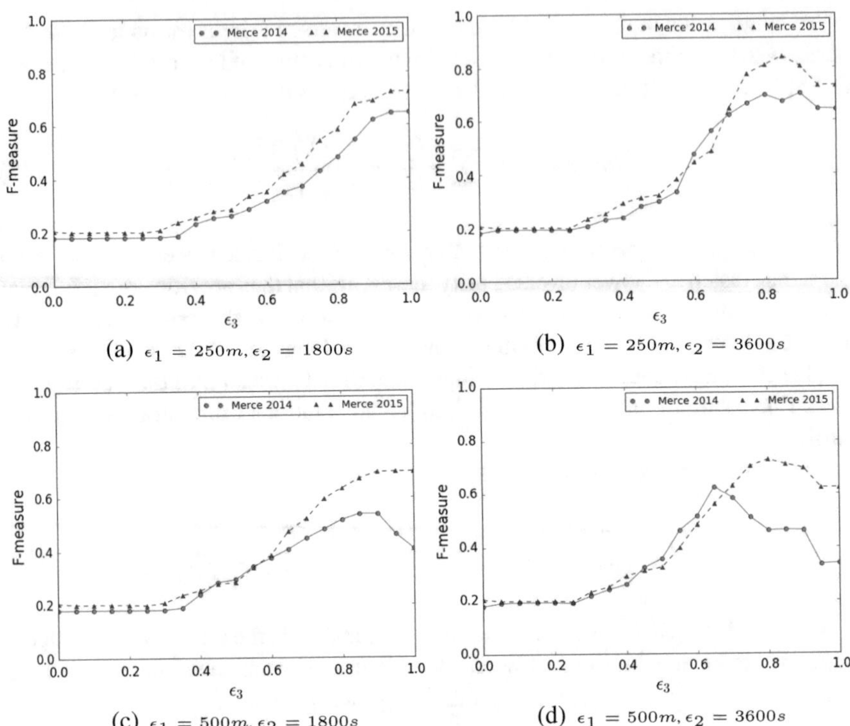

(a) $\epsilon_1 = 250m, \epsilon_2 = 1800s$

(b) $\epsilon_1 = 250m, \epsilon_2 = 3600s$

(c) $\epsilon_1 = 500m, \epsilon_2 = 1800s$

(d) $\epsilon_1 = 500m, \epsilon_2 = 3600s$

Fig. 7 F-measure for different ϵ_1, ϵ_2, ϵ_3 and *MinPts* = 10, $\mu = 0.5$

A Tweet-SCAN configuration for short distances in time and space ($\epsilon_1 = 250$ m, $\epsilon_2 = 1800$ s) optimizes F-measure for $\epsilon_3 = 1$, see Fig. 7a. This means that Tweet-SCAN disregards the textual component for this spatio-temporal setup and it can be explained by the fact that these $\epsilon_1\epsilon_2$-neighborhoods are too narrow for the tagged events.

For larger temporal neighborhoods ($\epsilon_1 = 250$ m, $\epsilon_2 = 3600$ s), the optimum value for ϵ_3 is achieved within the range 0.8–0.9 in both data sets, see Fig. 7b. Now, we can also see that this spatio-temporal configuration performs the best.

If we increase the spatial component, but we keep the temporal short ($\epsilon_1 = 500$ m, $\epsilon_2 = 1800$ s), F-measure score is lower in both data sets, but the optimum value for ϵ_3 is attained within 0.8–0.9 in "La Mercè" 2014, and $\epsilon_3 = 1$ in "La Mercè" 2015, see Fig. 7c.

Last, we increase both dimensions to ($\epsilon_1 = 500$ m, $\epsilon_2 = 3600$ s) as shown in Fig. 7d. Although the optimum F-measure score for this setup is lower than the best performing configuration, we observe that the textual component becomes more relevant. This is due to the fact that large $\epsilon_1\epsilon_2$-neighborhoods need textual discrimination to identify meaningful events.

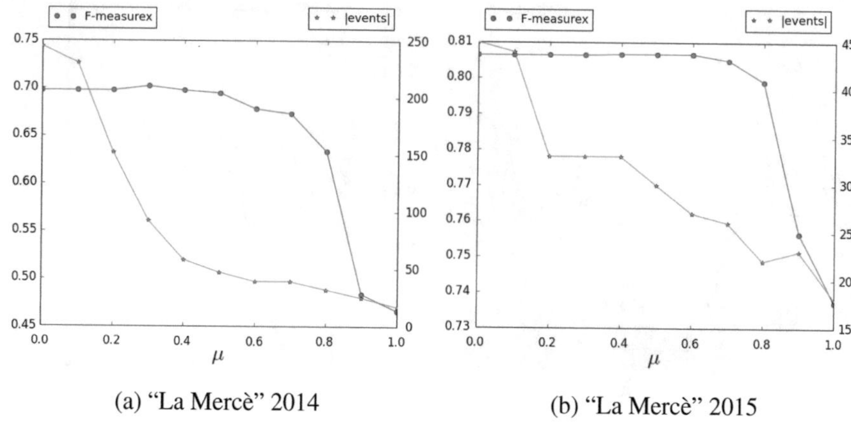

(a) "La Mercè" 2014

(b) "La Mercè" 2015

Fig. 8 F-measure for different μ values

Next, we examine the effect of different user diversity levels μ in terms of F-measure and number of discovered events. To do that, we fix the spatio-temporal and textual parameters to the best performing parameter set ($\epsilon_1 = 250\,\text{m}$, $\epsilon_2 = 3600\,\text{s}$, $\epsilon_3 = 0.8$, $MinPts = 10$) and we compute F-measure as function of the user diversity level μ. Low user diversity levels will cause that few users could generate an event in Tweet-SCAN, while higher values will entail that events are generated by many different users. Since different μ values influences the number of detected clusters by Tweet-SCAN, we will also add the number of events into the figure.

Figure 8 plots the F-measure and number of clusters as a function of μ for both data sets. It is clear from the figures that F-measure starts decreasing after a level of μ around 0.6. Similarly, the number of discovered clusters decreases but much faster and sooner than F-measure. We observe that a user diversity level of 50 % ($\mu = 0.5$) gives high figures of F-measure and reasonable number of events ($\approx$50 events for "La Mercè" 2014 and $\approx$30 events for "La Mercè" 2015). Given that the size of "La Mercè" 2015 data set is nearly four times smaller, make sense to obtain less number of events for the same μ level.

5.3.2 Learning WARBLE Background Component

In what follows, we learn the background component for the WARBLE model from "La Mercé" dataset. In particular, we consider all geo-located tweets from the 20th to the 23th of September 2014 to build the spatio-temporal backgrounds, L_B and T_B, to be used later in the 24th of September for event detection.

Figure 9a (left) shows the daily histogram of tweets in which we observe a valley during the early morning and a peak at night, indicating low and high tweeting activity during these hours, respectively. The 1-d histogram has been computed with

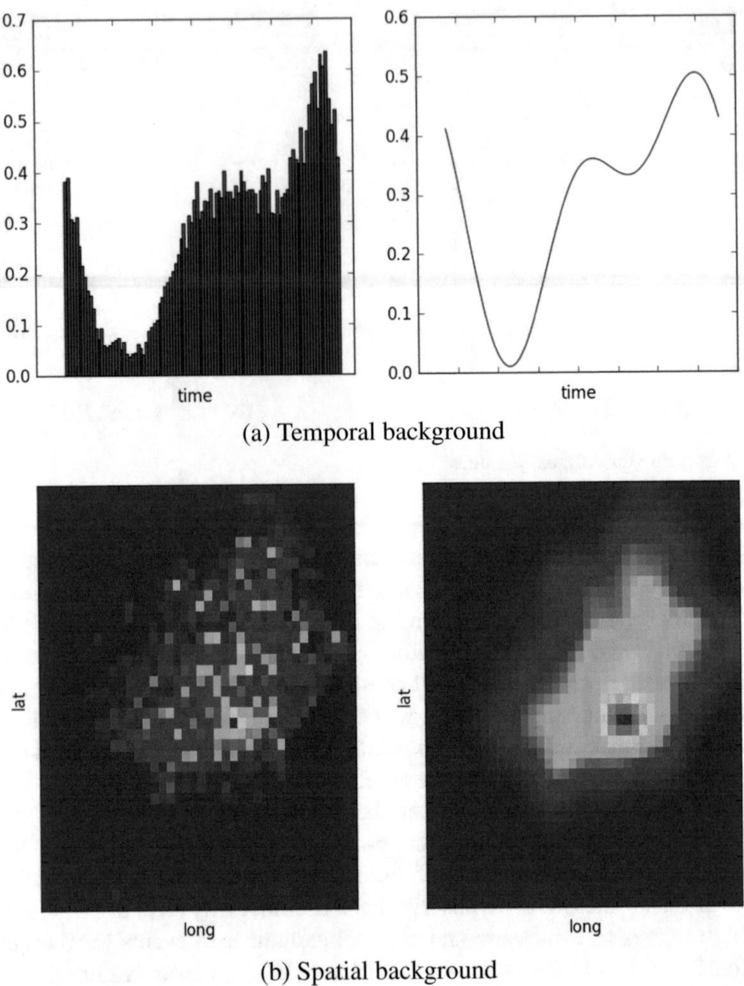

(a) Temporal background

(b) Spatial background

Fig. 9 Spatio-temporal backgrounds

$I_T = 100$ bins. Figure 9a (right) is the filtered histogram signal that will be used for setting the temporal background parameters $T_{B_1}, T_{B_2}, \ldots, T_{B_{I_T}}$.

Figure 9b (left) is the spatial histogram of all tweet locations. The smoothed version, Fig. 9b (right), provides the parameters for the spatial background $L_{B_1}, L_{B_2}, \ldots,$ $L_{B_{I_L}}$. The 2-d histogram has been computed with $I_L = 1600$ bins. We observe that the most likely areas in the filtered histogram (in red) correspond to highly dense regions of Barcelona like the city center, while city surroundings are colored in blue indicating lower density of tweets.

We note that WARBLE considers priors over most of model variables. We have considered non-informative priors and we have not experimented substantial differences in the results when varying its hyper parameters.

5.3.3 Comparative Evaluation

Finally, we compare the detection performance of Tweet-SCAN and WARBLE tuned up as described in the previous sections during the main day of "La Mercè 2014", which was the 24th of September. During that day, 7 events happened in the city of Barcelona: a music concert at *Bogatell* beach area and its revival the morning after, human towers exhibition at *Plaça Sant Jaume*, open day at *MACBA* museum, a food market at *Parc de la Ciutadella*, a wine tasting fair at *Arc de Triomf* and fireworks near *Plaça d'Espanya*.

Together with Tweet-SCAN and WARBLE, we will also consider McInerney & Blei model [29] and two WARBLE variants for comparison. Next, we enumerate event detection techniques under assessment,

(A) McInerney & Blei model [29], which does not consider background and does not perform simultaneous topic-event learning.
(B) The WARBLE model without simultaneous topic-event learning.
(C) The WARBLE model without modeling background.
(D) The WARBLE model.
(E) Tweet-SCAN.

For McInerney & Blei, WARBLE and its variants we consider the number of components K to be 8 so that the model is able to capture the 7 events occurring. Moreover, we also consider the number of topics T to be 30 for all models. Regarding those models that do not perform simultaneous topic-event learning (B and E), the Latent Dirichlet Allocation model [7] is separately trained with tweets aggregated by key terms as proposed earlier in Sect. 4.1.3.

Figure 10 shows the results for each event detection technique introduced earlier in terms of set matching metrics. Results show that the complete WARBLE model outperforms in terms of F-measure and purity. Moreover, by analyzing the results of models B and C we see a clear synergy between background modeling and simultaneous topic-event learning. Neither of them separately achieves a large increase of the F-measure, but when combined they do.

Figure 11 provides visual insight on the quality of the events detected by each of the alternatives, by drawing tweets in a 3-dimensional space corresponding to the spatial (lat, long) and temporal (time) features. Each tweet is colored with the maximum likelihood event assignment (c_n^*) for that tweet. Moreover, to improve visualization, the most populated cluster, which usually is the background, is plotted with tiny dots for all models, except model A, which fails to capture a clear background cluster. The figure shows that the similarity between hand-labeled data (F) and the WARBLE model (D) can only be compared to that of Tweet-SCAN (E).

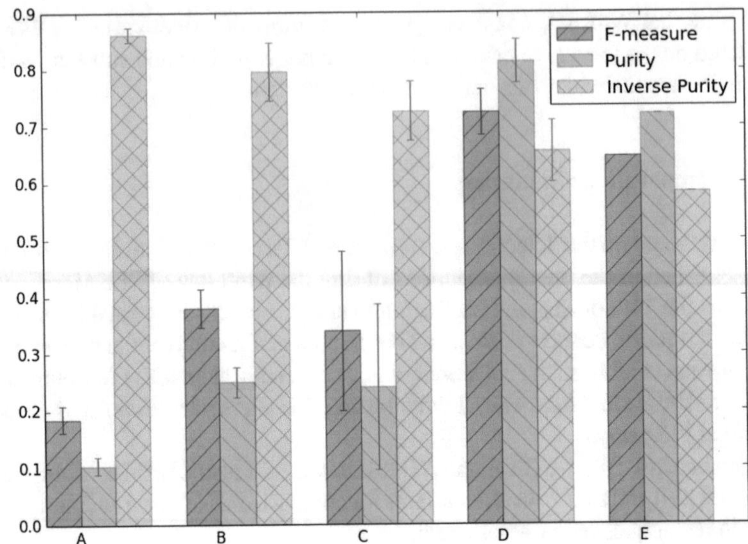

Fig. 10 Detection performance. (*A*) McInerney & Blei model (*B*) WARBLE w/o simultaneous topic-event learning (*C*) WARBLE w/o background model (*D*) WARBLE model (*E*) Tweet-SCAN

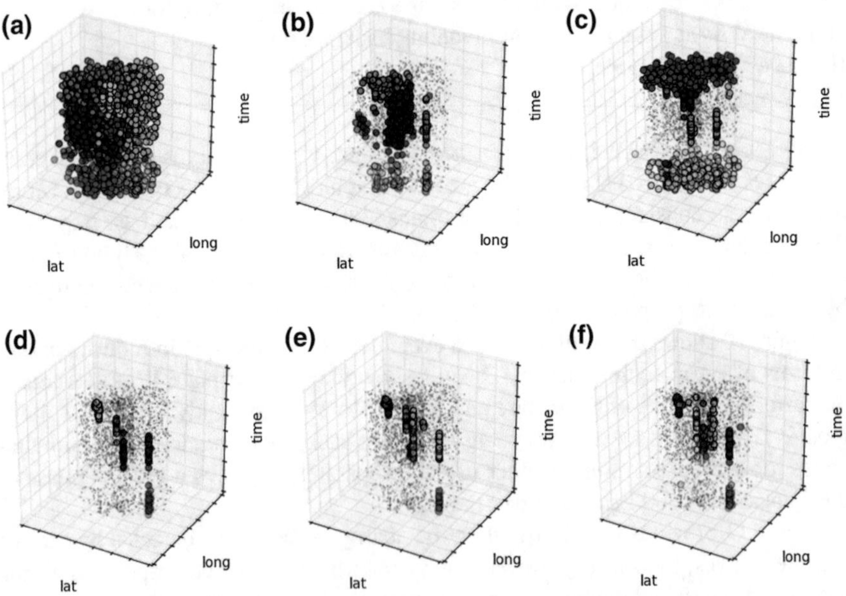

Fig. 11 Resulting real-world events. **a** McInerney & Blei model **b** WARBLE w/o simultaneous topic-event learning **c** WARBLE w/o background model **d** WARBLE model **e** Tweet-SCAN **f** Labeled events

6 Conclusions and Future Work

6.1 Conclusions

In this chapter, we have introduced the problem of event detection in location-based social networks and we have motivated a computational intelligent approach that combines probabilistic methods and learning theory to identify and characterize a set of interesting events from Twitter. Following this paradigm, we have presented a machine learning-based technique called WARBLE which is based on heterogeneous mixture models. To show the differences with the classical data mining approach, we have also presented a DBSCAN-like algorithm for event detection called Tweet-SCAN. Both approaches are inspired on the anomaly-based event detection paradigm, in which events are groups of data points which are anomalous with respect to a baseline or background distribution.

On the one hand, the formulation of Tweet-SCAN within the framework of DBSCAN defines events as density-connected set of tweets in their spatial, temporal and textual dimension. This technique allows the discovery of arbitrary-shaped events, but restricts the definition of 'normality' to simply be sparse regions of tweets and has no notion of the data generation process. On the other hand, WARBLE can define richer background models and account for seasonality and uneven population densities, but the spatio-temporal shape for events is explicitly constrained to be Gaussian.

The experimental results show that both techniques performs similarly well, although WARBLE does slightly better. For Tweet-SCAN, we have also shown that the technique performs much better when incorporating the textual and user features. More importantly, we have shown that Tweet-SCAN and WARBLE significantly outperforms the geographical topic model presented by [29]. This result encourages explicitly modeling 'normality' in a separate clustering component, either in a data mining approach like DBSCAN or in probabilistic models like mixture models.

We have shown that both approaches can scale up to large data volumes by means of distributed processing frameworks such as Apache Spark. A parallel version of Tweet-SCAN splits data into separate partitions which might reside in separate computers and apply local event detection and subsequent merging to obtain the same results as the sequential Tweet-SCAN. The scaling of WARBLE benefits from stochastic optimization to avoid having all data cached in memory or in the same local machine. Moreover, general frameworks like Splash enable parallel and distributed learning on top of Apache Spark.

6.2 Future Work

Future work will put together larger Twitter datasets to corroborate our preliminary findings regarding the accuracy of both techniques and validate our approach to scale them up through the proposed parallel schemes and general purpose data processing engines, such as Apache Spark.

Moreover, we will consider non-parametric approaches for the proposed WARBLE model in which the number of events and topics can be automatically inferred from data. For instance, existing work in mixture models uses Dirichlet Process [8] as a prior distribution and that of topic modeling uses Hierarchically-nested Dirichlet process [41].

Probabilistic approaches to event detection also provide a mechanism to reason about unseen observations or partially observed data in a principled way. For example, posts that have not been geo-referenced, words that have been misspelled or pictures without captions, can be taken into account by these models.

Finally, online or prospective event detection has to be addressed in such a way that events can be detected as early and reliably as possible and deal with the fact that the 'normality' might change over time.

Our vision is that computational intelligent approaches that combine probabilistic modeling and learning theory can pave the way to build event detection systems which self-adapt to fast changing scenarios and that are capable to reason with partially observed and noisy data.

Acknowledgements This work is partially supported by Obra Social "la Caixa", by the Spanish Ministry of Science and Innovation under contract (TIN2015-65316), by the Severo Ochoa Program (SEV2015-0493), by SGR programs of the Catalan Government (2014-SGR-1051, 2014-SGR-118), Collectiveware (TIN2015-66863-C2-1-R) and BSC/UPC NVIDIA GPU Center of Excellence. We would also like to thank the reviewers for their constructive feedback.

References

1. Agarwal A, Duchi JC (2011) Distributed delayed stochastic optimization. In: Advances in Neural Information Processing Systems, pp 873–881
2. Akyildiz IF, Su W, Sankarasubramaniam Y, Cayirci E (2002) A survey on sensor networks. IEEE Communications Magazine 40(8):102–114, doi:10.1109/MCOM.2002.1024422
3. Atefeh F, Khreich W (2015) A survey of techniques for event detection in twitter. Computational Intelligence 31(1):132–164
4. Banfield JD, Raftery AE (1993) Model-based gaussian and non-gaussian clustering. Biometrics pp 803–821
5. Bishop CM (2013) Model-based machine learning. Philosophical Transactions of the Royal Society A: Mathematical, Physical and Engineering Sciences 371(1984):20120, 222
6. Blei DM (2012) Probabilistic topic models. Communications of the ACM 55(4):77–84
7. Blei DM, Ng AY, Jordan MI (2003) Latent dirichlet allocation. the Journal of machine Learning research 3:993–1022
8. Blei DM, Jordan MI, et al (2006) Variational inference for dirichlet process mixtures. Bayesian analysis 1(1):121–144
9. Capdevila J, Cerquides J, Nin J, Torres J (2016a) Tweet-scan: An event discovery technique for geo-located tweets. Pattern Recognition Letters pp –
10. Capdevila J, Cerquides J, Torres J (2016b) Recognizing warblers: a probabilistic model for event detection in twitter, iCML Anomaly Detection Workshop

11. Capdevila J, Cerquides J, Torres J (2016c) Variational forms and updates for the WARBLE model. Tech. rep., https://www.dropbox.com/s/0qyrkivpsxxv55v/report.pdf?dl=0
12. Capdevila J, Pericacho G, Torres J, Cerquides J (2016d) Scaling dbscan-like algorithms for event detection systems in twitter. In: Proceedings of 16th International Conference, ICA3PP, Granada, Spain, December 14–16, 2016, Springer, vol 10048
13. Cea D, Nin J, Tous R, Torres J, Ayguadé E (2014) Towards the cloudification of the social networks analytics. In: Modeling Decisions for Artificial Intelligence, Springer, pp 192–203
14. Chandola V, Banerjee A, Kumar V (2009) Anomaly detection: A survey. ACM Comput Surv 41(3):15:1–15:58, doi:10.1145/1541880.1541882, URL http://doi.acm.org/10.1145/1541880.1541882
15. Cordova I, Moh TS (2015) Dbscan on resilient distributed datasets. In: High Performance Computing Simulation (HPCS), 2015 International Conference on, pp 531–540, doi:10.1109/HPCSim.2015.7237086
16. Endres DM, Schindelin JE (2003) A new metric for probability distributions. IEEE Transactions on Information theory
17. Escobar MD, West M (1995) Bayesian density estimation and inference using mixtures. Journal of the american statistical association 90(430):577–588
18. Ester M, Kriegel HP, Sander J, Xu X (1996) A density-based algorithm for discovering clusters in large spatial databases with noise. In: Kdd, vol 96, pp 226–231
19. Fox CW, Roberts SJ (2012) A tutorial on variational Bayesian inference. Artificial Intelligence Review 38(2):85–95, doi:10.1007/s10462-011-9236-8
20. Fraley C, Raftery AE (2002) Model-based clustering, discriminant analysis, and density estimation. Journal of the American statistical Association 97(458):611–631
21. He Y, Tan H, Luo W, Feng S, Fan J (2014a) Mr-dbscan: a scalable mapreduce-based dbscan algorithm for heavily skewed data. Frontiers of Computer Science 8(1):83–99, doi:10.1007/s11704-013-3158-3, URL http://dx.doi.org/10.1007/s11704-013-3158-3
22. He Y, Tan H, Luo W, Feng S, Fan J (2014b) Mr-dbscan: a scalable mapreduce-based dbscan algorithm for heavily skewed data. Frontiers of Computer Science 8(1):83–99, doi:10.1007/s11704-013-3158-3, URL http://dx.doi.org/10.1007/s11704-013-3158-3
23. Hoffman M, Bach FR, Blei DM (2010) Online learning for latent dirichlet allocation. In: advances in neural information processing systems, pp 856–864
24. Hoffman MD, Blei DM, Wang C, Paisley J (2013) Stochastic variational inference. The Journal of Machine Learning Research 14(1):1303–1347
25. Hong L, Davison BD (2010) Empirical study of topic modeling in twitter. In: Proceedings of the First Workshop on Social Media Analytics, ACM, pp 80–88
26. Jordan MI, Ghahramani Z, Jaakkola TS, Saul LK (1999) An introduction to variational methods for graphical models. Machine learning 37(2):183–233
27. Koller D, Friedman N (2009) Probabilistic graphical models: principles and techniques. MIT press
28. Li L, Goodchild MF, Xu B (2013) Spatial, temporal, and socioeconomic patterns in the use of twitter and flickr. Cartography and Geographic Information Science 40(2):61–77
29. McInerney J, Blei DM (2014) Discovering newsworthy tweets with a geographical topic model. NewsKDD: Data Science for News Publishing workshop Workshop in conjunction with KDD2014 the 20th ACM SIGKDD Conference on Knowledge Discovery and Data Mining
30. McMinn AJ, Moshfeghi Y, Jose JM (2013) Building a large-scale corpus for evaluating event detection on twitter. In: Proceedings of the 22nd ACM international conference on Information & Knowledge Management, ACM, pp 409–418
31. Murphy KP (2012) Machine learning: a probabilistic perspective. MIT press
32. Newman D, Smyth P, Welling M, Asuncion AU (2007) Distributed inference for latent dirichlet allocation. In: Advances in neural information processing systems, pp 1081–1088
33. Newman N (2011) Mainstream media and the distribution of news in the age of social discovery. Reuters Institute for the Study of Journalism, University of Oxford
34. Panagiotou N, Katakis I, Gunopulos D (2016) Detecting events in online social networks: Definitions, trends and challenges. Solving Large Scale Learning Tasks: Challenges and Algorithms

35. Quan X, Kit C, Ge Y, Pan SJ (2015) Short and sparse text topic modeling via self-aggregation. In: Proceedings of the 24th International Conference on Artificial Intelligence, AAAI Press, pp 2270–2276
36. Rajasegarar S, Leckie C, Palaniswami M (2008) Anomaly detection in wireless sensor networks. IEEE Wireless Communications 15(4):34–40, doi:10.1109/MWC.2008.4599219
37. Salton G, Buckley C (1988) Term-weighting approaches in automatic text retrieval. Inf Process Manage 24(5):513–523
38. Salton G, Wong A, Yang CS (1975) A vector space model for automatic indexing. Communications of the ACM 18(11):613–620
39. Sander J, Ester M, Kriegel HP, Xu X (1998) Density-based clustering in spatial databases: The algorithm gdbscan and its applications. Data Mining and Knowledge Discovery 2(2):169–194
40. Stelter B, Cohen N (2008) Citizen journalists provided glimpses of mumbai attacks. URL http://www.nytimes.com/2008/11/30/world/asia/30twitter.html
41. Teh YW, Jordan MI, Beal MJ, Blei DM (2006) Hierarchical dirichlet processes. Journal of the american statistical association 101(476)
42. Van Rijsbergen CJ (1974) Foundation of evaluation. Journal of Documentation 30(4):365–373
43. Wong WK, Neill DB (2009) Tutorial on event detection. In: KDD
44. Yang Y, Pierce T, Carbonell J (1998) A study of retrospective and on-line event detection. In: Proceedings of the 21st annual international ACM SIGIR conference on Research and development in information retrieval, ACM, pp 28–36
45. Yuan Q, Cong G, Ma Z, Sun A, Thalmann NM (2013) Who, where, when and what: discover spatio-temporal topics for twitter users. In: Proceedings of the 19th ACM SIGKDD international conference on Knowledge discovery and data mining, ACM, pp 605–613
46. Zaharia M, Chowdhury M, Franklin MJ, Shenker S, Stoica I (2010) Spark: cluster computing with working sets. In: Proceedings of the 2nd USENIX conference on Hot topics in cloud computing, vol 10, p 10
47. Zhang Y, Jordan MI (2015) Splash: User-friendly programming interface for parallelizing stochastic algorithms. arXiv preprint arXiv:150607552
48. Zheng Y (2012) Tutorial on location-based social networks. In: Proceedings of the 21st international conference on World wide web, WWW, ACM

Part II
Applications

Using Computational Intelligence for the Safety Assessment of Oil and Gas Pipelines: A Survey

Abduljalil Mohamed, Mohamed Salah Hamdi and Sofiène Tahar

Abstract The applicability of intelligent techniques for the safety assessment of oil and gas pipelines is investigated in this study. Crude oil and natural gas are usually transmitted through metallic pipelines. Working under unforgiving environments, these pipelines may extend to hundreds of kilometers, which make them very susceptible to physical damage such as dents, cracks, corrosion, etc. These defects, if not managed properly, can lead to catastrophic consequences in terms of both financial losses and human life. Thus, effective and efficient systems for pipeline safety assessment that are capable of detecting defects, estimating defects sizes, and classifying defects are urgently needed. Such systems often require collecting diagnostic data that are gathered using different monitoring tools such as ultrasound, magnetic flux leakage, and Closed Circuit Television (CCTV) surveys. The volume of the data collected by these tools is staggering. Relying on traditional pipeline safety assessment techniques to analyze such huge data is neither efficient nor effective. Intelligent techniques such as data mining techniques, neural networks, and hybrid neuro-fuzzy systems are promising alternatives. In this paper, different intelligent techniques proposed in the literature are examined; and their merits and shortcomings are highlighted.

Keywords Oil and gas pipelines · Safety assessment · Big data · Computational intelligence · Data mining · Artificial neural networks · Hybrid neuro-fuzzy systems · Artificial intelligence · Defect sizing · Magnetic flux leakage

---

A. Mohamed (✉) · M.S. Hamdi
Department of Information Systems, Ahmed Bin Mohamed Military College, P. O. Box 22713, Doha, Qatar
e-mail: ajamoham@abmmc.edu.qa

M.S. Hamdi
e-mail: mshamdi@abmmc.edu.qa

S. Tahar
Department of Electrical and Computer Engineering, Concordia University, 1515 St. Catherine W, Montreal, Canada
e-mail: tahar@ece.concordia.ca

W. Pedrycz and S.-M. Chen (eds.), *Data Science and Big Data: An Environment of Computational Intelligence*, Studies in Big Data 24,
DOI 10.1007/978-3-319-53474-9_9

1 Introduction

Oil and gas are the leading sources of energy the world relies on today; and pipelines are viewed as one of the most cost efficient ways to move that energy and deliver it to consumers. The latest data, in 2015, gives a total of more than 3.5 million km of pipeline in 124 countries of the world. Many other thousands of kilometers of pipelines are planned and under construction. Pump stations, along the pipeline, move oil and gas through the pipelines. Because the pipeline walls are under constant pressure, tiny cracks may arise in the steel. Under the continuous load, they can then grow into critical cracks or even leaks. Pipelines conveying flammable or explosive material, such as natural gas or oil, pose special safety concerns; and various accidents have been reported [1]. Damage to the pipeline may cause the occurrence of large and enormous human and economic losses. Moreover, damaged pipelines obviously represent an environmental hazard. Therefore, pipeline operators must identify and remove pipeline failures caused by corrosion and other types of defects as early as possible.

Today, inspection tools, called “Pipeline Inspection Gauges” or “Smart Pigs”, employ complex measuring techniques such as ultrasound and magnetic flux leakage. They are used for the inspection of such pipelines, and have become major components to pipeline safety and accident prevention. These smart pigs are equipped with hundreds of highly tuned sensors that produce data that can be used to locate and determine the thickness of cracks, fissures, erosion and other problems that may affect the integrity of the pipeline. In each inspection passage, huge amounts of data (several hundred gigabytes) are collected. A team of experts will look at these data and assess the health of the pipeline segments.

Because of the size and complexity of pipeline systems and the huge amounts of data collected, human inspection alone is neither feasible nor reliable. Automating the inspection process and the evaluation and interpretation of the collected data have been an important goal for the pipeline industry for a number of years. Significant progress has been made in that regard, and we currently have a number of techniques available that can make the highly challenging and computationally-intensive task of automating pipeline inspection possible. These techniques range from analytical modeling, to numerical computations, to methods employing artificial intelligence techniques such as artificial neural networks. This paper presents a survey of the state-of-the-art in methods used to assess the safety of the oil and gas pipelines, with emphasis on intelligent techniques. The paper explains the principles behind each method, highlights the settings where each method is most effective, and shows how several methods can be combined to achieve higher accuracy.

The rest of the paper is organized as follows. In Sect. 2, we review the five stages of the pipeline reliability assessment process. The theoretical principals behind the intelligent techniques surveyed in this study are discussed in Sect. 3. In

Sect. 4, the pipeline safety assessment approaches using the intelligent techniques reported in Sect. 3 are presented and analyzed. We conclude with final remarks in Sect. 5.

2 Safety Assessment in Oil and Gas Pipelines

The pipeline reliability assessment process is basically composed of five stages, namely data processing, defect detection, determination of defect size, assessment of defect severity, and repair management. Once a defect is detected, the defect assessment unit proceeds by determining the size (the defect's depth and length) of the defect. This is really an important step as the severity of the defect is based on its physical characteristics. Based on the severity level of the detected defect, an appropriate action is taken by the repair management. These five stages of the pipeline assessment process are summarized in the following subsections.

2.1 Big Data Processing

The most common nondestructive evaluation (NDE) method of scanning oil and gas pipelines for possible pipeline defects utilizes magnetic flux leakage (MFL) technology [2], in which autonomous devices containing magnetic sensors are sent on periodic basis into the pipeline under inspection. The magnetic sensors are used to measure MFL signals every three-millimeters along the pipeline length. Figure 1 shows a rolled-out representation of a pipeline wall. The MFL sensors are equally distributed around the circumference of the pipeline and move parallel to the axis of the pipeline.

For pipelines that extend hundreds of kilometers, the data sets collected by the MFL sensors are so big and complex that traditional data processing techniques to analyze such data are inadequate. To reduce the quantity of the data, redundant and

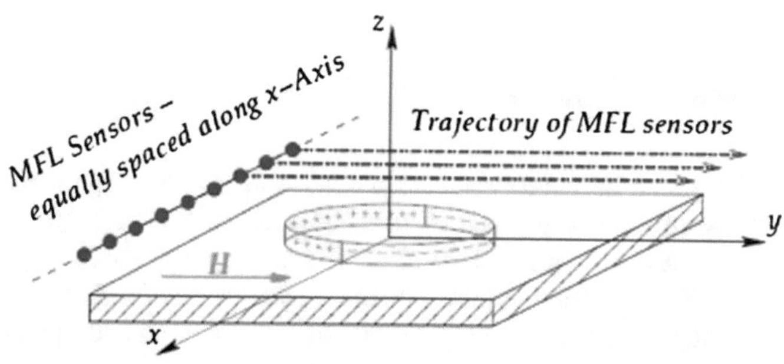

Fig. 1 Rolled-out representation of pipeline wall

irrelevant data are removed using feature extraction and selection techniques. The most relevant features are selected, and then used to determine the depth and length of the detected defect.

2.2 Defect Detection

In this stage, the diagnostic data are examined for the existence of possible defects in the pipeline. To detect and identify the location of potential defects, wavelet techniques are widely used [3]. They are very powerful mathematical methods [4–6]. They were reported in many applications such as data compression [7], data analysis and classification [8], and de-noising [9–11].

2.3 Determination of Defect Size

To determine the severity level of the detected defect, the defect's depth and length are calculated. However, the relationship between the given MFL signals and particular defect type and shape is not well-known. Hence, it is very difficult to derive an analytical model to describe this relationship. To deal with this problem, researchers resort to intelligent techniques to estimate the required parameters. One of these intelligent tools is the Adaptive Neuro-Fuzzy Inference System (ANFIS).

2.4 Assessment of Defect Severity

Based on the defect parameters (i.e., depth and length) obtained in the previous stage, an industry standard known as ASME B31G is often used to assess the severity level of the defect [12]. It specifies the pipeline stress under operating pressure and what defect parameters that may fail the hydro pressure test [13].

2.5 Repair Management

In order to determine an appropriate maintenance action, the repair management classifies the severity level of pipeline defects into three basic categories, namely: severe, moderate, and acceptable. Severe defects are given the highest priority and an immediate action is often required. The other two severity levels are not deemed critical, thus, a repair action can be scheduled for moderate and acceptable defects.

3 Computational Intelligence

As mentioned in the previous section, MFL signals are widely used to determine the depth and length of potential defects. From recorded data, it has been observed that the magnitude of MFL signals varies from one defect depth and length to another. In the absence of analytical models that can describe the relationship between the amplitude of MFL signals and their corresponding defect dimensions, computational intelligence provides an alternative approach. Given sufficient MFL data, there are different computational techniques such as data mining techniques, artificial neural networks, and hybrid neuro-fuzzy systems that can be utilized to learn such relationships. In the following, the theoretical principals behind each of these techniques are summarized.

3.1 Data Mining

The k-nearest neighbor (k-NN) and support vector machines (SVM) are widely used in data mining to solve classification problems. Within the context of the safety assessment in oil and gas pipelines, these two techniques can be employed to assign detected defects to a certain severity level.

3.1.1 K-Nearest Neighbor (KNN)

The KKN is a non-parametric learning algorithm as it does not make any assumptions on the underlying data distribution. This may come in handy since many real world problems do not follow such assumptions. The KNN learning algorithm is also referred to as a lazy algorithm because it does not use the training data points to do any generalization. Thus, there is no training stage in the learning process, but rather KNN makes its decision based on the entire training data set. The learning algorithm assumes that all instances correspond to points in the n-dimensional space. The nearest neighbors of an instance are identified using the standard Euclidean distance. Let us assume that a given defect x is characterized by a feature vector:

$$\langle a_1(x), a_2(x), \ldots, a_n(x) \rangle, \tag{1}$$

where $a_r(x)$ denotes the value of the rth attribute of instance x. Thus, the distance d between two instances x_i and x_j is calculated as follows:

$$d(x_i, x_j) = \sqrt{\sum_{r=1}^{n} \left(a_r(x_i) - a_r(x_j)\right)^2}, \tag{2}$$

For the safety assessment in oil and gas pipeline application, the target function is discrete. That is, it assigns the feature vector of the detected defect to one of the three severity levels severe, moderate, or acceptable. If we suppose $k = 1$, then the 1-nearest neighbor assigns the feature vector to the severity level where the training instance of that severity level is nearest to the feature vector. For larger values of k, the algorithm assigns the most common severity level among the k nearest training examples. e only assumption made is that the data is in a feature space.

3.1.2 Support Vector Machine (SVM)

The SVM is a discriminant classifier defined by a separating hyperplane. Given labeled training data, the SVM algorithm outputs an optimal hyperplane that can categorize new examples. Support vector machines are originally designed for binary classification problems. For a linearly separable set of 2D-points, there will be multiple straight lines that may offer a solution to the problem. However, a line is considered bad if it passes too close to the points because it will be susceptible to noise. The task of the SVM algorithm is to find the hyperplane that gives the largest minimum distance (i.e., margin) to the training examples.

To solve multi-class classification problems, the SVM should be extended. The training algorithms of SVMs look for the optimal separating hyperplane which has a maximized margin between the hyperplane and the data, which in turn, minimizes the classification error. The separating hyperplane is represented by a small number of training data, called support vectors (SVs). However, the real data cannot be separated linearly, thus the data are mapped into a higher dimensional space. Practically, a kernel function is utilized to calculate the inner product of the transformed data. The efficiency of the SVM depends mainly on the kernel.

Formally, the hyperplane is defined as follows:

$$f(x) = \beta_0 + \beta^T x, \tag{3}$$

where β is known as the *weight vector* and β_0 as the bias. The optimal hyperplane can be represented in an infinite number of different ways by scaling of β and β_0. The hyperplane chosen is:

$$\left|\beta_0 + \beta^T x\right| = 1, \tag{4}$$

where x symbolizes the training examples closest to the hyperplane, which are called support vectors. The distance between a point x and a hyperplane (β,β_0) can be calculated as:

$$distance = \frac{\left|\beta_0 + \beta^T x\right|}{\|\beta\|}, \tag{5}$$

For the canonical hyperplane, the numerator is equal to one, thus,

$$distance = \frac{|\beta_0 + \beta^T x|}{||\beta||} = \frac{1}{||\beta||}, \tag{6}$$

The margin (*M*) is twice the distance to the closest examples:

$$M = \frac{2}{||\beta||}, \tag{7}$$

Now, maximizing *M* is equivalent to the problem of minimizing a function $L(\beta)$ subject to some constrains as follows:

$$\min_{\beta, \beta_0} L(\beta) = \frac{1}{2}|\beta|^2, \tag{8}$$

subject to:

$$y_i = \left(\beta^T x_i + \beta_0\right) \geq 1 \forall_i, \tag{9}$$

where y_i represents each of the labels of the training examples.

3.2 *Artificial Neural Networks*

Artificial neural networks (ANN) are suitable for the safety assessment in oil and gas pipelines as they are capable of solving ill-defined problems. Essentially they attempt to simulate the neural structure of the human brain and its functionality.

The multi-layer perceptron (MLP) with the back propagation learning algorithm is considered the most common neural network and being widely used in a large number of applications. A typical MLP neural network of one hidden layer is depicted in Fig. 2. There are *d* inputs (example, *d* dimensions of input pattern *X*), h hidden nodes, and c outputs nodes.

The output of the *j*th hidden node is $z_j = f_j(a_j)$, where $a_j = \sum_{i=0}^{d} w_{ji} x_i$, and $f_j(.)$ is an activation function associated with hidden node *j*. w_{ji} is the connection weight from the input node *i* to *j*, and w_{j0} denotes the bias for the hidden node *j*. For an input node *k*, its output is $y_k = f_k(a_k)$, where $a_k = \sum_{j=0}^{h} w_{kj} z_j$, and $f_k(.)$ is the activation function associated with output node *k*. w_{kj} is the connection weight from hidden node *j* to output node *k*. w_{k0} denotes the bias for output node *k*. The activation function is often chosen as the unipolar sigmoidal function:

$$f(a) = \frac{1}{1 + \exp(-\gamma a)}, \tag{10}$$

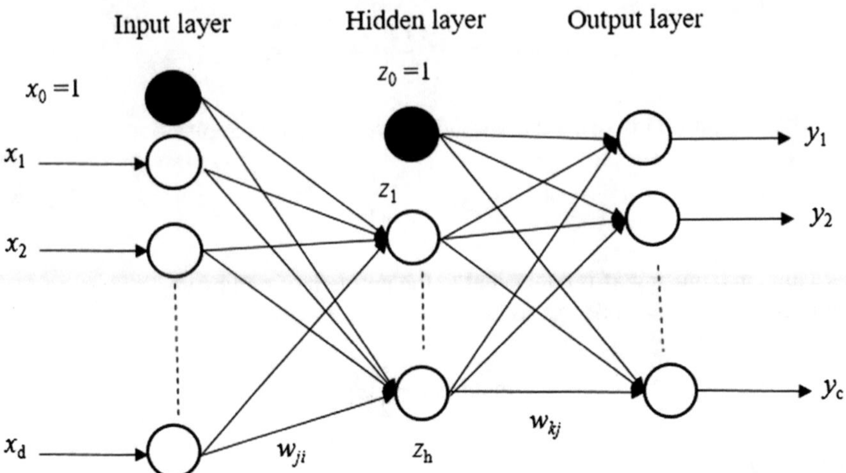

Fig. 2 A multi-layer perceptron neural network

In MLP, the back propagation learning algorithm is used to update weights so as to minimize the following squared error function:

$$J(w) = \frac{1}{2}\sum_{k=1}^{c}(e_k - y_k(X))^2, \tag{11}$$

3.3 *Hybrid Neuro-Fuzzy Systems*

The focus of intelligent hybrid systems in this study will be on the combination of neural networks and fuzzy inference systems. One of these systems is the adaptive neuro-fuzzy inference system (ANFIS), which will be used as an illustrative example of such hybrid systems. ANFIS, as introduced by Jang [14], utilizes fuzzy IF-THEN rules, where the membership function parameters can be learned from training data, instead of being obtained from an expert [15–23]. Whether the domain knowledge is available or not, the adaptive property of some of its nodes allows the network to generate the fuzzy rules that approximate a desired set of input-output pairs. In the following, we briefly introduce the ANFIS architecture as proposed in [14]. The structure of the ANFIS model is basically a feedforward multi-layer network. The nodes in each layer are characterized by their specific function, and their outputs serve as inputs to the succeeding nodes. Only the parameters of the adaptive nodes (i.e., square nodes in Fig. 3) are adjustable during the training session. Parameters of the other nodes (i.e., circle nodes in Fig. 3) are fixed.

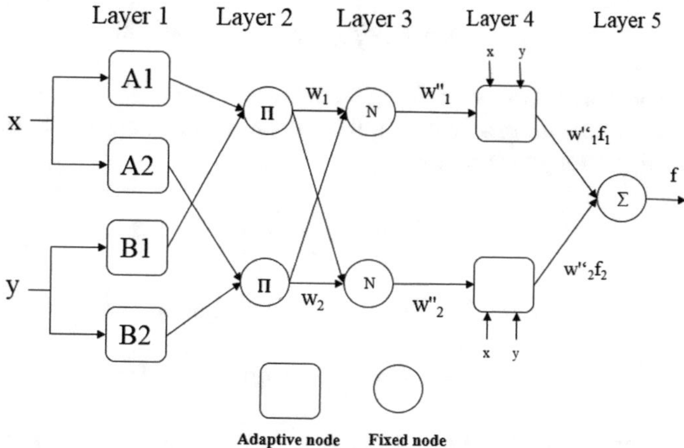

Fig. 3 The architecture of ANFIS

Suppose there are two inputs x, y, and one output f. Let us also assume that the fuzzy rule in the fuzzy inference system is depicted by one degree of Sugeno's function [14].

Rule 1: if x is A_1 and y is B_1 then $f = p_1x + q_1y + r_1$
Rule 2: if x is A_2 and y is B_2 then $f = p_2x + q_2y + r_2$

where p_i, q_i, r_i are adaptable parameters.
The node functions in each layer are described in the sequel.

Layer 1: Each node in this layer is an adaptive node and is given as follows:

$$o_i^1 = \mu_{Ai}(x), \quad i = 1, 2$$
$$o_i^1 = \mu_{Bi-2}(y), \quad i = 3, 4$$

where x and y are inputs to the layer nodes, and A_i and B_{i-2} are linguistic variables. The maximum and minimum of the bell-shaped membership function are 1 and 0, respectively. The membership function has the following form:

$$\mu_{Ai}(x) = \frac{1}{1 + \left\{\left(\frac{x - c_i}{a_i}\right)^2\right\}^{bi}}, \tag{12}$$

where the set $\{a_i, b_i, c_i\}$ represents the premise parameters of the membership function. The bell-shaped function changes according to the change of values in these parameters.

Layer 2: Each node in this layer is a fixed node. Its output is the product of the two input signals as follows:

$$o_i^2 = w_i = \mu_{Ai}(x)\mu_{Bi}(y), \quad i = 1, 2, \tag{13}$$

where w_i refers to the firing strength of a rule.

Layer 3: Each node in this layer is a fixed node. Its function is to normalize the firing strength as follows:

$$o_i^3 = w_i^{''} = \frac{w_i}{w_1 + w_2}, \quad i = 1, 2 \tag{14}$$

Layer 4: Each node in this layer is adaptive and adjusted as follows:

$$o_i^4 = w_i^{''} f_i = w_i^{''}(p_i x + q_i y + r_i), \quad i = 1, 2 \tag{15}$$

where $w_i^{''}$ is the output of layer 3 and $\{p_i + q_i + r_i\}$ is the consequent parameter set.

Layer 5: Each node in this layer is fixed and computes its output as follows:

$$o_i^5 = \sum_{i=1}^{2} w_i^{''} f_i = \frac{\left(\sum_{i=1}^{2} w_i f_i\right)}{w_1 + w_2}, \tag{16}$$

The output of layer 5 sums the outputs of nodes in layer 4 to be the output of the whole network. If the parameters of the premise part are fixed, the output of the whole network will be the linear combination of the consequent parameters, i.e.,

$$f = \frac{w_1}{w_1 + w_2} f_1 + \frac{w_2}{w_1 + w_2} f_2, \tag{17}$$

The adopted training technique is hybrid, in which, the network node outputs go forward till layer 4, and the resulting parameters are identified by the least square method. The error signal, however, goes backward till layer 1, and the premise parameters are updated according to the descent gradient method. It has been shown in the literature that the hybrid-learning technique can obtain the optimal premise and consequent parameters in the learning process [14].

4 Pipeline Safety Assessment Using Intelligent Techniques

In this section, pipeline safety assessment approaches using the above intelligent techniques that are reported in the literature are presented and analyzed. Most of these have been proposed for either predicting pipeline defect dimensions or detecting and classifying defect types [24].

4.1 Data Mining-Based Techniques

A recognition and classification of pipe cracks using images analysis and a neuro-fuzzy algorithm is proposed [25]. In the preprocessing step the scanned images of the pipe are analyzed and crack features are extracted. In the classification step the neuro-fuzzy algorithm is developed that employs a fuzzy membership function and an error back-propagation algorithm. The classification of underground pipe defects is carried out using the Euclidean distance method, a fuzzy-KNN algorithm, a conventional back-propagation neural network, and a neuro-fuzzy algorithm. The theoretical backgrounds of all classifiers are presented and their relative advantages are discussed. In conventional recognition methods, the Euclidean distance has been commonly used as a distance measure between two vectors. The Euclidean distance is defined by Eq. 2.

The fuzzy k-NN algorithm assigns class membership to a sample observation based on the observation distance from its k-nearest neighbors and their membership. The neural network universal approximation property guarantees that any sufficiently smooth function can be approximated using a two-layer network. Neuro-fuzzy systems belong to hybrid intelligent systems. Neural networks are good for numerical knowledge (data sets), fuzzy logic systems are good for linguistic information (fuzzy sets). The proposed neuro-fuzzy algorithm is a mixture, where the input and the output of the ANN is a fuzzy entity. Fuzzy neural networks such as the ones proposed in this study provide more flexibility in representing the input space by integrating vagueness usually associated with fuzzy patterns with learning capabilities of neural networks. In fact, by using fuzzy variables as input to the neural network structure, the boundaries of the decision space become represented in a less restrictive manner (unlike the conventional structure of neural networks where the input are required to be crisp), and permits the representation of data possibly belonging to overlapping boundaries. As such more information could be represented without having recourse to the storage of a huge amount of data, which are usually required for the training and testing of conventional "crisp-based data training" neural networks.

The main disadvantage of the KNN algorithm, in addition to determining the value of the parameter k, is that, for a large number of images or MFL data, the computation cost is high because we need to compute the distance of each instance to all training samples. Moreover, it takes up a lot of memory to store all the image properties and features of MFL samples. However, it is simple and effective due to the large data.

SVM-based approaches are reported in [26–28]. In [26], the proposed approach aims at detecting, identifying, and verifying construction features while inspection the condition of underground pipelines. The SVM is used to classify features extracted from the signals of a NDE sensor. The SVM model to be trained for this work uses the RFT data and the ground truth labels to learn how to separate construction features (CF) from other data (non-CF) from CCTV images. The CFs represent pipeline features such as joints, flanges, and elbows. The learned SVM model is later employed to detect CF in unseen data. In [27], the authors propose an SVM method to reconstruct defects shape features. To create a defect feature picture, a large number of samples are collected for each defect. The SVM model reconstruction error is below 4%. For the analysis of magnetic flux leakage images in pipeline inspection, the authors in [28] apply support vector regression among other techniques. In this paper, the focus is on the binary detection problem of classifying anomalous image segments into one of two classes: the first class is the one which consists of injurious or non-benign defects such as various crack-like anomalies and metal losses in girth welds, long-seam welds, or in the pipe wall itself, which if left untreated, could lead to pipeline rupture. The second class consists of non-injurious or benign objects such as noise events, safe and non-harmful pipeline deformations, manufacturing irregularities, etc.

Although finding the right kernel for the SVM classifier is a challenge, but once obtained, it can work well despite the fact that the MFL data is not linearly separable. The main disadvantage is that it is fundamentally a binary classifier; thus, there is no particular way for dealing with multi-defect pipeline problems.

4.2 Neural Network-Based Techniques

Artificial neural networks have been used extensively in safety assessment in oil and gas pipelines [29–33]. In [29], Carvalho et al. propose an artificial neural network approach for detection and classification of pipe weld defects. These defects were manufactured and deliberately implanted. The ANN was able to distinguish between defect and non-defect signals with great accuracy (94.2%). For a particular type of defect signals, the ANN recognized them 92.5% of the time. In [29], a Radial Basis Function Neural Network (RBFNN) is deemed to be a suitable technique and a corrosion inspection tool to recognize and quantify the corrosion characteristics. An Immune RBFNN (IRBFNN) algorithm is proposed to process the MFL data to determine the location and size of the corrosion spots on the pipeline. El Abbasy et al. in [31] propose an artificial neural network models to evaluate and predict the condition of offshore oil and gas pipelines. The inspection data for selected factors are used to train the ANN in order to obtain ANN-based condition prediction models. The inspection data points were divided randomly into three sets: (1) 60% for training; (2) 20% for testing; and (3) 20% for validation. The training set is used to train the network whereas the testing set is used to test the network during the development/training and also to continuously correct it by adjusting the weights of network links. The authors in [32] propose a machine learning approach for big data in oil and gas pipelines, in which three different network architectures are examined, namely static feedforward neural networks (static FFNN), cascaded FFNN, and dynamic FFNN as shown in Figs. 4, 5, and 6, respectively.

In the static FFNN architecture, the extracted feature vector is fed into the first hidden layer. Weight connections, based on the number of neurons in each layer,

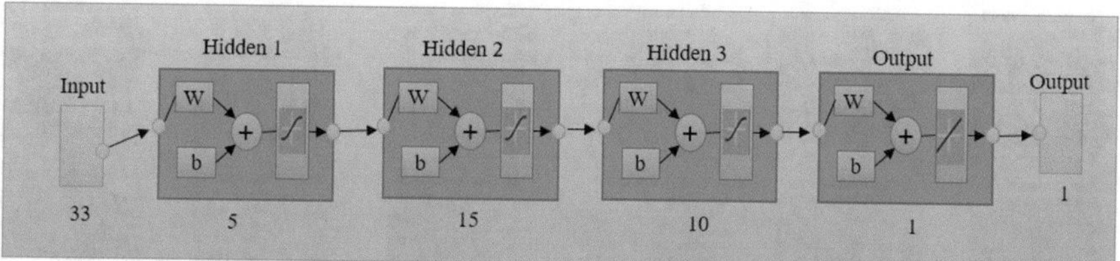

Fig. 4 Architecture of static FFNN

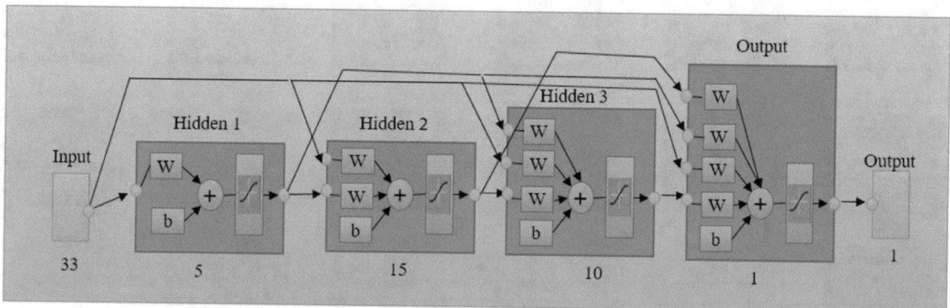

Fig. 5 Architecture of cascaded FFNN

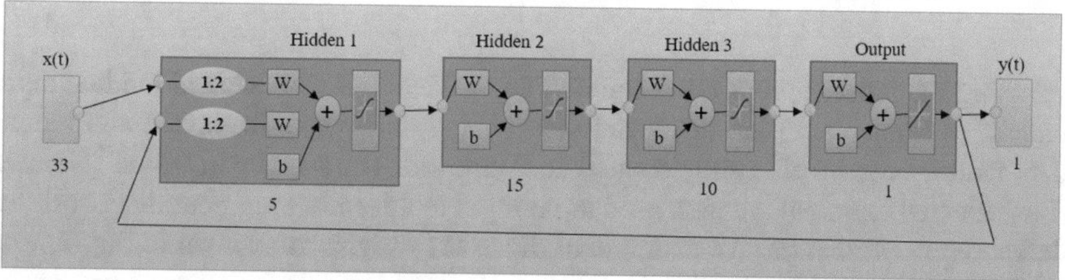

Fig. 6 Architecture of dynamic FFNN

are assigned between every adjacent layers. While in the cascaded FFNN architecture, include a weight connection from the input layer to each other layer, and from each layer to the successive layers. In the dynamic architecture, the network outputs depend not only on the current input feature vector, but also on the previous inputs and outputs of the network. Compared with the performance of pipeline inspection techniques reported by service providers such as GE and ROSEN, the results obtained using the method we proposed are promising. For instance, within ±10% error-tolerance range, the obtained estimation accuracy is 86%, compared to only 80% reported by GE; and within ±15% error-tolerance range, the achieved estimation accuracy is 89% compared to 80% reported by ROSEN.

Mohamed et al. propose a self-organizing map-based feature visualization and selection for defect depth estimation in oil and gas pipelines in [33]. The authors use the self-organizing maps (SOMs) as feature visualization tool for the purpose of

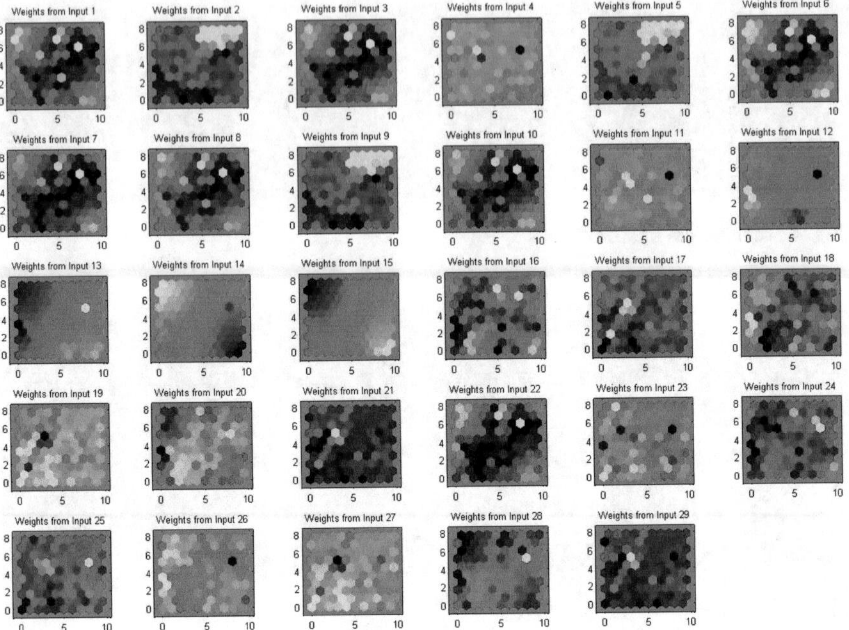

Fig. 7 SOM weights for each input feature [33]

selecting the most appropriate features. The SOM weights for each individual input feature (weight plane) are displayed then visually analyzed. Irrelevant and redundant features can be efficiently spotted and removed. The remaining “good” features (i.e., selected features) are then used as an input to a feedforward neural network for defect depth estimation. An example of the SOM weights are shown in Fig. 7. The 21 features selected by the SOM approach are used to evaluate the performance of the three FFNN structures. Experimental work has shown the effectiveness of the proposed approach. For instance, within ±5% error-tolerance range, the obtained estimation accuracy, using the SOM-based feature selection, is 93.1%, compared to 74% when all input features are used (i.e., no feature selection is performed); and within ±10% error-tolerance range, the obtained estimation accuracy, using the SOM-based feature selection, is 97.5%, compared to 86% when all the input features are used (i.e., no feature selection is performed).

The disadvantage of using neural networks is that the neural network structure (i.e., number of neurons, hidden layers, etc.) is determined by trial and error approach. Moreover, the learning process can take very long due to the large number of MFL samples. The main advantage is that there is no need to find a mathematical model that describes the relationship between MFL signals and pipeline defects.

4.3 Hybrid Neuro-Fuzzy Systems-Based Techniques

Several approaches that utilize hybrid systems have been reported in the literature. In [34], the authors propose a neuro-fuzzy classifier for the classification of defects by extracting features in segmented buried pipe images. It combines a fuzzy membership function with a projection neural network where the former handles feature variations and the latter leads to good learning efficiency as illustrated in Fig. 8. Sometimes the variation of feature values is large, in which case it is difficult to classify objects correctly based on these feature values. Thus, as shown in the figure, the input feature is converted into fuzzified data which are input to the projection neural network. The projection network combines the utility of both the restricted coulomb energy (RCE) network and backpropagation approaches. A hypersphere classifier such as RCE places hyper-spherical prototypes around training data points and adjusts their radii. The neural network inputs are projected onto a hypersphere in one higher dimension and the input and weight vectors are confined to lie on this hypersphere. By projecting the input vector onto a hypersphere in one higher dimension, prototype nodes can be created with closed or open classification surfaces all within the framework of a backpropagation trained feedforward neural network. In general, a neural network passes through two phases: training and testing. During the training phase, supervised learning is used to assign the output membership values ranging in [0,1] to the training input vectors. Each error in membership assignment is fed back and the connection weights of the network are appropriately updated. The back-propagated error is computed with respect to each desired output, which is a membership value denoting the degree of belongingness of the input vector to a certain class. The testing phase in a fuzzy network is equivalent to the conventional network.

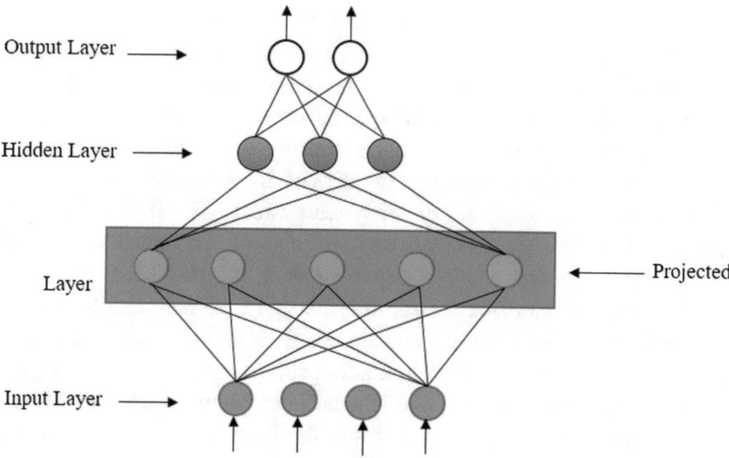

Fig. 8 A hybrid neuro-fuzzy classifier

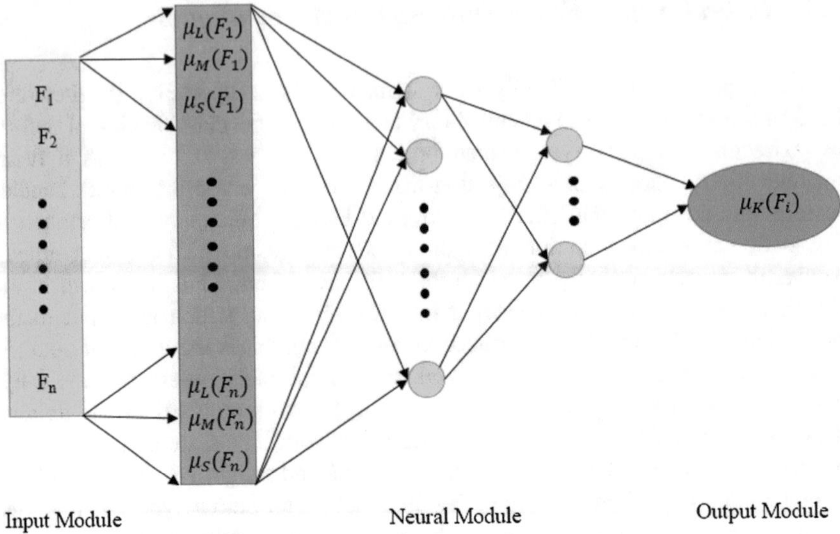

Fig. 9 Neuro-fuzzy neural network architecture

In [35], a classification of underground pipe scanned images using feature extraction and neuro-fuzzy algorithm is proposed. The concept of the proposed fuzzy input and output module and neural network module is illustrated in Fig. 9. The fuzzy ANN model has three modules: the fuzzy input module, the neural network module, and the fuzzy output module. The neural network module is aconventional feedforward artificial neural network. A simple three-layer network with a backpropagation training algorithm is used in this study. To increase the rate of convergence, a momentum term and a modified backpropagation training rule called the delta–delta rule are used. The input layer of this network consists of 36 nodes (because of the use of fuzzy sets to screen the 12 input variables; and the output layer consists of seven nodes (trained with fuzzy output values). As shown in Fig. 9, the input layer of this fuzzy ANN model is actually an output of the input module. On the other hand, the output layer becomes an input to the output module. The input and output modules, for preprocessing and post-processing purposes, respectively, are designed to deal with the data of the ANN using fuzzy sets theory.

In [36], an adaptive neuro-fuzzy inference system (ANFIS)-based approach is proposed to estimate defect depths from MFL signals. To reduce data dimensionality, discriminant features are first extracted from the raw MFL signals. Representative features that characterize the original MFL signals can lead to a better performance for the ANFIS model and reduce the training session. The following features are extracted: maximum magnitude, peak-to-peak distance, integral of the normalized signal, mean average, and standard deviation. Moreover, MFL signals can be approximated by polynomial series of the form, $a_nX^n + \ldots + a_1X + a_0$. The proposed approach is tested for different levels of error-tolerance. At the levels of

±15, ±20, ±25, ±30, ±35, and ±40%, the best defect depth estimates obtained by the new approach are 80.39, 87.75, 91.18, 95.59, 97.06, and 98.04%, respectively.

The advantages of using ANFIS is that the MFL data can be exploited to learn the fuzzy rules required to model the pipeline defects, and it converges faster than typical feedforward neural networks. However, the number of rules extracted is exponential with the number of used MFL features, which may prolong the learning process.

5 Conclusion

In this paper, the applicability of computational intelligence in the safety assessment in oil and gas pipelines is surveyed and examined. The survey covers safety assessment approaches that utilize data mining techniques, artificial neural networks, and hybrid neuro-fuzzy systems, for the purpose of detecting pipeline defects, estimating their dimensions, and identifying (classifying) their severity level. Obviously, techniques of computational intelligence offer an attractive alternative to traditional approaches as they can cope with complexity resulting from the uncertainty accompanying the collected diagnostic data, as well from the large size of the collected data. For intelligent techniques such as KNN, SVM, neural networks, and ANFIS, there is no need to derive a mathematical model that describes the relationship between pipeline defects and the diagnostic data (i.e., MFL and ultra sound signals, images, etc.). For typically large MFL data, KNN and SVM classifiers perform well and can provide optimal results. However, KNN may require large memory to store MFL samples. Obtaining suitable kernel functions for the SVM model has proven to be difficult. While, large MFL data may effectively be used to train different types and structures of neural networks, the learning process may take long time. Moreover, appropriate fuzzy rules can be extracted from the MFL data for the ANFIS model, which has the advantage of converging much faster than regular neural networks. The number of rules extracted, however, may increase exponentially with the number of the used MFL features.

References

1. http://www.ntsb.gov/investigations/AccidentReports/Pages/pipeline.aspx
2. K. Mandal and D. L. Atherton, A study of magnetic flux-leakage signals. Journal of Physics D: Applied Physics, 31(22): 3211, 1998.
3. K. Hwang, et al, Characterization of gas pipeline inspection signals using wavelet basis function neural networks. NDT & E International, 33(8): 531-545, 2000.
4. I. Daubechies, Ten Lectures on Wavelets. SIAM: Society for Industrial and Applied Mathematics, 1992.
5. S. Mallat, A Wavelet Tour of Signal Processing. Academic Press, 2008.

6. M. Misiti, Y. Misiti, G. Oppenheim, and J. M. Poggi, Wavelets and their Applications. Wiley-ISTE, 2007.
7. J. N. Bradley, C. Bradley, C. M. Brislawn, and T. Hopper, FBI wavelet/scalar quantization standard for gray-scale fingerprint image compression. In: SPIE Procs, Visual Information Processing II, 1961: 293–304, 1993.
8. M. Unser and A. Aldroubi, A review of wavelets in biomedical applications. Proceedings of the IEEE, 84 (4): 626–638, 1996.
9. S. Shou-peng and Q. Pei-wen, Wavelet based noise suppression technique and its application to ultrasonic flaw detection. Ultrasonics, 44(2): 188–193, 2006.
10. A. Muhammad and S. Udpa, Advanced signal processing of magnetic flux leakage data obtained from seamless gas pipeline. Ndt & E International, 35(7): 449–457, 2002.
11. S. Mukhopadhyay and G. P. Srivastava, Characterization of metal loss defects from magnetic flux leakage signals with discrete wavelet transform. NDT & E International, 33(1): 57–65, 2000.
12. American Society of Mechanical Engineers, ASME B31G Manual for Determining Remaining Strength of Corroded Pipelines, 1991.
13. A. Cosham and M. Kirkwood, Best practice in pipeline defect assessment. Paper IPC00–0205, International Pipeline Conference, 2000.
14. J. R. Jang, ANFIS: Adaptive-network-based fuzzy inference system. IEEE Transactions on Systems, MAN, and Cybernetics, 23: 665–685, 1993.
15. N. Roohollah, S. Safavi, and S. Shahrokni, A reduced-order adaptive neuro-fuzzy inference system model as a software sensor for rapid estimation of five-day biochemical oxygen demand. Journal of Hydrology, 495: 175–185, 2013.
16. K. Mucsi, K. Ata, and A. Mojtaba, An Adaptive Neuro-Fuzzy Inference System for estimating the number of vehicles for queue management at signalized intersections. Transportation Research Part C: Emerging Technologies, 19(6): 1033–1047, 2011.
17. A. Zadeh, et al, An emotional learning-neuro-fuzzy inference approach for optimum training and forecasting of gas consumption estimation models with cognitive data. Technological Forecasting and Social Change, 91: 47–63, 2015.
18. H. Azamathulla, A. Ab Ghani, and S. Yen Fei, ANFIS-based approach for predicting sediment transport in clean sewer. Applied Soft Computing, 12(3): 1227–1230, 2012.
19. A. Khodayari, et al, ANFIS based modeling and prediction car following behavior in real traffic flow based on instantaneous reaction delay. IEEE Intelligent Transportation Systems Conference, 599–604, 2010.
20. A. Kulaksiz, ANFIS-based estimation of PV module equivalent parameters: application to a stand-alone PV system with MPPT controller. Turkish Journal of Electrical Engineering and Computer Science, 21(2): 2127–2140, 2013.
21. D. Petković, et al, Adaptive neuro-fuzzy estimation of conductive silicone rubber mechanical properties. Expert Systems with Applications, 39(10): 9477–9482, 2012.
22. M. Chen, A hybrid ANFIS model for business failure prediction utilizing particle swarm optimization and subtractive clustering. Information Sciences, 220: 180–195, 2013.
23. M. Iphar, ANN and ANFIS performance prediction models for hydraulic impact hammers. Tunneling and Underground Space Technology, 27(1): 23–29, 2012.
24. Layouni, Mohamed, Sofiene Tahar, and Mohamed Salah Hamdi, "A survey on the application of neural networks in the safety assessment oil and gas pipelines." Computational Intelligencefor Engineering Solutions", 2014 IEEE Symposium on. IEEE, 2014.G. S. Park, and E. S. Park, Improvement of the sensor system in magnetic flux leakage-type nod-destructive testing. IEEE Transactions on Magnetics, 38(2): 1277–1280, 2002.
25. Sinha, Sunil K., and Fakhri Karray. "Classification of underground pipe scanned images using feature extraction and neuro-fuzzy algorithm." IEEE Transactions on Neural Networks 13.2 (2002): 393–401.R. K. Amineh, et al, A space mapping methodology for defect characterization from magnetic flux leakage measurement. IEEE Transactions on Magnetics, 44(8): 2058-2065, 2008.

26. Vidal-Calleja, Teresa, et al. "Automatic detection and verification of pipeline construction features with multi-modal data." 2014 IEEE/RSJ International Conference on Intelligent Robots and Systems. IEEE, 2014.
27. Lijian, Yang, et al. "Oil-gas pipeline magnetic flux leakage testing defect reconstruction based on support vector machine." Intelligent Computation Technology and Automation, 2009. ICICTA'09. Second International Conference on. Vol. 2. IEEE, 2009.
28. Khodayari-Rostamabad, Ahmad, et al. "Machine learning techniques for the analysis of magnetic flux leakage images in pipeline inspection." IEEE Transactions on magnetics 45.8 (2009): 3073–3084.
29. Carvalho, A. A., et al. "MFL signals and artificial neural networks applied to detection and classification of pipe weld defects." Ndt & E International 39.8 (2006): 661–667.
30. Ma, Zhongli, and Hongda Liu. "Pipeline defect detection and sizing based on MFL data using immune RBF neural networks." Evolutionary Computation, 2007. CEC 2007. IEEE Congress on. IEEE, 2007.
31. El-Abbasy, Mohammed S., et al. "Artificial neural network models for predicting condition of offshore oil and gas pipelines." Automation in Construction 45 (2014): 50–65.
32. A. Mohamed, M. S. Hamdi, and S. Tahar, A machine learning approach for big data in oil and gas pipelines. 3rd International Conference on Future Internet of Things and Cloud (FiCloud), IEEE, 2015.
33. A. Mohamed, M. S. Hamdi, and S. Tahar, Self-organizing map-based feature visualization and selection for defect depth estimation in oil and gas pipelines. 19th International Conference on Information Visualization (iV), IEEE, 2015.
34. Sinha, Sunil K., and Paul W. Fieguth. "Neuro-fuzzy network for the classification of buried pipe defects." Automation in Construction 15.1 (2006): 73–83.
35. Sinha, Sunil K., and Fakhri Karray. "Classification of underground pipe scanned images using feature extraction and neuro-fuzzy algorithm." IEEE Transactions on Neural Networks 13.2 (2002): 393–401.
36. A. Mohamed, M. S. Hamdi, and S. Tahar, An adaptive neuro-fuzzy inference system-based approach for oil and gas pipeline defect depth estimation. SAI Intelligent Systems Conference. IEEE, 2015.

Big Data for Effective Management of Smart Grids

Alba Amato and Salvatore Venticinque

Abstract The Energy industry is facing a set of changes. The old grids need to be replaced, alternative energy market is increasing and consumers want more control of their consumption. On the other hand, the ever-increasing pervasiveness of technology together with the smart paradigm, are becoming the reference point of anyone involved in innovation, and energy management issues. In this context, the information that can potentially be made available by technological innovation is obvious. Nevertheless, in order to turn it into better and more efficient decisions, it is necessary to keep in mind three sets of issues: those related to the management of generated data streams, those related to the quality of the data and finally those related to their usability for human decision-maker. In smart grid, large amounts of and various types of data, such as device status data, electricity consumption data, and user interaction data are collected. Then, as described in several scientific papers, many data analysis techniques, including optimization, forecasting, classification and other, can be applied on the large amounts of smart grid big data. There are several techniques, based on Big Data analysis using computational intelligence techniques, to optimize power generation and operation in real time, to predict electricity demand and electricity consumption and to develop dynamic pricing mechanisms. The aim of the chapter is to critically analyze the way Big Data is utilized in the field of Energy Management in Smart Grid addressing problems and discussing the important trends.

Keywords Big Data · Smart Grid · Computational intelligence · Dynamic energy management · Predictive analytics · Peer to Peer · SCADA · Interoperability, NOSQL, learning theory

---

A. Amato (✉) · S. Venticinque
Department of Industrial and Information Engineering, Second University of Naples, Caserta, Italy
e-mail: alba.amato@unina2.it

S. Venticinque
e-mail: salvatore.venticinque@unina2.it

W. Pedrycz and S.-M. Chen (eds.), *Data Science and Big Data: An Environment of Computational Intelligence*, Studies in Big Data 24,
DOI 10.1007/978-3-319-53474-9_10

1 Introduction

Smart grids are an evolution of the existing power distribution networks. They respond to the growing demand for energy, and the availability of several solutions of renewable energy sources that have stimulated the formulation of plans aiming at expanding and upgrading existing power grids in several countries. A fundamental characteristic of the smart grid is also the ability to manage, via protocols and information flows, generators and loads active in the network, coordinating them to perform certain functions in real-time as, for example, to cope with a peak, balance the load of a power supply or make up for a sudden drop in voltage by drawing more districts where there is a surplus. By linking information technologies with the electric power grid to provide "electricity with a brain" the smart grid promises many benefits, including increased energy efficiency, reduced carbon emissions, and improved power reliability, but there is an urgent need to establish protocols and standards [1].

Nevertheless, in order to turn it into better and more efficient decisions, it is necessary to keep in mind three sets of issues: those related to the management of generated data streams, those related to the quality of the data and finally those related to their usability for human decision-maker.

The recent introduction of smart meters and the creation of the Smart Grid, the first of which constitutes one of the fundamental elements, has completely revolutionized the utility system. From a processing point of view of a smart grid is inexhaustible and valuable data source in order to analyse time series, crossing them with weather data to make predictions about the electricity consumption of long or very short time; Distribute more efficiently the supply on the territory; Evaluate the quality of service provided in real time; Analyse time series to prevent potential failures and to intervene promptly; Produce more accurate bills making more conscious consumer. Smart Grid therefore generates a stream of data that must be captured, processed and analyzed efficiently from all business areas. So the data obtained, together with the data coming from more traditional sources, can be used to perform several types of analysis such as churn analysis, or research and development of new tariff plans depending on the type of consumer or fraud detection.

In fact, in smart grid, large amounts of and various types of data, such as device status data, electricity consumption data, and user interaction data are collected. Then, as described in several scientific papers [2], many data analysis techniques, including optimization, forecasting, classification and other, can be applied on the large amounts of smart grid big data [3]. There are several techniques, based on Big Data analysis using computational intelligence techniques, to optimize power generation and operation in real time, to predict electricity demand and electricity consumption and to develop dynamic pricing mechanisms [4]. Computational Intelligence can provide effective and efficient decision support for all of the producers, operators, customers and regulators in smart grid enabling all those stakeholders to have more control over the energy utilization. The aim of the chapter is to critically analyze the way Big Data analysis using computational intelligence techniques is utilized in the field of Energy Management in Smart Grid addressing problems and discussing the important trends. In particular computational intelligence techniques can be used to overcome the challenges posed by large and complex software systems. Computational intelligence aims to produce good solutions to problems in a reasonable amount of time and it is widely used for several real world applications and problems that are widespread in Smart Grid field, e.g., routing problems; assignment and scheduling problems; forecasting problems; etc.

The first part of the chapter presents overview, background and real life applications of Internet of Energy with particular emphasis on Smart Grid. Successively the state of the art of Big Data Analytics issues in IoE are discussed together with the current solutions and future trends and challenges.

The paper is organized as follows: in Sect. 2 smart energy concepts are presented, Sect. 3 introduces big data properties of smart grid and some important information in order to better understand the problem. Section 4 presents an overview of research lines and research project dealing with energy management in Smart Grid. Conclusions are drawn in Sect. 5.

2 Smart Grids and Smart Micro-Grids

Smart grids is now part of a wider smart energy concepts that includes not only the provisioning of intelligences to the power grid by an ICT solution, but also management of smart buildings, the monitoring and analysis of user's information, user's devices, environmental parameters and others [5].

In the context of smart energy, it is increasingly spreading the idea that the road leading to the reduction of global energy consumption depends of the capability to deliver usable information in the hands of energy managers, users and consumers for effective decision making. This would require lower investment than advancing the power grid or green restructuring buildings. In fact an immediate and effective solution to obtain improvements in fuel consumption and emissions appears to ensure that the existing infrastructure is at its maximum efficiency. To reach this goal it can certainly be useful the creation of a data analysis system, which is a convenient solution also from an investment perspective of their occupants. Such capability relies on innovative ICT solutions which can extract the information potentially contained in data collected by advanced metering technologies and deliver effectively it to applications and users.

The logic architecture of a smart grid is shown in Fig. 1 that is also introduced in [6]. The fabric layer of smart grids is composed of a number of smart meters and controllers, which collect data, perform simple elaboration and feed directly SCADA and other types of systems. The more representative example is the network of downstream detection instruments, which connect the user's power grid to the distribution network, and are used to measure the consumption on the various branches of a network, in particular for billing purpose. They can be also used to check and control power peek, but do not allow for improving energy efficiency, in fact usually measures are not collected and processed in real time for management purpose. These

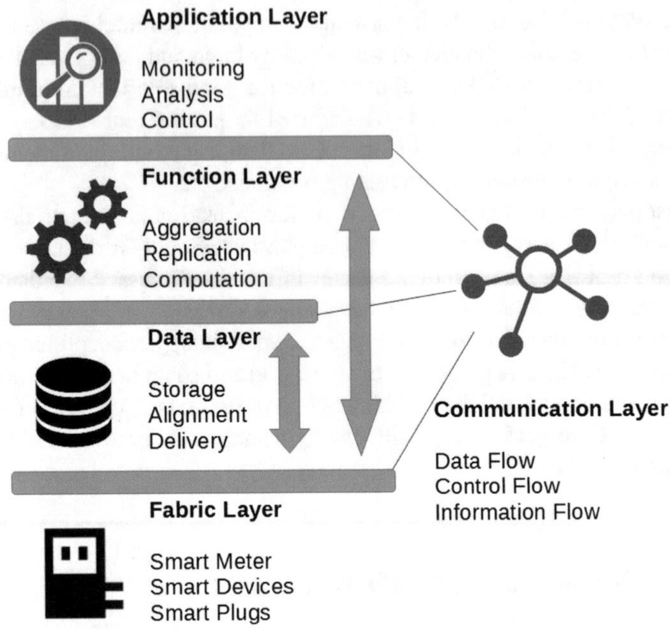

Fig. 1 Layered model of smart energy

usually represent the leaves of an energy provider and provide the finest grain of information.

On the other hand the spread of smart metering systems is growing as the information technology is assuming a central role in the technologies for energy management sector also at the building dimension. The instruments to detect energy consumption are now reaching affordable and their market penetration appears rapidly growing. For this reason the dimension of the fabric layer is growing, but it usually feeds, at building level, proprietary systems, which implement all the remaining stack of the layered architecture, caring of data communication, collection, monitoring, analysis and control. These solutions are usually provided to users which aim at increasing the degree of automation of their home, in this case to improve energy efficiency. Such kind of grid, limited to an apartment or to a building is called smart micro-grid.

The communication layer allows for transmission of data and control signals using heterogeneous technologies and across different kinds of area networks. For example data can be collected locally using embedded systems that are hosted at the user's home, in the Cloud, or directly to the user's smartphone, according to the complexity of applications, the amount of data and the provided functionalities. Hybrid solutions exists [7].

At the next layer data are collected in centralized or distributed repositories. Data flowing from the fabric layer are characterized here by complexity complexity of different type, that make challenging the extraction of relevant information they provide as a whole. First we must consider different solutions, because data are heterogeneous as they come from very different sources or they are representative of different magnitudes and phenomena (the sources can be rather than utility meter sensors that detect environmental quantities or human phenomena). Sometimes the same technology cannot adequately manage the characteristics of data from different sources, and for example, the integration of the electrical consumption data with those of other energy sources such as fuel or water is seldom supported. But also the data representation make it difficult the correlation between different kinds of energy related information or environmental variables. Furthermore, virtually all the data come from field measurements and need to be consolidated into a single data structure, there is the problem of managing the transport and ensure their integrity. Lies also the problem of data sets to be processed keeping in mind that interoperability with NoSQL database, it is not a very common feature among the energy data management systems, although this, like other technologies that allow you to work with big date, could be a very important resource to support processing in real time, ensuring accurate and timely analysis capabilities and deliver it to decision-makers with the latest information.

The collective layer services orchestrate services provided by the underlying layers to reduce the complexity due to distribution, heterogeneity, computational load, distribution etc. Effort spent to increase innovation could be spent at this layer to integrate available technologies such as in the Cloud field, to meet computational requirements of big data processing.

At highest level we found application which implement monitoring, analysis and sometimes business intelligence to decision makers for an effective management of the infrastructure. Building Management Systems (BMS) are implemented and Energy Information Systems (EIS) are implemented at this layer. BMS are conceived to perform operational tasks (maintenance of electrical and mechanical systems, the occupant complaints management ...), as well as the energy and the operation of the individual subsystems such as the heating or air conditioning. EIS allow for recognition and fault diagnosis, alarm management and energy management. In those three areas, if implemented properly, they are able to provide a very important support by speeding up the procedures of intervention in case of failures, focusing the attention of engineers on the most critical events that are happening in real time, integrating and consolidating data flows from environmental or energy meter and presenting the results of calculations with usable mode.

Available technological solutions implement all or part of the discussed architecture. For example in Fig. 2 we can see some example of implementation for smart micro-grids and smart grid:

- A commercial solution[1] that provides a wifi gateway from a sensor network of smart meters to an app that can be used for monitoring and management of the smart home. The apps implements all the layers of the architecture.

---

[1]http://www.4-noks.com.

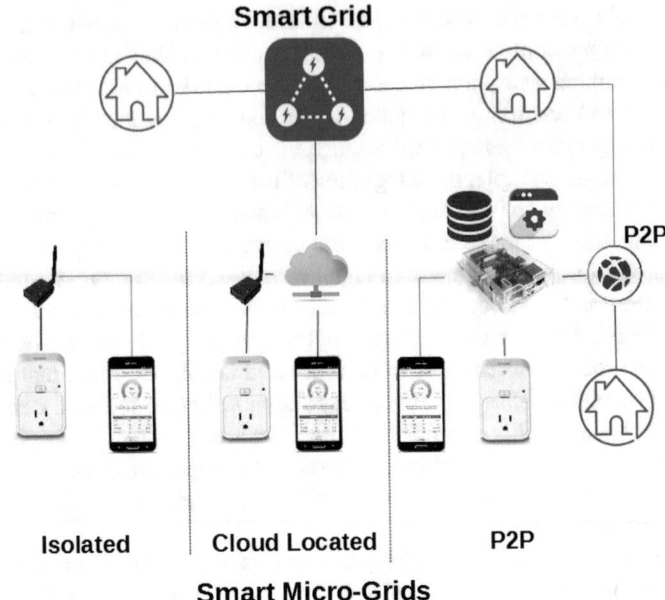

Fig. 2 Technological implementations of smart micro-grids

- A different technology[2] that use a resident gateway to forward collected data to the cloud, where they are stored and analyzed, monitoring and control dashboard are provided to customers by application as a service (AaaS). In this case the data, the application and the business logic are hosted in Cloud.
- The CoSSMic [8] research project prototype implements a peer to peer network of collaborating smart micro-grids, each of them collecting and storing data locally. All layers are hosted at home, in an PC or an embedded system with enough computational resources.
- Finally smart grid solutions are provided by some big commercial companies, which usually provide ad hoc solution which are not integrated with the micro-grids, a part of the monitoring and total energy flow between the household and the distribution network.

Figure 3 shows how the variation in energy consumption patterns between buildings in a neighbourhood can be optimized by coordinating load shifting and the use of storage capacities. The investigation of a scalable, at least at neighborhood level, is investigated in [6] by a Peer to Peer (P2P) collaboration of smart micro-grids [5, 7].

Unfortunately fragmentation of the smart grid is observed both across layers and within the same layers, because of different reasons which range from interoperability to security. In the current technological scenario more and more micro-grids are

[2]http://smartbeecontrollers.com/.

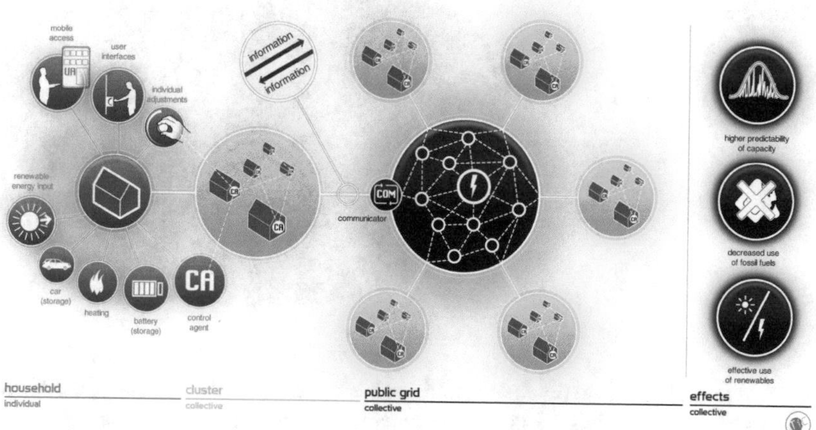

Fig. 3 Collaborating smart grids

growing, which are all connected to the power grid, but isolated from the smart grid. Open issues affect different layer of the presented architectures. Some of them can be addressed independently such as growing supply of services in mobility, flexibility and customization of user interfaces. Other ones rely on interoperability and security, that are drivers to deliver and integrate data to upper levels, for improving the application awareness, for aggregate processing and information fusion.

But this process has started and more and more data are collected and must be stored, aggregated and processed. For this reason the smart energy energy is already now a big data problem. In the following sections we will focus on the data and the application layer discussing issues, research efforts and technological solution related to big data properties of smart energy.

3 Big Data Properties of Smart Grid

"Not only will the integration of Big Data technologies help make the grid more efficient, it will fundamentally change who sells electric power; how it gets priced; and how regulator, utilities grid operators and end user interact" is claimed in [9].

An interesting view of what are the Big Data has been exposed to Gartner that defines Big Data as "high volume, velocity and/or variety information assets that demand cost-effective, innovative forms of information processing that enable enhanced insight, decision making, and process automation" [10]. In fact the huge size is not the only property of Big Data. Only if the information has the characteristics of Volume,Velocity and/or Variety we can talk about Big Data [11] as shown in Fig. 4. Volume refers to the fact that we are dealing with ever-growing data expand-

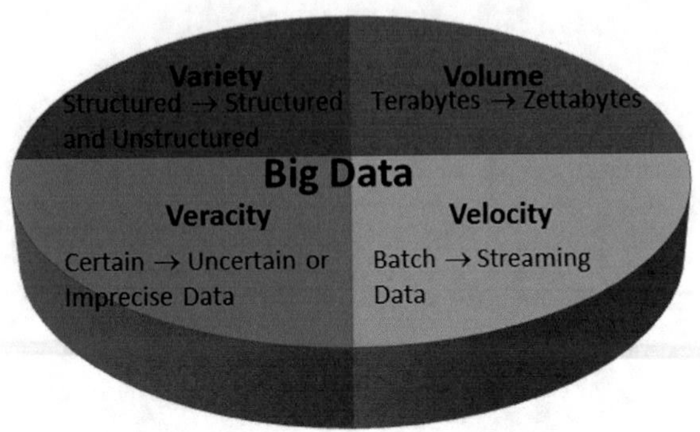

Fig. 4 Big Data characteristics

ing beyond terabytes into petabytes, and even exabytes (1 million terabytes). Variety refers to the fact that Big Data is characterized by data that often come from heterogeneous sources such as machines, sensors and unrefined ones, making the management much more complex. Finally, the third characteristic, that is velocity that, according to Gartner [12], "means both how fast data is being produced and how fast the data must be processed to meet demand". In fact in a very short time the data can become obsolete. Dealing effectively with Big Data "requires to perform analytics against the volume and variety of data while it is still in motion, not just after" [11]. IBM [13] proposes the inclusion of veracity as the fourth Big Data attribute to emphasize the importance of addressing and managing the uncertainty of some types of data. Striving for high data quality is an important Big Data requirement and challenge, but even the best data cleansing methods cannot remove the inherent unpredictability of some data, like the weather, the economy, or a customer's actual future buying decisions. The need to acknowledge and plan for uncertainty is a dimension of Big Data that has been introduced as executives seek to better understand the uncertain world around them.

In the case of smart grids most of information are provided by *Machine-generated data* coming from smart meters to measure the consumption of electricity, but we have also human-generated data. Data frequency will varie a lot. In fact the plans will be update by the user at low frequency at not regular intervals. Monitoring must be available on demand, as with social media data. On the other hand prediction and measures could came periodically and at higher frequency. Even if data analysis can be performed in batch mode, on the other hand negotiation and scheduling should take place in near real time. In any case we handle structured data.

About data complexity we have to consider that volume of data increases continuously, either because of the continuous production of metering information and because the foreseen connection of new users and devices. Such issue pose a challenge about using traditional methods, such as relational database engines to store, search, share, analyze, and visualize using. Data processing can make use of massive parallel processing power on available hardware because of an application model based on collaborating autonomous agents, however location of data and privacy is an open issue in current big data solutions.

Data velocity can obviously increase and response adaptation may be required.

The quality of the data made available by an energy information system is certainly another area of research on which it is important to reflect. Quality can be understood from the point of view of fairness and integrity of the data source to user, but also from the perspective of adherence data to the needs of decision-makers and their proper context. The first part of the problem may seem simple solution, but taking into account the heterogeneity of the sources and the absence of specific reference models, combined with the specific nature of the data collected in each real application context, you realize that it can be a longer process complicated than expected. It must be added that not always the transport of energy data takes place in an ideal way, since the communication networks built for this purpose are, in many cases the result of compromises due to the need to make them coexist with the existing classical structures of the buildings which are the subject Business management Energy. Standardize the data in a consistent and functional structure to the needs of the organization that will exploit them, it is also a prerequisite for people and structures who base their decisions on these data. The second part of the reasoning on data quality has instead to do with both the design that with the implementation of these systems. The fact that the data provided by a Smart Grid are more or less appropriate in decision-making that are intended to support, depends both on the instrument's ability to offer the depth and breadth of analysis, that the skills of those who are planning to deploy to a specific application context. By focusing on the power of the analysis instruments, the question regards the minimum level of granularity provided, and it is evident that this can only be a trade-off between the complexity of the system and the large number of listening points (the meters and sensors) it has. It is inevitably an issue that involves aspects concerning the scalability of the systems, the complexity of the necessary data structures and performance processing; identifying the right balance between these three dimensions is surely one of the most significant issues to be addressed, and will be both a mobile equilibrium influenced by the evolution of the technologies that make up the energy management systems.

Big data is changing the way of energy production and the pattern of energy consumption. Energy big data have brought opportunities and challenges at the same time for us. Some of the primary and urgent challenges include: (a) how to effectively collect, store and manage the energy big data; (b) how to efficiently analyze and mine the energy big data; (c) how to use the energy big data to support more effective and efficient decision makings; (d) how to get insights and obtain values from the energy big data; and (e) how to effectively prevent risks and protect privacy while utilizing the energy big data [14].

So, the challenge is to find a way to transform raw data into valuable information. To capture value from Big Data, it is necessary an innovation in technologies and techniques that will help individuals and organizations to integrate, analyze, visualize different types of data at different spatial and temporal scales. Based on Big Data Characteristics, [14] describes the characteristics of energy big data and proposes a mapping of the two. So the 4 V characteristics became the 3E (energy, exchange and empathy) characteristics of energy big data. Energy (data-as-an-energy) means that energy savings can be achieved by big data analytics. Energy big data with its easy to transport properties, and in the course of constantly refining and value-added. Under the premise of can protect the interests of users, in each link of the low energy. By saving energy, the process of energy big data is the process of water and electricity energy release in a sense, through the energy big data analysis to achieve the purpose of energy saving, is the largest investment in energy infrastructure [15].

Exchange (data-as-an-exchange) refers to that the big data in energy system need to exchange and integrate with the big data from other sources to better realize its value. Energy big data has a value that reflect the entire national economy, social progress and development of all walk of life and other aspects of innovation and to play a greater value of its premise and the key is to interact with external data. In fact big data energy fusion and a full range of mining, analysis and presentation on this basis, can effectively improve the current industry [15].

Empathy (data-as-an-empathy) means that better energy services can be provided, users needs can be better satisfied, and consumer satisfaction can be improved based on energy big data analytics. Enterprise's fundamental purpose is to create customers, create demand. Energy big data natural contact households, factories and enterprises to promote power industry by the electric power production as the center to take the customer as the center which is the essence of ultimate concern for power users. Through the power user needs to fully tap and met to establish an emotional connection, for the majority of electricity users to provide more high-quality, safe and reliable electric service. In the process of energy industry contribution to maximize the value of energy industry also found a new source of power often changed frequently, empathy can benefit [15].

In conclusion, big data have a large impact on the management of power utilities as smart grid operation and future energy management will be hugely data-intensive. There are many challenges which affect the success of big data applications in smart grid. It is necessary to gain practical experience in integrating big data with smart grid together with more effort to develop more advanced and efficient algorithms for data analysis. With this aim, in the next section, we introduce several research contribution in the field of big data analytics for smart grid.

4 Research Lines and Contribution

In this section we present an overview of research contribution for big data analytics applications to the smart grid. In [16] authors identify several future research directions about smart grid, we focus here on those which are related to big data.

An effort to improve interoperability needs to aggregate all available information from different sources, such as individual smart meters, energy consumption schedulers, solar radiation sensors, wind-speed meters and relays.

As it is strengthened in [17], it is relevant to improve data collection and management, starting from optimizing data sampling, choosing effective storage solutions.

The next open issue deals with real-time monitoring and forecasting, which requires effort in the application and development of data mining techniques and big data analytic. Techniques for categorization of the information and successful recognition of the different patterns are necessary here and at the business level.

Big data analytic is also relevant to provide at the right communication point necessary decision making support.

Research Focus	Requirements	Technology
Interoperability and standardization	Metrics, taxonomies, protocols	Ontologies, semantic
Big Data storages and Cloud availability	Performance, reliability, aggregation, processing, NOSQL	
Big data analytic	Categorization, pattern recognition, decision making, monitoring and forecasting	Data mining, business intelligence

4.1 Interoperability and Standardization

Smart Grid data are collected from different sources, such as individual smart meters, energy consumption schedulers, aggregators, solar radiation sensors, wind-speed meters. Some effort have been spent by widespread deployment of wireless communication power meters, availability of customer energy usage data, development of remote sensing for determining real-time transmission and distribution status, and protocols for electric vehicle charging.

National Institute of Standards and Technology (NIST) ongoing efforts aim at facilitating and coordinating smart grid interoperability standards development and smart grid-related measurement science and technology, including the evolving and continuing NIST relationship with the Smart Grid Interoperability Panel (SGIP[3]) public-private partnership. NIST developed an initial (now completed) three-phase plan whose first step is the identification and consensus on smart grid standards. In particular interoperability on data is addressed at technical level, focusing on syntactic interoperability, and at information layer, where it deals with semantic understanding business context. Syntactic interoperability means understanding of data structures, while semantic means understanding of concept contained in data structures. At the state of the art improving interoperability requires development and application of techniques and technologies to achieve such alignment on big data which already now are available in smart grids.

Table Definition Language (TDL), described in terms of the XML-based Exchange Data Language (EDL) that can be used to constrain oft-utilized information into a well-known form, are proposed for data representation.

[3]http://www.sgip.org/.

Many standard and proposal are already available.[4]

They are related to information exchange among many components of a smart-grids such as energy usage in kilowatt hours from a meter, load profiles and control information, communications protocols for field devices and systems in SCADA Distribution Grid Management, Markets Energy Price and Transactions, Schedules and Calendars, Standard for Common Format for Event Data Exchange (COMFEDE) for Power Systems.

4.2 Big Data Storages

Widespread integration of grid-tied renewable along with attendant large-scale storage has been recognized as a key scientific and technological areas related to smart-grid [18].

Research effort at this layer deals with the effective utilization and integration of available technologies to address different kind of data of smart grid.

Big Data are so complex and large that it is really difficult and sometime impossible, to process and analyze them using traditional approaches. In fact traditional relational database management systems (RDBMS) can not handle Big Data sets in a cost effective and timely manner. These technologies may not be enabled to extract,from large data set, rich information that can be exploited across of a broad range of topics such as market segmentation, user behavior profiling, trend prediction, events detection, etc. and in many fields like public health, economic development and economic forecasting. Besides Big Data have a low information per byte, and, therefore, given the vast amount of data, the potential for great insight is quite high only if it is possible analyze the whole dataset [11].

The term NoSQL (meaning 'not only SQL') is used to describe a large class of databases which do not have properties of traditional relational databases and which are generally not queried with SQL (structured query language). NoSQL data stores are designed to scale well horizontally and run on commodity hardware. Also, the 'one size fit's it all' [19] notion does not work for all scenarios and it is a better to build systems based on the nature of the application and its work/data load [20]. They can provide solution both at data and collective layer of a smart grid architecture, in fact some functionalities can be:

- Sharding, also referred to as horizontal scaling or horizontal partitioning. It is a partitioning mechanism in which records are stored on different servers according to some keys. Data is partitioned in such a way that records, that are typically accessed/updated together, reside on the same node. Data shards may also be replicated for reasons of reliability and load-balancing.
- Consistent hashing [21] The idea behind consistent hashing is to use the same hash function, used to generate fixed-length output data that acts as a shortened refer-

[4]http://collaborate.nist.gov/twiki-sggrid/bin/view/SmartGrid/SGIPCoSStandardsInformation Library.

ence to the original data, for both the object hashing and the node hashing. This is advantageous to both objects and machines. The machines will get an interval of the hash function range and the neighboring machines can take over portions of the interval of their adjacent nodes if they leave and can assign parts of their interval if some new member node joins and gets mapped to a nearby interval. Another advantage of consistent hashing is that clients can easily determine the nodes to be contacted to perform read or write operations.

- Advanced computation capabilities such as MapReduce [22] for processing large data sets. A popular open source implementation is Apache Hadoop [23], a framework that allows for the distributed processing of large data sets across clusters of computers using simple programming models. It is designed to scale up from single servers to thousands of machines, each offering local computation and storage and to execute queries and other batch read operations against massive datasets that can be tens or hundreds of terabytes and even petabytes in size. It supports both high performant stream and batch processing, which are respectively used for smart grids to detect in real time alerts, error or for business analytics.

On the other hand, to chose the right technological solution we have to take into account the type of data of the smart energy domain and the kind of requirements for processing them. In particular format of the content for the energy measure is of course composed of real or integer values. Hence we have to store, communicate and process time series which should be available for off-line analysis and reporting but also for on-line monitoring and learning. Moreover we need to handle transaction for negotiation and historical data and prediction.

Key-value data stores (KVS) typically provide replication, versioning, locking, transactions, sorting, and/or other features. The client API offers simple operations including puts, gets, deletes, and key lockups. Notable examples include: Amazon DynamoDB [24], Project Voldemort [25], Memcached [26], Redis [27] and RIAK [28]. They can be exploited to organize information. Many of them offer in-memory solution and workload distribution [26], which can be used to improve performance for stream processing, when it needs to elaborate on line high frequency measures which are received from smart meters.

Document data stores (DDS). DDS typically store more complex data than KVS, allowing for nested values and dynamic attribute definitions at runtime. Unlike KVS, DDS generally support secondary indexes and multiple types of documents (objects) per database, as well as nested documents or lists. Notable examples include Amazon SimpleDB [29], CouchDB [30], Membase/Couchbase [31], MongoDB [32] and RavenDB [33]. They could be exploited at business level to store report, result of analysis, billing documents.

Extensible record data stores (ERDS). ERDS store extensible records, where default attributes (and their families) can be defined in a schema, but new attributes can be added per record. ERDS can partition extensible records both horizontally (per-row) or vertically (per-column) across a datastore, as well as simultaneously using both partitioning approaches. Notable examples include Google BigTable [34],

HBase [35], Hypertable [36] and Cassandra [37]. In smart grid they are the solution for building datawarehouse and supporting business analytics.

Another important category is constituted by Graph data stores. They [38] are based on graph theory and use graph structures with nodes, edges, and properties to represent and store data. Different Graph Store products exist in the market today. Some provide custom API's and Query Languages and many support the W3C's RDF standard. Notable examples include neo4j [39], AllegroGraph [40] and InfiniteGraph [41]. The graph data model fits better to model domain problems that can be represented by graph as ontologies, relationship, maps etc. Particular query languages allow querying the data bases by using classical graph operators as neighbor, path, distance etc. Unlike the NoSQL systems we presented, these systems generally provide ACID transactions. In smart grid they can be exploited to record relationships among events, users, devices and in general to represent the knowledge of the smart energy domain. They allow to apply at application level effective inference models to recognize situation, deduct reactions and predict future events.

We cannot forget a Time Series Databases which are optimized for handling time series data. Meters output are mainly represented as an array of numbers indexed by time. They are used also to represent profiles, curves, or traces in the smart energy domain. Some available technologies are Graphite,[5] InfluxDB[6] and OpenTSDB.[7] Some of them are not really conceived to store big data, but to effectively manage time series in window time-frames.

Table 1 provides a comparison of all the examples given in terms of Classification, Licence and Storage System. Comparison based on several issues are available at [42].

Exploitation of big data storages in Cloud is another scientific and technological challenge.

In [43] authors discuss about about how Cloud computing model can be used for developing Smart Grid solutions. They propose to exploit to use advantages of Cloud computing to achieve the most important future goals of a large-scale Smart Grid, such as energy savings, two-way communication, and demand resource management. In [44] smart grid data management is based on specific characteristics of cloud computing, such as distributed data management for real-time data gathering, parallel processing for real-time information retrieval, and ubiquitous access.

4.3 Big Data Analytic

In order to build an accurate real-time monitoring and forecasting system, it is necessary to integrate all available information from different sources, such as individual smart meters, energy consumption schedulers, aggregators, solar radiation

[5]http://graphite.net.

[6]https://influxdata.com/.

[7]opentsdb.net/.

Table 1 Data store comparison

Name	Classification	License	Data storage
Dynamo	KVS	Proprietary	Plug-in
Voldemort	KVS	Open Source	RAM
Memcached	KVS	Open Source	RAM
Redis	KVS	Open Source	RAM
RIAK	KVS	Open Source	Plug-in
SimpleDB	DDS	Proprietary	S3 (Simple Storage Service)
CouchDB	DDS	Open Source	Disk
Couchbase	DDS	Open Source	Disk
MongoDB	DDS	Open Source	Disk
RavenDB	DDS	Open Source	Disk
Google BigTable	ERDS	Proprietary	GFS
HBase	ERDS	Open Source	Hadoop
Hypertable	ERDS	Open Source	Disk
Cassandra	ERDS	Open Source	Disk
Neo4J	Graph	Open Source	Disk
AllegroGraph	Graph	Proprietary	Disk
InfiniteGraph	Graph	Proprietary	Disk
RRDtool	TSDB	Open Source	Disk (Circular Buffer)
Graphite	TSDB	Open Source	Disk
Druid	TSDB	Open Source	NFS
Riak TS	TSDB	Open Source	Hadoop
OpenTSDB	TSDB	Open Source	Hadoop, Hbase
kdb+	TSDB	Commercial	Disk, column oriented
splunk	TSDB	Commercial	Oracle DB

sensors [45]. Moreover [45] also individuate two important issues. the appropriate forecasting system should rely on effective data sampling, improved categorization of the information and successful recognition of the different patterns. Second, suitable adaptive algorithms and profiles for effective dynamic, autonomous, distributed, self-organized and fast multi-node decision-making have to be designed. This requires to invest effort in big data analytics, as it will give utilities and grid operators insights on smart grid data for grid management, reliability and efficiency. Analytics application can utilize various big data options for administering smart meter gateways and meter data processing for market participants. Energy theft and overloaded distribution equipment detections through grid load analysis, grid incident analysis and end-customer consumption load analysis are examples of information that can be extracted. The big data option will also allow the creation of load forecasts for different levels in the distribution grid as well as an analysis of distributed energy resources.

Descriptive smart meter analytics have already proved to be quite valuable for utilities that are searching for ways to use data to understand root causes and provide awareness for better decision making.

Diagnostic analytic models is closely coupled with descriptive ones a diagnostic model will further analyze the data to look for trends and patterns in the monitoring stage. Thus, the successful model will use drill-downs, factor analytics, and advanced statistical exploration.

Predictive big data analytics is a powerful tool to expose risks, uncover opportunities, and reveal relationships among myriad variables.

Relevant use case are the need for load balancing according to the day-to-day and hour-to-hour costs of power. The goal is saving both money and energy by predicting the costs of power and demand based on a constant flow of signals, allowing the distributors to buy and sell accordingly while shaving load during peak hours. Predictive analytics, applied to big-data, allows to create an interaction layer between the bulk power system and the distribution systems.

Decision making support can be implemented, exploiting Big Data Analytics Strategies for producing recommendations to find the best decision in a particular context making a more informed guess about the most high-value action.

Research contribution have proposed how to use computing resources for applying such techniques. A hierarchical structure of cloud computing centers to provide different types of computing services for information management and big data analysis is described in [46]. Security issues are addressed by a solution based on identity-based encryption, signature and proxy re-encryption.

In [47] it has been shown that the performance of multi-node load forecasting is clearly better than that of single-node forecasting. [47] proposes a load data hierarchical and partitioned processing method, establishes a formula to reflect their mutual restraint and relation, creates a model to describe transmission system multi-node load dynamic characteristic on the basis of top layer forecasting using recursive least square support vector machines algorithm, and constructs an ultra-short term load forecasting overall frame of adaptive dynamic model. As shown in [48], the designed algorithms should be based on realistic consensus functions or voting by incorporating probability terms models, where the large computations can be parallelized. The algorithmic results are the state estimation, the estimated production and consumption, and the STLF in SGs.

For the most efficient pattern-recognition and state estimation in the SGs environment, the following methodologies and technologies can be used:

- *Feature Selection and Extraction*. The traditional factors include the weather conditions, time of the day, season of the year, and random events and disturbances. On the other hand, the smart grid factors include the electricity prices, demand response, distributed energy sources, storage cells and electric vehicles.
- *Online Learning*. In contrast to statistical machine learning, online learning algorithms do not make stochastic assumptions about the observed data. Some application use a streaming processing of incoming data o recognize already known shape-lets and classify events.

- *Randomized Model Averaging*. It is concerned with the design and development of algorithms that allow computers to evolve behaviours based on empirical data. A major focus of research is to automatically learn to recognize complex patterns such as the features of smart grids, and make intelligent decisions based on data.

Even if many contributions focused on short term load forecasting, using regression models, linear time-series-based, state-space models, and nonlinear time-series modeling, on the other hand they have been seldom used at meter aggregate levels, such as distribution feeders and substation. Moreover, very little progress has been made in the field of the very-short-term load forecasting, which could be very useful to compensate aggregate power fluctuation of co-located photovoltaic panels, when storages are full or do not exist.

4.4 Research Projects Networked with Companies

Power utilities are cooperating with IT companies to develop big data analytics for smart grid applications. There are several research project, networked with big companies, aimed at investigate big data analytics applications to the smart grid. The aim of those research project networked with companies is to achieve also competence and results that can be used as a springboard for the companies' own research and development projects.

Global technology company Siemens announced on February 2016 the integration of a big data option in its smart grid application. They are providing more and more standard Business Intelligence (BI) reports to their products and supporting applications into the smart grid solution. Also European Commission is funding several European Project based on Big Data driven Smart Grid solution. Some important project aiming at solve the issue related to big data analytics, data transformation and management, demand response, settlement end forecasting are the following.

The EU-funded SPARKS[8] (Smart Grid Protection Against Cyber Attacks) project aims to provide innovative solutions in a number of ways, including approaches to risk assessment and reference architectures for secure smart grids. The project will make recommendations regarding the future direction of smart grid security standards. Furthermore, key smart grid technologies will be investigated, such as the use of big data for security analytics in smart grids, and novel hardware-supported approaches for smart meter (gateway) authentication. All of these contributions and technologies will be assessed from a societal and economic impact perspective, and evaluated in real-world demonstrators.

RealValue[9] is an European energy storage project funded by Horizon 2020, the largest Research and Innovation Programme in Europe. RealValue will use a combination of physical demonstrations in Ireland, Germany and Latvia along with innovative modelling techniques, in order to demonstrate how local small-scale energy

[8]https://project-sparks.eu/.

[9]http://www.realvalueproject.com/.

storage, optimised across the whole EU energy system, with advanced ICT, could bring benefits to all market participants. Using Big Data technologies, the smart system will also interpret data to forecast heat demand and consumer comfort preferences, and it will interact with the grid to manage supply and congestion.

IES (Integrating the Energy System)[10] is an European Project that combines Big Data technologies and cyber security in order design a modular process chain to achieve interoperability of standards, a specification of a normalised use of these standards in interoperability profiles and a demonstration of the processes for testing interoperability. The description of the processes and the practical implementation is to be vendor-neutral, in order to ensure long-term interoperability and acceptance in energy domain. The transparency of the method and the open database for technical specifications and profiles should be accessible for technology providers for interoperable products and services. In this way interoperability will create increasing competition with decreasing prices for better products on binding security level.

Secure Operation of Sustainable Power Systems (SOSPO)[11] is a project whose main goal is to carry out research and development for the purpose of methods for a real-time assessment of system stability and security, as well as methods for intelligent wide-area prosumption control that can ensure stable and secure operation of the future power system. The research in the SOSPO project focuses on methods that enable system stability and security assessment in real-time and on methods for automatically determining control actions that regain system security when an insecure operation has been detected.

5 Conclusion

In this chapter we have presented a survey of the various research issues, challenges, technical and technological solutions and analyzed critically the utilization of Big Data to manage smart grid introducing several research problems and describing several research project trying to bring out the weaknesses and strengths of the different solutions. The main goal of the research was to identify the critical issues and highlight strengths and potential on the basis of direct experience accumulated. The main challenges are [49]:

- the difficulty in collecting the data by itself due the existence of multiple sources with different formats and types and different usage and access policies;
- the difficulty in categorize and organize and an easily accessible way for applications to use the data itself due to the unstructured nature of the data;
- the difficulty in create a unified understanding of data semantics and a knowledge base in order to extract new knowledge based on specific real-time data;
- the difficulty in retrieve and transform the data automatically and universally into a unified data source for useful analysis;

[10] http://www.offis.de/en/offis_in_portrait/structure/structure/projekte/ies-austria.html.

[11] http://www.sospo.dk/.

- the data uncertainty and trustworthiness;
- security and privacy issues, as databases may include confidential information, so it is necessary to protect this data against unauthorized use and malicious attacks;
- the size of generated data that rapidly grows according to the population grows. So Smart Grid applications need to evolve quickly and extend efficiently to handle the growing volume and variety of big data.

Nevertheless, the success of the new energy transition relies on the ability to adopt Big Data analysis using computational intelligence techniques. Computational Intelligence provides solutions for such complex real-world problems to which traditional modeling can be useless for a few reasons: the processes might be too complex and it contains some uncertainties during the process. Computational Intelligence represents a powerful and versatile methodology for a wide range of data analysis problems, so to effectively and efficiently overcome the challenges it is necessary to pay more attention to these techniques.

References

1. NIST: Nist smart grid. http://www.nist.gov/smartgrid/ (2012)
2. Jin, X., Wah, B., Cheng, X., Wang, Y.: Significance and challenges of big data research. Big Data Research **2** (2015) 59–64
3. Diamantoulakis, P., Kapinas, V., Karagiannidis, G.: Big data analytics for dynamic energy management in smart grids. Big Data Research **2** (2015) 94–101
4. Pakkanen, P., Pakkala, D.: Reference architecture and classification of technologies, products and services for big data systems. Big Data Research **2** (2015) 166–186
5. Amato, A., Aversa, R., Di Martino, B., Scialdone, M., Venticinque, S., Hallsteinsen, S., Horn, G.: Software agents for collaborating smart solar-powered micro-grids. Volume 7. (2015) 125–133
6. Amato, A., Di Martino, B., Scialdone, M., Venticinque, S.: Design and evaluation of p2p overlays for energy negotiation in smart micro-grid. Computer Standards and Interfaces **44** (2016) 159–168
7. Amato, A., Di Martino, B., Scialdone, M., Venticinque, S., Hallsteinsen, S., Jiang, S.: A distributed system for smart energy negotiation. Lecture Notes in Computer Science (including subseries Lecture Notes in Artificial Intelligence and Lecture Notes in Bioinformatics) **8729** (2014) 422–434
8. Horn, G., Venticinque, S., Amato, A.: Inferring appliance load profiles from measurements. Lecture Notes in Computer Science (including subseries Lecture Notes in Artificial Intelligence and Lecture Notes in Bioinformatics) **9258** (2015) 118–130
9. Stimmel, C.L.: Big Data Analytics Strategies for the Smart Grid. CRC Press (2014)
10. Gartner: Hype cycle for big data, 2012. Technical report (2012)
11. IBM, Zikopoulos, P., Eaton, C.: Understanding Big Data: Analytics for Enterprise Class Hadoop and Streaming Data. 1st edn. McGraw-Hill Osborne Media (2011)
12. Gartner: Pattern-based strategy: Getting value from big data. Technical report (2011)
13. Schroeck, M., Shockley, R., Smart, J., Romero-Morales, D., Tufano, P.: Analytics: The real-world use of big data. Ibm institute for business value - executive report, IBM Institute for Business Value (2012)
14. Zhou, K., Fu, C., Yang, S.: Big data driven smart energy management: From big data to big insights. Renewable and Sustainable Energy Reviews **56** (2016) 215–225

15. Wan, X., Wang, B.: Key technology research based on big data era hydroelectric energy. In: Advances in Energy Equipment Science and Engineering. CRC Press (2015) 81–85
16. Diamantoulakis, P.D., Kapinas, V.M., Karagiannidis, G.K.: Big data analytics for dynamic energy management in smart grids. Big Data Research **2** (2015) 94–101
17. Yang, P., Yoo, P., Fernando, J., Zhou, B., Zhang, Z., Zomaya, A.: Sample subset optimization techniques for imbalanced and ensemble learning problems in bioinformatics applications. IEEE Transactions on Cybernetics **44** (2014) 445–455
18. Amato, A., Venticinque, S.: Big data management systems for the exploitation of pervasive environments. Studies in Computational Intelligence **546** (2014) 67–89
19. Stonebraker, M., Cetintemel, U.: "one size fits all": An idea whose time has come and gone. In: Proceedings of the 21st International Conference on Data Engineering. ICDE '05, Washington, DC, USA, IEEE Computer Society (2005) 2–11
20. Gajendran, S.K.: A survey on nosql databases. Technical report (2012)
21. Karger, D., Lehman, E., Leighton, T., Panigrahy, R., Levine, M., Lewin, D.: Consistent hashing and random trees: distributed caching protocols for relieving hot spots on the world wide web. In: Proceedings of the twenty-ninth annual ACM symposium on Theory of computing. STOC '97, New York, NY, USA, ACM (1997) 654–663
22. Dean, J., Ghemawat, S.: Mapreduce: simplified data processing on large clusters. Commun. ACM **51** (2008) 107–113
23. Apache: Hadoop (2012) http://hadoop.apache.org/, [Online; 10-July-2016].
24. DeCandia, G., Hastorun, D., Jampani, M., Kakulapati, G., Lakshman, A., Pilchin, A., Sivasubramanian, S., Vosshall, P., Vogels, W.: Dynamo: amazon's highly available key-value store. SIGOPS Oper. Syst. Rev. **41** (2007) 205–220
25. Sumbaly, R., Kreps, J., Gao, L., Feinberg, A., Soman, C., Shah, S.: Serving large-scale batch computed data with project voldemort. (2009)
26. Memcached: Memcached (2012) http://memcached.org/, [Online; 7-July-2016].
27. Redis: (2012) http://redis.io/documentation, [Online; 10-July-2016].
28. Riak: (2012) http://basho.com/riak/ [Online; 6-July-2016].
29. Amazon: Simpledb (2012) http://aws.amazon.com/simpledb/, [Online; 6-July-2016].
30. Apache: Couchdb (2012) http://couchdb.apache.org/, [Online; 6-July-2016].
31. Couchbase: (2012) http://www.couchbase.com/, [Online; 6-July-2016].
32. MongoDB: (2012) http://www.mongodb.org/, [Online; 6-July-2016].
33. RavenDB: (2012) http://ravendb.net/, [Online; 6-July-2016].
34. Chang, F., Dean, J., Ghemawat, S., Hsieh, W.C., Wallach, D.A., Burrows, M., Chandra, T., Fikes, A., Gruber, R.E.: Bigtable: A distributed storage system for structured data. ACM Trans. Comput. Syst. **26** (2008) 4:1–4:26
35. HBase: Hbase (2012) [Online; 6-July-2016].
36. Hypertable: (2012) http://hypertable.com/documentation/, [Online; 6-July-2016].
37. : Cassandra (2012) http://cassandra.apache.org/, [Online; 6-July-2016].
38. Robinson, I., Webber, J., Eifrem, E.: Graph Databases. O'Reilly Media, Incorporated (2013)
39. Neo Technology, I.: Neo4j, the world's leading graph database. (2012) http://www.neo4j.org/, [Online; 7-July-2016].
40. AllegroGraph: (2012) http://www.franz.com/agraph/allegrograph/, [Online; 6-July-2016].
41. InfiniteGraph: (2012) http://www.objectivity.com/infinitegraph, [Online; 6-July-2016].
42. findthebest.com: Compare nosql databases (2012) [Online; 6-July-2016].
43. Markovic, D.S., Zivkovic, D., Branovic, I., Popovic, R., Cvetkovic, D.: Smart power grid and cloud computing. Renewable and Sustainable Energy Reviews **24** (2013) 566–577
44. Rusitschka, S., Eger, K., Gerdes, C.: Smart grid data cloud: A model for utilizing cloud computing in the smart grid domain. In: Smart Grid Communications (SmartGridComm), 2010 First IEEE International Conference on. (2010) 483–488
45. Diamantoulakis, P.D., Kapinas, V.M., Karagiannidis, G.K.: Big data analytics for dynamic energy management in smart grids. CoRR **abs/1504.02424** (2015)
46. Baek, J., Vu, Q.H., Liu, J.K., Huang, X., Xiang, Y.: A secure cloud computing based framework for big data information management of smart grid. IEEE Transactions on Cloud Computing **3** (2015) 233–244

47. Han, L., Han, X., Hua, J., Geng, Y.: A hybrid approach of ultra-short term multinode load forecasting. (2007) 1321–1326
48. Hajdu, A., Hajdu, L., Jns, ., Kovacs, L., Tomn, H.: Generalizing the majority voting scheme to spatially constrained voting. IEEE Transactions on Image Processing **22** (2013) 4182–4194
49. Al Nuaimi, E., Al Neyadi, H., Mohamed, N., Al-Jaroodi, J.: Applications of big data to smart cities. Journal of Internet Services and Applications **6** (2015) 1–15

Distributed Machine Learning on Smart-Gateway Network Towards Real-Time Indoor Data Analytics

Hantao Huang, Rai Suleman Khalid and Hao Yu

Abstract Computational intelligence techniques are intelligent computational methodologies such as neural network to solve real-world complex problems. One example is to design a smart agent to make decisions within environment in response to the presence of human beings. Smart building/home is a typical computational intelligence based system enriched with sensors to gather information and processors to analyze it. Indoor computational intelligence based agents can perform behavior or feature extraction from environmental data such as power, temperature, and lighting data, and hence further help improve comfort level for human occupants in building. The current indoor system cannot address dynamic ambient change with a real-time response under emergency because processing backend in cloud takes latency. Therefore, in this chapter we have introduced distributed machine learning algorithms (SVM and neural network) mapped on smart-gateway networks. Scalability and robustness are considered to perform real-time data analytics. Furthermore, as the success of system depends on the trust of users, network intrusion detection for smart gateway has also been developed to provide system security. Experimental results have shown that with a distributed machine learning mapped on smart-gateway networks real-time data analytics can be performed to support sensitive, responsive and adaptive intelligent systems.

Keywords Computational intelligence · Smart home · Indoor positioning · Distributed machine learning · Network intrusion detection · Support vector machine · Neural network

H. Huang (✉) · R.S. Khalid · H. Yu
Nanyang Technological University, 50 Nanyang Avenue, Block S3.2, Level B2, Singapore 639798, Singapore
e-mail: HHUANG013@e.ntu.edu.sg

R.S. Khalid
e-mail: RAIS0001@e.ntu.edu.sg

H. Yu
e-mail: haoyu@ntu.edu.sg

W. Pedrycz and S.-M. Chen (eds.), *Data Science and Big Data: An Environment of Computational Intelligence*, Studies in Big Data 24,
DOI 10.1007/978-3-319-53474-9_11

1 Introduction

1.1 Computational Intelligence

Computational intelligence is the study of the theory, design and application of biologically and linguistically motivated computational paradigms [1–3]. Computational intelligence is widely applied to solve real-world problems which traditional methodologies can neither solve fficiently nor model feasibly. A typical computational intelligence based system can sense data from real-world, use this information to reason the environment and then performed desired actions. In a computational intelligent system such as smart building/home, collecting environmental data, reasoning the accumulated data and then selecting actions can further help to improve comfort level for human occupants. The intelligence of the systems comes from appropriate actions by reasoning the environmental data, which is mainly based on computational intelligence such as fuzzy logic and machine learning. To have a real-time response to the dynamic ambient change, a distributed system is preferred since a centralized system suffers long latency of processing in the back end [4]. Computational intelligence techniques (machine learning algorithms) have to be optimized to utilize the distributed yet computational resource limited devices.

1.2 Distributed Machine Learning

To tackle the challenge of high training complexity and long training time of machine learning algorithms, distributed machine learning is developed to utilize computing resources on sensors and gateway. Many recent distributed learning algorithms are developed for parallel computation across a cluster of computers by applying MapReduce software framework [5]. MapReduce shows a high capacity in handling intensive data and Hadoop is a popular implementation of MapReduce [6]. A prime attractive feature of MapReduce framework is its ability to take good care of data/code transport and nodes coordination. However, MapReduce services always have a high hardware requirement such as large processing memory in order to achieve good performance. However, IoT platforms, such as smart gateways, are with limited resources to support MapReduce operations.

Another kind of approaches is Message Passing Interface (MPI) based algorithms [7, 8]. MPI-based distributed machine learning has very low requirement for hardware and memory sources and it is very suitable for implementation and application in smart gateway environment. However, distribution schedulers for traditional machine learning algorithms are naive and ineffective to utilize computational loads among nodes. Therefore, learning algorithms should be optimized to map on the distributed computational platform.

1.3 Indoor Positioning

GPS provides excellent outdoor services, but due to the lack of Line of Sight (LoS) transmissions between the satellites and the receivers, it is not capable of providing positioning services in indoor environment [9]. Developing a reliable and precise indoor positioning system (IPS) has been deeply researched as a compensation for GPS services in indoor environment. Wi-Fi based indoor positioning is becoming very popular these days due to its low cost, good noise immunity and low set-up complexity [10, 11]. Many WiFi-data based positioning systems have been developed recently for indoor positioning based on received signal strength indicator (RSSI) [12]. As the RSSI parameter can show large dynamic change under environmental change (such as obstacles) [13–15], the traditional machine-learning based WiFi data analytic algorithms can not adapt to the environment change because of the large latency. This is mainly due to the centralized computational system and the high training complexity [16], which will introduce large latency and also cannot be adopted on the sensor network directly. Therefore, in this chapter, we mainly focus on developing distributing indoor positioning algorithm targeting to computational resource limited devices.

1.4 Network Intrusion Detection

Any successful penetration is defined to be an intrusion which aims to compromise the security goals (i.e. integrity, confidentiality or availability) of a computing and networking resource [17]. Intrusion detection systems (IDSs) are security systems used to monitor, recognize, and report malicious activities or policy violations in computer systems and networks. They work on the hypothesis that an intruder's behavior will be noticeably different from that of a legitimate user and that many unauthorized actions are detectable [18, 19]. Anderson et al. [17] defined the following terms to characterize a system prone to attacks:

- **Threat**: The potential possibility of a deliberate unauthorized attempt to access information, manipulate information or render a system unreliable or unusable.
- **Risk**: Accidental and unpredictable exposure of information, or violation of operations integrity due to malfunction of hardware or incomplete or incorrect software design.
- **Vulnerability**: A known or suspected flaw in the hardware or software design or operation of a system that exposes the system to penetration of its information to accidental disclosure.
- **Attack**: A specific formulation or execution of a plan to carry out a threat.
- **Penetration**: A successful attack in which the attacker has the ability to obtain unauthorized/undetected access to files and programs or the control state of a computer system.

The aim of cyber physical security techniques such as network intrusion detection system (NIDS) is to provide a reliable communication and operation of the whole system. This is especially necessary for network system such as Home Area Network (HAN), Neighborhood Area Network (NAN) and Wide Area Network (WAN) [20, 21]. Network intrusion detection system can be placed at each network to detect network intrusion.

In this chapter, we will develop algorithms on distrusted gateway networks to detect network intrusions to provide system security.

1.5 Chapter Organizations

This chapter will be organized as follows. Firstly, we introduce a distributed computational platform on smart gateways for smart home management system in Sect. 2. Then, in Sect. 3, an indoor positioning system by support vector machine (SVM) and neural network is discussed and mapped on distributed gateway networks. In the following Sect. 4, a machine learning based network intrusion detection system (NIDS) is designed to provide system security. Finally, in Sect. 5, conclusion is drawn that distributed machine learning can utilize the limited computing resources and boost the performance of smart home.

2 Distributed Data Analytics Platform on Smart Gateways

2.1 Smart Home Management System

Smart Home Management System (SHMS) is an intelligent system built for residents to benefit from automation technology. By collecting environmental data including temperature, humidity and human activities, a system can react towards residents' best experience [22–25]. Figure 1 depicts the basic components and working strategies in our SHMS test bed:

- **Smart gateways** to be the control center, harboring the ability in storage and computation. Our smart gateway will be BeagleBoard-xM.
- **Smart sensors** to collect environmental information on light intensity, temperature, humidity, and occupancy.
- **Smart sockets** to collect current information of home appliances.
- **Smart devices** with GUI to interact with users; residents have access to environmental information and can control home appliances through a smart phone or tablet.

To ensure high quality performance of SHMS, a robust indoor positioning system (IPS) is indispensable because knowledge about occupants of a building and their

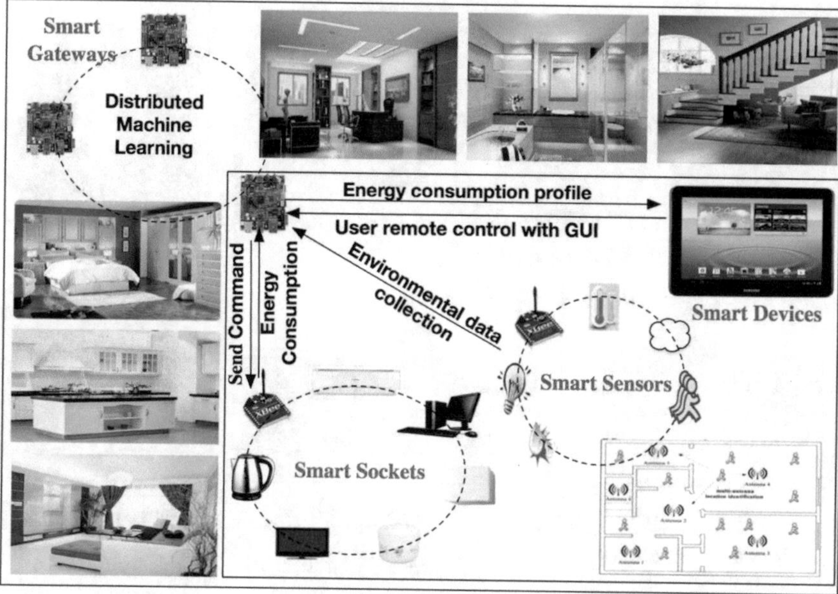

Fig. 1 The overview of smart home management system

movements is essential [26]. Applications can include the scenarios when a resident comes back and enters his room, SHMS automatically powers on air conditioner, heater, humidifier and sets the indoor environment to suit the fitness condition; when nobody is in the house, the system turns off all appliances except for fridge and security system for energy saving issue.

2.2 Distributed Computation Platform

Figure 2 shows our computation platform for real time data analytics. The major computation is performed on smart gateways in a distributed fashion. Data communication between gateways is performed through Wi-Fi using message passing interface (MPI). This distributed computation platform can perform real-time data analytics and store data locally for privacy purpose. Also, shared machine learning engine is developed in the smart gateway to perform real-time feature extraction and learning. Therefore, these learnt features can support indoor positioning services and provide network security protection.

An indoor positioning system (IPS) by WiFi-data consists of at least two hardware components: a transmitter unit and a measuring unit. Here we use smart gateways to collect WiFi signal emitted from other smart devices (phone, pad) of moving occupants inside the building. The IPS determines the positioning with WiFi-data analyzed from the smart gateway network [27]. The central unit in SHMS is BeagleBoard-xM as shown in Fig. 2b, which is also utilized in our positioning

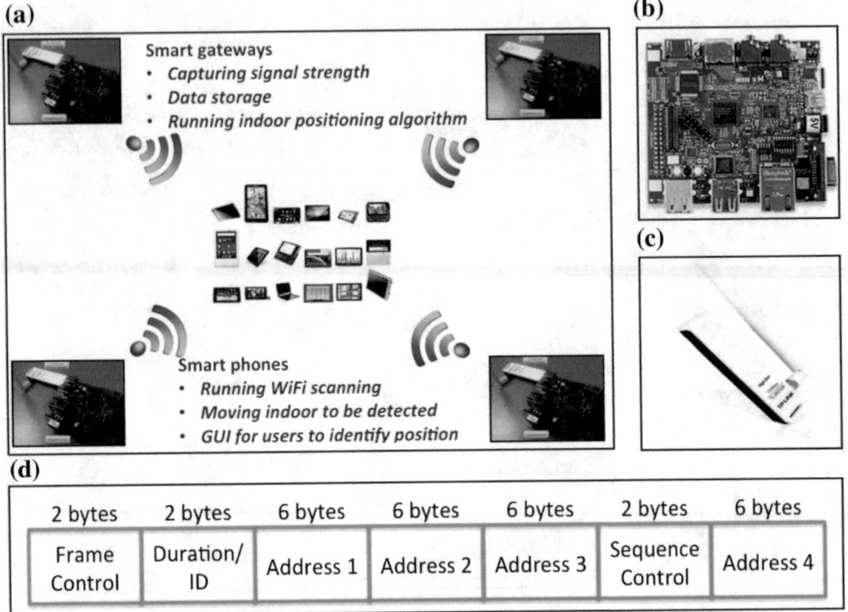

Fig. 2 **a** Distributed computation platform. **b** BeagleBoard xM. **c** TL-WN722N. **d** MAC frame format in Wi-Fi header field

systems. In Fig. 2c, TL-WN722N wireless adapter is our Wi-Fi sensor for wireless signals capturing. BeagleBoard-xM runs Ubuntu 14.04 LTS with all the processing done on board, including data storage, Wi-Fi packet parsing, and positioning algorithm computation. TL-WN722N works in monitor mode, capturing packets according to IEEE 802.11. They are connected with a USB 2.0 port on BeagleBoard-xM.

As depicted in Fig. 2d, Wi-Fi packet contains a header field (30 bytes in length), which contains information about Management and Control Address (MAC). This MAC address is unique to identify the device where the packet came from. Another useful header, which is added to the Wi-Fi packets when capturing frames, is the radio-tap header, which is added by the capturing device (TL-WN722N). This radio-tap header contains information about the RSSI, which reflects the information of distance [28].

Received Signal Strength Indicator (RSSI) is the input of indoor positioning system. It represents the signal power received at a destination node when signal was sent out from a source passing through certain space. RSSI has a relationship with distance, which can be given as:

$$RSSI = -KlogD + A; \tag{1}$$

where K is the slope of the standard plot, A is a fitting parameter and D is the distance [13]. So RSSI, as a basic measurement for distance, has been widely applied in Wi-Fi indoor positioning.

As described in Fig. 2, our SHMS is heavily dependent upon different entities communicating with each other over different network protocols (Wi-Fi, Ethernet, Zigbee). This in turn is prone to network intrusions. Some of the security violations that would create abnormal patterns of system usage include:

- Remote-to-Local Attacks (R2L): Unauthorized users trying to get into the system.
- User-to-Root Attacks (U2R): Legitimate users doing illegal activities and having unauthorized access to local superuser (root) privileges.
- Probing Attacks: Unauthorized gathering of information about the system or network.
- Denial of Service (DOS) Attacks: Attempt to interrupt or degrade a service that a system provides to its intended users.

Therefore, the ability to detect a network intrusion is crucial for intelligent system to provide data and communication security.

3 Distributed Machine Learning Based Indoor Positioning Data Analytics

3.1 Problem Formulation

The primary objective is to locate the target as accurate as possible considering the scalability and complexity.

Objective 1: Improve the accuracy of positioning subject to the defined area.

$$\begin{aligned} \min e &= \sqrt{(x_e - x_0)^2 + (y_e - y_0)^2} \\ s.t.\ label&(x_e, y_e) \in \mathbf{T} \end{aligned} \tag{2}$$

where (x_e, y_e) is the system estimated position belongs to the positioning set $\mathbf{T}$ and (x_0, y_0) is the real location coordinates. Therefore, a symbolic model based positioning problem can be solved using training set Ω to develop neural-network.

$$\Omega = \{(s_i, t_i), i = 1, \dots, N, s_i \in R^n,\ t_i \in \mathbf{T}\} \tag{3}$$

where N represents the number of datasets and n is the number of smart gateways, which can be viewed as the dimension of the signal strength space. s_i is the vector containing RSSI values collected in i_{th} dataset, $t_i \in \{-1, 1\}$ is a discrete value and denoted as a label to represent the indoor positioning coordinates. Note that $\mathbf{T} \in \{-1, 1\}$ labels the physical position and by changing the sequence of -1 and 1, different labels can be represented. The more labels are used, the more accurate the positioning service is.

Objective 2: Reduce the training time of machine learning on Hardware Beagle Board-xM. To distribute training task on gateways with n number, the average training time should be minimized to reflect the reduced complexity on such gateway system.

$$\min \frac{1}{n} \sum_{i=1}^{n} t_{train,i} \\ s.t.\ e < \epsilon \tag{4}$$

where $t_{train,i}$ is the training time on ith smart gateway, e is the training error and ϵ is the tolerable maximum error.

3.2 *Indoor Positioning by Distributed SVM*

Support vector machine (SVM) is one robust machine learning algorithm and can be viewed as a special form of neural network [29]. Due to its reliability and high accuracy in classification, it is one of the most popular algorithms in positioning. The basic idea behind SVM in indoor positioning is to build a classification model based on a training set [30]. To obtain the decision function in the classification problem, LIBSVM [31], one of the most popular SVM solver, is chosen for our multi-category positioning problem.

For indoor positioning application, different positioning zones mean different classes, so multi-category classification is required here. Multi-category classification in LIBSVM uses one-versus-one (1-v-1) SVM, i.e., if there are k classes, the solution follows binary problems solved by 1 versus 2, 1 versus 3, …, 1 versus k, 2 versus 3, …, 2 versus k, …, k−1 versus k, with a total number of k(k −1)/2. This will generate k(k −1)/2 decision functions of where F:$R^n \rightarrow \{C_i, C_j\}, i, j = 1, 2, ..., k, i \neq j$ where C_i means i_{th} class.

3.2.1 Workload Scheduling for DSVM

Experimental results show that iteration time, the time spent on generating binary decision functions, is the major part of time consumed in training (more than 95%). The balanced training time can be expressed as (5):

$$\sum_{i=1}^{n} t_{train,i} = \sum_{p=1}^{K(K-1)/2} t_p \tag{5}$$

where t_p means the time consumed for subtask p and $t_{train,i}$ is the training time on i_{th} smart gateway, Thus, we can ease the load on a single gateway by simply distributing iteration tasks. This section will focus on evenly distributing all the sub tasks onto multiple gateways for processing in order to improve performance. In order to elaborate our task distribution strategies with more details, we first introduce two necessary definitions:

Definition 1 The task workload is the factor to measure each task time in computation. The higher workload a task has, the more computational time it needs for this sub-task. The workloads for all sub-tasks are stored in a matrix **L**, where a L_p stores the workload of sub-task p. As the computation time in SVM training is quadratic in terms of the number of training instances, task workloads are represented by square of dataset size of sub-training tasks.

Definition 2 The decision matrix **B** is an $M \times P$ matrix to define whether sub-task p is allocated to gateway m, where M is the number of gateways and P is the number of sub-tasks.

$$\mathbf{B}_{m,p} = \begin{cases} 1, & \text{sub task p is with gate m} \\ 0, & \text{sub task p not with gate m} \end{cases} \tag{6}$$

Based on the task workload distribution and decision matrix, we can have the final workload allocation on each gate:

$$\begin{aligned} \mathbf{G}_m &= \mathbf{B}_m \mathbf{L}, \quad m = 1, 2, \dots, M \\ &= [\mathbf{B}_{m,1}, \mathbf{B}_{m,2}, \dots, \mathbf{B}_{m,p}][L_1, L_2, \dots, L_p]^T \end{aligned} \tag{7}$$

where **L** represents all sub-tasks stored, where L_p stores the workload of sub-task p. $\mathbf{B}_m$ represents the decision matrix for gateway m and $\mathbf{G}_m$ is the total workload allocated to m_{th} gateway. With these definitions, an optimization WL-DSVM model can be described as (8)

$$min[max(\mathbf{G}_m) + \sqrt{\sum_{m=1}^{M} (\mathbf{G}_m - \overline{\mathbf{G}_m})^2 / M}] \tag{8}$$

$$s.t. \begin{cases} \mathbf{G}_m = \mathbf{B}_m \mathbf{L}, & m = 1, 2, \dots, M \\ \mathbf{B}_{m,p} = 0 \text{ or } 1, & m = 1, 2, \dots, M; p = 1, 2, \dots, P \end{cases} \tag{9}$$

The conventional allocation scheme in [30] is following a computational sequence, and ignoring the difference in computation complexity between sub-problems. However, we take workload of each sub-task into consideration and reorder task allocation. The distribution schemes comparison is described in Fig. 3. If we have a 5-class problem, which will generate 10 sub-tasks (or binary-class classifications), sequential DSVM (SQ-DSVM) will allocate the sub-tasks between 2 gateways by following the computational sequence as shown in Fig. 3: tasks 1–4 allocated to gateway 1; tasks 5–7 allocated to gateway 2; tasks 8–9 allocated to gateway 1; tasks 10 allocated

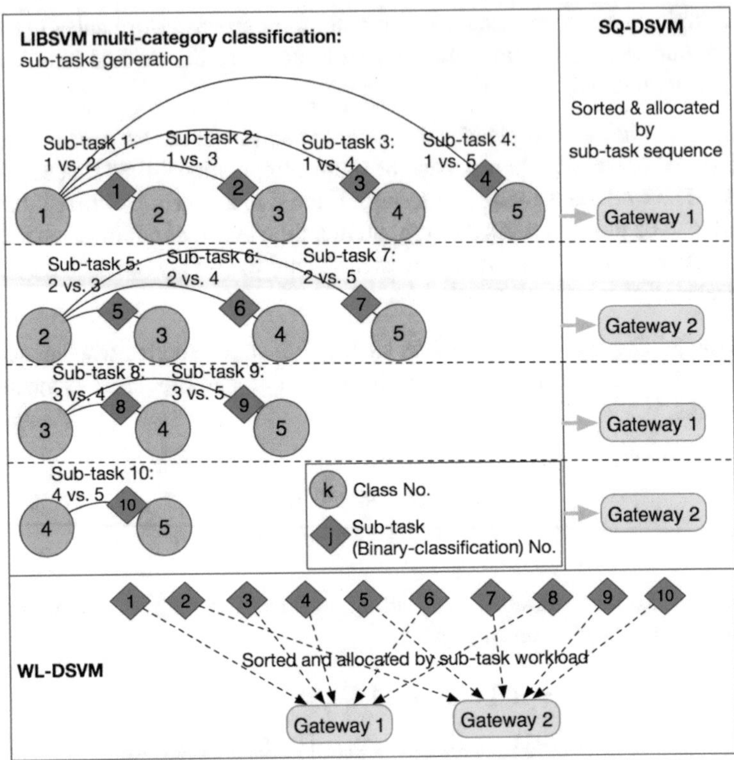

Fig. 3 Working scheduling schemes of sequential based distributed SVM (SQ-DSVM) and workload based distributed SVM (WL-DSVM)

to gateway 2. To achieve an even distribution, we need to rearrange the sub-tasks according to workload distribution and allocate them in an even manner to gateways. In order to realize it, we propose an allocation scheme based on a greedy rule that always allocates the heaviest workload to the gateway with the lightest workload.

As a summary, the pseudo-code and initial settings in the proposed workload-based DSVM are shown in Algorithm 1. Initial values are presented as input. In line 1 we sort the workloads of gates from large ones to small ones. The updated sub-task index in l_s is with line 2. In line 3 we randomize gate index in case that some gates are always allocated more loads than others. Loop line 4–9 is the main part of Workload-based WL-DSVM computation load allocation. Finally in line 8, decision matrix will be updated and the iteration moves to the next loop.

Algorithm 1: WL-DSVM Workload Allocation

Input : Sub-task number n
Computational load vector for each sub-task L
Index vector for each sub-task $l_s = [1, 2, \ldots, P]$
Working load vector for each gate G = 0
Index vector for each gateway $l_g = [1, 2, \ldots, M]$
Decision Matrix B = 0

Output: Working load allocation for each sub-task B

SortFromLargeToSmall(a)
Update order in l_s based on L
RandGateIndex (I_g)
for $p \leftarrow 1$ **to** *P //load allocation for each gate* **do**
  SortFromSmallToLarge(G)
  Update order in I_g based on G
  Update gateway load $G_1 \leftarrow G_1 + L_p$
  Update decision matrix $B(l_{g,1,p}) \leftarrow 1$
end

3.2.2 Working Flow for DSVM-Based IPS

With a distributed fashion applied in support vector machine (SVM), an ideal result is that working load can be distributed and training time can be reduced to:

$$T_{train} = t_{run} + \sum_{p=1}^{P} t_p/P \tag{10}$$

For predicting phase, as only a very short time is needed to predict a user's position, it's not necessary to distribute one single prediction task among different gateways. However, for plenty of incoming RSSI arrays, to predict position of all data on one gateway will have the working load unevenly allocated and there will be one node put under too much load. The solution is to introduce a distributed data storage system. RSSI series stored in mth gateway can be determined using the following formula:

$$S_i = \{s | RSSI_m = max(s), RSSI_{m'} \neq max(s), m' < m\} \tag{11}$$

which means the maximum RSSI value of the data stored in m_{th} gateway is exactly $RSSI_m$; s represents a RSSI array. But if there existing case of $RSSI_1 = RSSI_2$, assuming that we have stored the dataset in gateway 1, Eq. (11) will avoid storing the same dataset in gateway 2 again ($m = 2$, $m' = 1$ as described in (11)). As such, working flow of proposed DSVM indoor positioning can be described as Fig. 4.

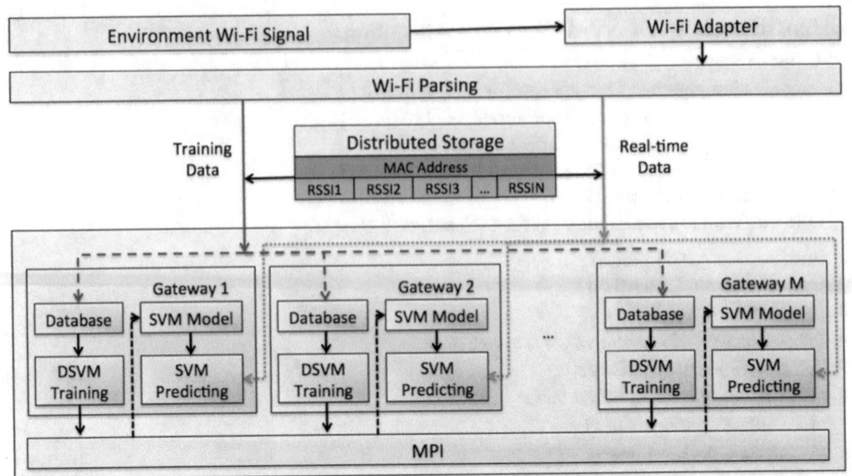

Fig. 4 Working scheduling schemes of SQ-DSVM and WL-DSVM

3.2.3 Experiment Result

System and Simulation Setup

Before verifying the effectiveness of distributed support vector machine, we test the computational capacity of PC and gateway with 4800 × Dup (Dup = 1 to 10 means size of datasets are 4800 × *Dup*, separately, the duplicate count in Fig. 5) datasets for 8 classes, 70% for training, 30% for testing; each dataset has 5 features, i.e. RSSI1, RSSI2, ..., RSSI5. Figure 5 shows the time for training/testing on BeagleBoard-xM (BB)/PC. As is shown in Fig. 5, training on smart gateway (AM37x 1GHz ARM

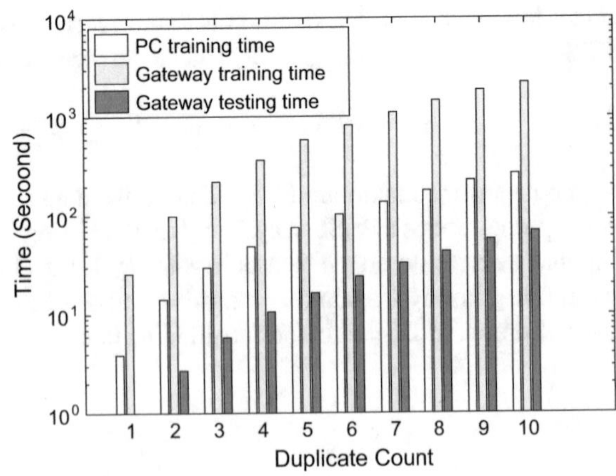

Fig. 5 Runtime comparison for PC and BB based SVMs

processor) takes a rather long time compared with PC (Intel® CoreTM i5 CPU 650 3.20 GHz), and the divergence is about 10 times. Conversely, the predicting time is low enough that it can be quickly handled on the board.

To verify the effectiveness of distributed support vector machine (DSVM) in indoor positioning system (IPS), we performed tests with our indoor environment. For comparison, we implemented conventional SVM algorithm on PC (Baseline 1) and BeagleBoard-xM (Baseline 2) separately, sequential based DSVM (SQ-DSVM) on board (Baseline 3), and finally workload-based DSVM (WL-DSVM) on board (Baseline 4).

Baseline 1: Centralized SVM algorithm. In this method, data (RSSI arrays and position) are collected by different gateways but data processing is performed on a single PC, denoted as SVM-PC.

Baseline 2: Centralized SVM algorithm. In this method, data are collected by different gateways and data processing is performed on Gateway 1, denoted as SVM-BB.

Baseline 3: SQ-DSVM algorithm. In this method, data collection and algorithm computation are both paralleled among 5 gateways. In this method, training phase is performed with sequential DSVM, it is denoted as SQ-DSVM-BB.

Baseline 4: WL-DSVM algorithm. In this method, data collection and algorithm computation are both paralleled among 5 gateways. In this method, training phase is performed with workload-based DSVM, denoted as WL-DSVM-BB.

Indoor test-bed environment for positioning is presented in Fig. 6, with total area being about 80 m^2 (8 m at width and 10 m at length) separated into 48 regular blocks, each block represents a research cubicle in the lab, and the cubicles are the position-

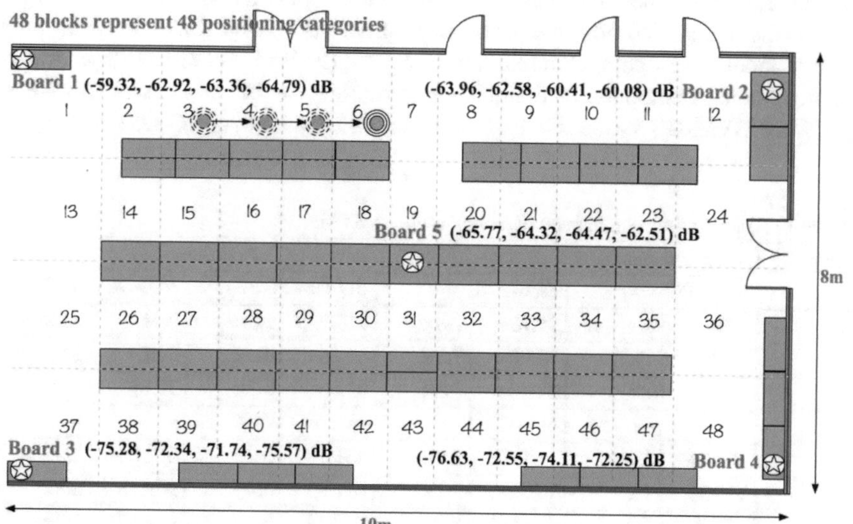

Fig. 6 An example of building floor with position tracking

ing areas in this chapter. 5 gateways, with 4 locations at 4 corners of the map, 1 in the center of the map, are set up for experiment.

To quantify our environment setting, here the positioning accuracy is defined as r, representing radius of target area. It is generated from $S = \pi r^2$, where S is the square of the whole possible positioning area.

Besides, positioning precision is defined as the probability that the targets are correctly positioned within certain accuracy. The definition is as follow:

$$Precision = \frac{N_{pc}}{N_p} \tag{12}$$

where N_{pc} is the number of correct predictions and N_p is the number of total predictions.

Performance Comparison

To verify the effectiveness of distributed support vector machine (DSVM) in indoor positioning system (IPS), in this part, performance comparison between WL-DSVM and other SVMs will be performed within our indoor environment. For comparison, we implemented centralized SVM algorithm on PC (SVM-PC) and BeagleBoard-xM (SVM-BB) separately, and distributed SVMs are tested with workload-based DSVM on board (WL-DSVM-BB) and sequential DSVM on board (SQ-DSVM-BB).

Table 1 (Dup = 1, 2, ..., 10 means size of datasets are 4800 × Dup, separately) mainly elaborates the advantages of distributed machine learning over centralized data analytics. (1) Without a distributed fashion, centralized SVM on board only shows about 1/8 in computational ability of PC. (2) With DSVM on 5 gateways, improvement of computational efficiency in runtime can be 2.5–3.5x of SVM-BB. (3) WL-DSVM-BB shows a higher efficiency in runtime than SQ-DSVM-BB; (4)

Table 1 Training time comparison of among different SVMs

Dup	Time consumed (s)							
	SVM-PC		SVM-BB		SQ-DSVM-BB		WL-DSVM-BB	
1	3.30	1×	25.78	7.8×	11.58	3.5×	9.41	2.9×
2	13.23	1×	99.09	7.5×	44.26	3.3×	36.78	2.8×
3	29.75	1×	221.84	7.5×	95.38	3.2×	79.85	2.7×
...	...	...	...	...	...	...	...	...
7	137.44	1×	1101.98	8.0×	451.18	3.3×	362.84	2.6×
8	179.33	1×	1452.87	8.1×	611.85	3.4×	509.18	2.8×
9	228.08	1×	1842.62	8.1×	770.86	3.4×	614.57	2.7×
10	266.68	1×	2195.79	8.2×	859.90	3.2×	679.27	2.5×

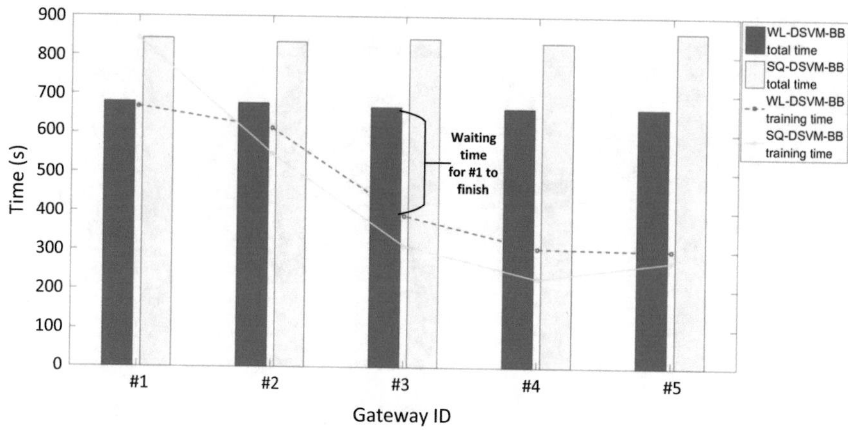

Fig. 7 Runtime comparison on each gateway for WL-DSVM and SQ-DSVM. Total time includes the training time, communication time and waiting time for other beagleboards

Increased size of dataset for training doesn't show an increase in relative time with distributed machine learning; it means that computational load is much more time consuming than data communication and needs more attention.

Figure 7 shows the running time comparison of each BeagleBoard with WL-DSVM-BB and SQ-DSVM-BB methods. The total time of each BeagleBoard includes the training time, communication time and waiting time for others. It clearly shows that WL-DSVM-BB has relative fair load distribution. Therefore, WL-DSVM-BB performs a higher training efficiency with total time 679 s ($max(G_m)$ in (8)), comparing with total time 860 s on SQ-DSVM-BB, achieving an improvement of 27% for the case of Dup = 10.

WL-DSVM-BB only needs 2.5× training time of PC, and results in 3.2× improvement in training time when compared with centralized SVM on board (SVM-BB), which means WL-DSVM has made sense in reducing the working load of a single node, which is useful when gateway nodes are in a large amount, such as hundreds of gateways, which will be very promising in real-time positioning for a changing environment. Conventional way in training phase is off-line training on PC and then sending the training predictor to sensor nodes. But with an ability to efficiently compute training phase on board, real-time data analysis can be performed on board so that we can get rid of server that requires extra cost and implementation.

In order to test the positioning precision improvement with WL-DSVM, we simulate an indoor environment where RSSI values vary according to Gaussian distribution every half an hour. WL-DSVM can update its predictor automatically while SVM-BB applies the initially generated predictor. This is a reasonable setting since WL-DSVM is much faster to perform training than SVM-BB. Due to the long training time SVM-BB is not likely to perform online updates while WL-DSVM is favorable to perform online updates. Here, we take accuracy of 2.91 m for an example. Results in Fig. 9 show that WL-DSVM can maintain the precision of prediction,

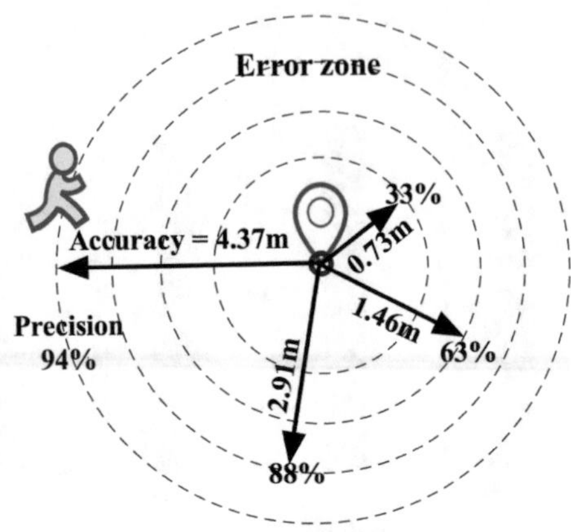

Fig. 8 Indoor positioning accuracy and precision by WL-DSVM

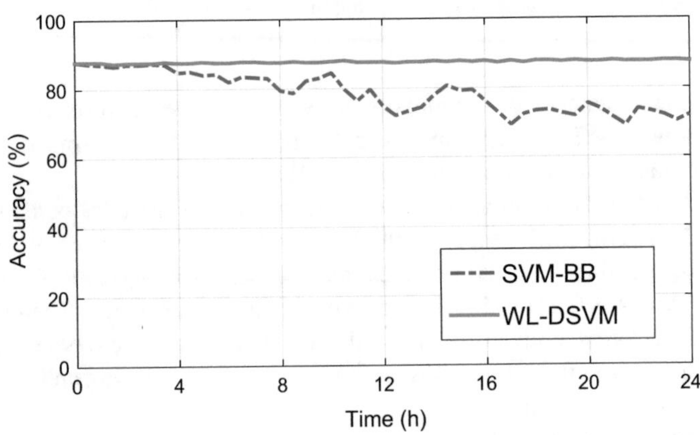

Fig. 9 Positioning precision comparison of changing environment for WL-DSVM and SVM-BB

while SVM shows a decreasing and unstable performance of precision. It is because WL-DSVM will update the predictor whenever environment changes. But traditionally, training phase is only done once so that the precision will decrease with a changing environment. Indoor positioning accuracy and its precision by WL-DSVM are shown in Fig. 8.

In conclusion, instead of dealing with data storage and analytics on one central server, the proposed distributed real-time data analytics is developed on networked gateways with limited computational resources. By utilizing the distributed support vector machine (DSVM) algorithm, the data analytics of real-time RSSI values of Wi-Fi data can be mapped on each individual gateway. The experimental results have shown that the proposed WL-DSVM can achieve a 3.2x improvement in runtime in comparison to a single centralized node and can achieve a performance improvement of 27% in runtime in comparison to conventional DSVM.

3.3 Indoor Positioning by Distributed-neural-network

In this section, we introduce a low computational complexity machine-learning algorithm that can perform WiFi—data analytics for positioning on smart gateway network. A distributed-neural-network (DNN) machine learning algorithm is introduced with the maximum posteriori probability based soft-voting. Experiment results have shown significant training and testing speed-up comparing to SVM.

3.3.1 Machine Learning Algorithm on Gateway

Single-Hidden-Layer Neural-Network

Our neural-network with two sub-systems is shown as Fig. 10, which is inspired by extreme learning machine and compressed sensing [32, 33]. Unlike previous work [14], the input weight is only connecting nearyby hidden nodes. The input weight in our proposed neural-network is connected to every hidden node and is randomly generated independent of training data [34]. Therefore, only the output weight is calculated from the training process. Assume there are N arbitrary distinct training samples $\mathbf{X} \in \mathbf{R}^{N \times n}$ and $\mathbf{T} \in \mathbf{R}^{N \times m}$, where $\mathbf{X}$ is training data representing scaled RSSI values from each gateway and $\mathbf{T}$ is the training label indicating its position respectively. In our indoor positioning cases, the relation between the hidden neural-node and input training data is addictive as

$$\mathbf{preH} = \mathbf{XA} + \mathbf{B}, \ \mathbf{H} = \frac{1}{1 + e^{-\mathbf{preH}}} \tag{13}$$

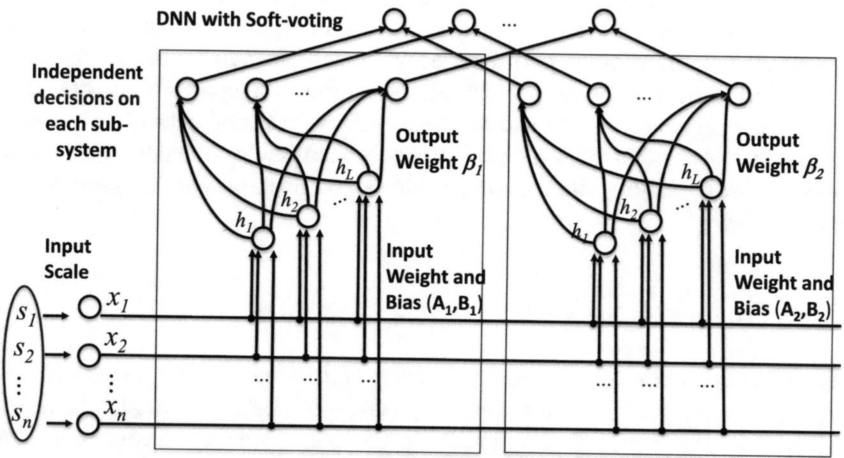

Fig. 10 Soft-voting based distributed-neural-network with 2 sub-systems

Algorithm 2: Learning Algorithm for Single Layer Network

Input : Training Set $(x_i, t_i), x_i \in \mathbf{R}^n, t_i \in \mathbf{R}^m, i = 1, ...N$, activation function $H(a_{ij}, b_{ij}, x_i)$, maximum number of hidden neural node L_{max} and accepted training error ϵ.

Output: Neural-network output weight $\boldsymbol{\beta}$

1 Randomly assign hidden-node parameters
2 $(a_{ij}, b_{ij}),\ a_{ij} \in \mathbf{A},\ b_{ij} \in \mathbf{B}$
3 Calculate the hidden-layer pre-output matrix $\mathbf{H}$
4 $\mathbf{preH} = \mathbf{XA} + \mathbf{B},\ \ \mathbf{H} = 1/(1 + e^{-\mathbf{preH}})$
5 Calculate the output weight
6 $\boldsymbol{\beta} = (\mathbf{H}^T\mathbf{H})^{-1}\mathbf{H}^T\mathbf{T}$
7 Calculate the training error *error*
8 $error = ||\mathbf{T} - \mathbf{H}\boldsymbol{\beta}||$
9 **if** $(L \leq L_{max}$ *and* $e > \epsilon)$ **then**
10   Increase number of hidden node
11     $L = L + 1$, repeat from Step 1
12 **end**

where $\mathbf{A} \in \mathbf{R}^{n \times L}$ and $\mathbf{B} \in \mathbf{R}^{N \times L}$. $\mathbf{A}$ and $\mathbf{B}$ are randomly generated input weight and bias formed by a_{ij} and b_{ij} between $[-1, 1]$. $\mathbf{H} \in \mathbf{R}^{N \times L}$ is the result from sigmoid function for activation. In general cases, the number of training data is much larger than the number of hidden neural nodes (i.e. $N > L$), to find the output weight $\boldsymbol{\beta}$ is an overdetermined system. Therefore, estimating the output weight is equivalent to minimize $||\mathbf{T} - \mathbf{H}\boldsymbol{\beta}||$, the general solution can be found as

$$\boldsymbol{\beta} = (\mathbf{H}^T\mathbf{H})^{-1}\mathbf{H}^T\mathbf{T},\ \mathbf{H} \in \mathbf{R}^{N \times L} \quad (14)$$

where $\boldsymbol{\beta} \in \mathbf{R}^{L \times m}$ and m is the number of symbolic classes. $(\mathbf{H}^T \times \mathbf{H})^{-1}$ exits for full column rank of $\mathbf{H}$ [32]. However, such method is computationally intensive. Moreover, as the number of hidden neural nodes can not be explicit from the training data to have small training error, [35] suggests to increase the number of hidden neural node L during the training stage, which will reduce the training error but at the cost of increasing computational cost and required memory for neural-network. Therefore, an incremental solution for (14) is needed to adjust the number of hidden node L with low complexity. The algorithm of single hidden layer neural-network is summarized in Algorithm 2.

Incremental Least-Square Solver

The key difficulty for solving training problem is the least square problem of minimizing $||\mathbf{T} - \mathbf{H}\boldsymbol{\beta}||$. This could be solved by using SVD, QR and Cholesky decomposition. The computational cost of SVD, QR and Cholesky decomposition is $O(4NL^2 - \frac{4}{3}L^3)$, $O(2NL^2 - \frac{2}{3}L^3)$ and $O(\frac{1}{3}L^3)$ respectively [36]. Therefore, we use Cholesky decomposition to solve the least square problem. Moreover, its incremental and symmetric property reduces the computational cost and saves half memory

Algorithm 3: Incremental L_2 Norm Solution

Input : Activation matrix $\mathbf{H}_L$, target matrix $\mathbf{T}$ and number of hidden nodes L
Output: Neural-network output weight $\boldsymbol{\beta}$

for $l \leftarrow 2$ **to** L **do**
  Calculate new added column
  $\mathbf{v}_l \leftarrow \mathbf{H}_{l-1}^T h_l$
  $g \leftarrow h_l^T * h_l$
  Calculate updated Cholesky matrix
  $\mathbf{z}_L \leftarrow \mathbf{Q}_{L-1}^{-1}\mathbf{v}_L,\ p \leftarrow \sqrt{g - \mathbf{z}_L^T\mathbf{z}_L}$
  Form new Cholesky Matrix $\mathbf{Q}_L \leftarrow \begin{pmatrix} \mathbf{Q}_{L-1} & 0 \\ \mathbf{z}_L^T & p \end{pmatrix}$
  Calculate output weight using forward and backward substitution
  $\mathbf{Q}_L\mathbf{Q}_L^T\boldsymbol{\beta} \leftarrow \mathbf{H}_L^T\mathbf{T}$
end

required [36]. Here, we use H_L to represent the matrix with L number of hidden neural nodes ($L < N$), which decomposes the symmetric positive definite matrix $\mathbf{H}^T\mathbf{H}$ into

$$\mathbf{H}_L^T\mathbf{H}_L = \mathbf{Q}_L\mathbf{Q}_L^T \tag{15}$$

where $\mathbf{Q}_L$ is a low triangular matrix and T represents transpose operation of the matrix.

$$\begin{aligned} \mathbf{H}_L^T\mathbf{H}_L &= \left[\mathbf{H}_{L-1}\ h_L\right]^T\left[\mathbf{H}_{L-1}\ h_L\right] \\ &= \begin{pmatrix} \mathbf{H}_{L-1}^T\mathbf{H}_{L-1} & \mathbf{v}_L \\ \mathbf{v}_L^T & g \end{pmatrix} \end{aligned} \tag{16}$$

where h_L is the new added column by increasing the size of L, which can be calculated from (13). The Cholesky matrix can be expressed as

$$\begin{aligned} &\mathbf{Q}_L\mathbf{Q}_L^T \\ &= \begin{pmatrix} \mathbf{Q}_{L-1} & \mathbf{0} \\ \mathbf{z}_L^T & p \end{pmatrix} \begin{pmatrix} \mathbf{Q}_{L-1}^T & \mathbf{z}_L \\ \mathbf{0} & p \end{pmatrix} \end{aligned} \tag{17}$$

As a result, we can easily calculate the $\mathbf{z}_L$ and scalar p for Cholesky factorization as

$$\mathbf{Q}_{L-1}\mathbf{z}_L = \mathbf{v}_L,\ p = \sqrt{g - \mathbf{z}_L^T\mathbf{z}_L} \tag{18}$$

where $\mathbf{Q}_{L-1}$ is the previous Cholesky decomposition result and $\mathbf{v}_L$ is known from (16), which means we can continue to use previous factorization result and update only according part. Algorithm 3 gives details on each step since $l \geq 2$. Please note when $l = 1$, Q_1 is a scalar and equals to $\sqrt{H_1^T H_1}$. Such method will greatly reduce computational cost and allow the online training on smart gateway for positioning.

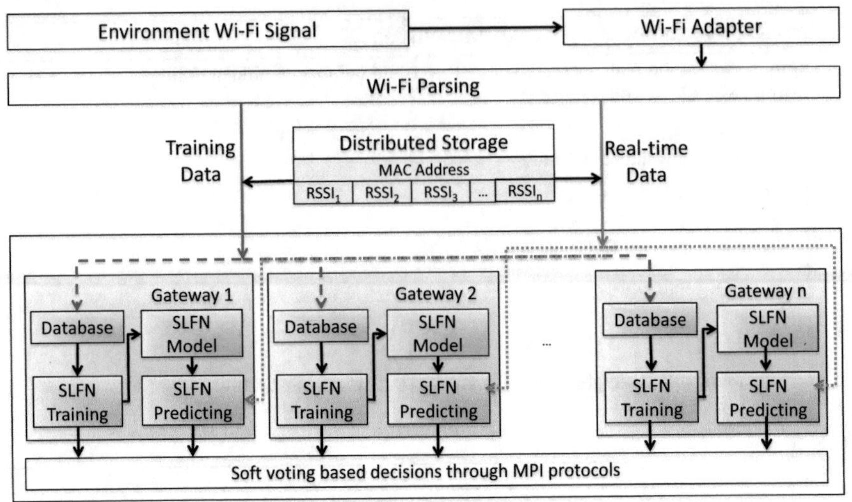

Fig. 11 Working flow of distributed-Neural-Network indoor positioning system

3.3.2 Distributed-Neural-Network with Soft-Voting

Distributed-Neural-Network for Indoor Positioning

The working flow of the distributed-neural-network (DNN) is described as Fig. 11. Environmental Wi-Fi signal is received from the Wi-Fi Adapter and through Wi-Fi parsing the data with MAC address and RSSI of each Wi-Fi adapter is stored. Such data is sent to gateway for training with label first. Please note that the training process is on the gateway. As we mentioned in Sect. 3.3.1, a single layer forward network (SLFN) is trained. A small data storage is required to store trained weight for the network. In the real time application, the same format data will be collected and sent into the well trained network to locate its position. In Fig. 11, the block for soft-voting is through message passing interface (MPI) protocols to collect all the testing result from each SLFN and soft-voting is processed in the central gateway. Note that n gateways together can form one or several SLFNs based on the accuracy requirement.

Soft-Voting

As we have discussed in Sect. 3.3.1, the input weight and bias $\mathbf{A}, \mathbf{B}$ are randomly generated, which strongly supports that each SLFN is an independent expert for indoor positioning. Each gateway will generate posteriori class probabilities $P_j(c_i|\mathbf{x}), i = 1, 2, ..., m, j = 1, 2, ..., N_{slfn}$, where $\mathbf{x}$ is the received data, m is the number of classes and N_{slfn} is the number of sub-systems for single layer network deployed on smart gateway. During the testing process, the output of single layer forward network (SLFN) will be a set of values $y_i, i = 1, 2, ..., m$. Usually, the maximum y_i is selected to represent its class i. However, in our case, we scale the training and testing input between $[-1, 1]$ and target labels are also formed using a set of $[-1, -1, ...1..., -1]$, where the only 1 represents its class and the target label has length m. The posteriori probability is estimated as

$$P_j(c_i|\mathbf{x}) = (y_i + 1)/2, \ j = 1, 2, ..., N_{slfn} \tag{19}$$

A loosely stated objective is to combine the posteriori of all sub-systems to make more accurate decisions for the incoming data $\mathbf{x}$. Under such case, information theory suggests to use a cross entropy (Kullback-Leibler distance) criterion [37], where we may have two possible ways to combine the decisions (Geometric average rule and Arithmetic average rule). The geometric average estimates can be calculated as

$$P(c_i) = \prod_{j=1}^{N_{slfn}} P_j(c_i|\mathbf{x}), \ i = 1, 2, ...m \tag{20}$$

and the arithmetic average estimate is shown as

$$P(c_i) = \frac{1}{N_{slfn}} \sum_{j=1}^{N_{slfn}} P_j(c_i|\mathbf{x}), \ i = 1, 2, ...m \tag{21}$$

where $P(c_i)$ is the posteriori probability to choose class c_i and will select the maximum posteriori $P(c_i)$ for both cases. In this chapter, we use arithmetic average as soft-voting of each gateway since [37] indicates that geometric average rule works poorly when the posteriori probability is very low. This may happen when the object to locate is far away from one gateway and its RSSI is small with low accuracy of positioning. The final decision is processed at the central gateway to collect the voting value from each sub-systems on other gateways. Such soft-voting will utilize the confidence of each sub-system and avoid the prerequisite that each sub-system maintains accuracy of more than 50% for hard-voting.

3.3.3 Experimental Results

Experiment Setup

Indoor test-bed environment for positioning is presented in Fig. 6, which is the same as Sect. 3.2.3. The summary for the experiment set-up is shown in Table 2. To avoid confusion, we use DNN to represent distributed neural network and SV-DNN represents soft-voting based DNN.

Table 2 Experimental set-up parameters

Parameter	Value
Traing date size	18056
Testing date size	2000
Data dimension	5
Number of labels	48
No. of gateway	5
Testing area	80 m^2

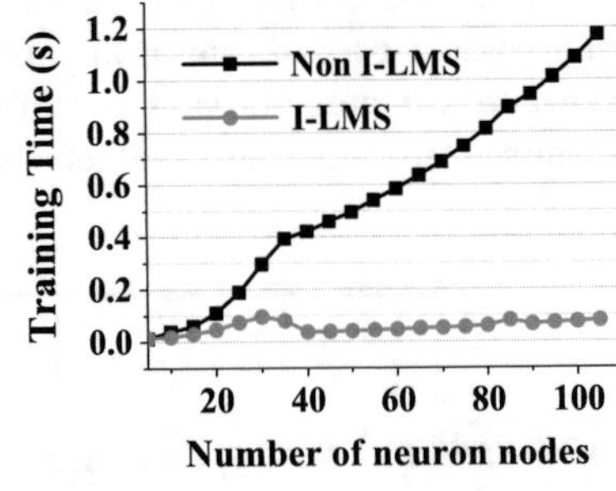

Fig. 12 Training time for SLFN by incremental Cholsky decomposition

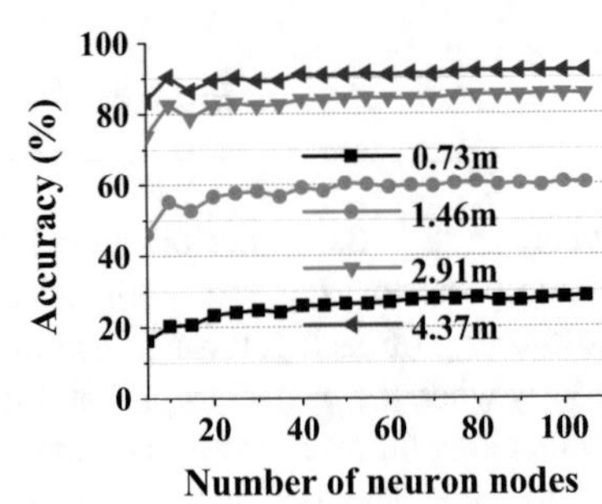

Fig. 13 Testing accuracy under different positioning scale

Real-Time Indoor Positioning Results

The result of the trained neural forward network is shown as Figs. 12 and 13. The training time can be greatly reduced by using incremental Cholesky decomposition. This is due to the reduction of least square complexity, which is the limitation for the training process. As shown in Fig. 12, training time maintains almost constant with increasing number of neural nodes when the previous training results are available. Figure 13 also shows the increasing accuracy under different positioning scales from 0.73 to 4.57 m. It also shows that increasing the number of neural nodes will increase the performance to certain accuracy and maintains almost flat at larger number of neural nodes.

Table 3 Comparison table with previous works

System/Solution	Precision
Proposed DNN	58% within 1.5 m, 74% within 2.2 m and 87% within 3.64 m
Proposed SV-DNN	62.5% within 1.5 m, 79% within 2.2 m and 91.2% within 3.64 m
Microsoft RADAR [38]	50% within 2.5 m and 90% within 5.9 m
DIT [39]	90% within 5.12 m for SVM; 90% within 5.40 m for MLP
Ekahau [40]	5 to 50 m accuracy (indoors)
SVM	63% within 1.5 m, 80% within 2.2 m and 92.6 % within 3.64 m

Performance Comparison

In Table 3, we can see that although single layer network cannot perform better than SVM but it outperforms other positioning algorithms proposed in [38–40]. Moreover, by using maximum posteriori probability based soft-voting, SV-DNN can be very close to the accuracy of SVM. Table 4 shows the detailed comparisons between proposed DNN positioning algorithm with SVM. Please note that the time reported is the total time for training data size 18056 and testing data size 2000. It shows more than 120× training time improvement and more than 54× testing time saving for proposed SLFN with 1 sub-network comparing to SVM. Even adding soft-voting with 3 sub-networks, 50x and 38x improvement in testing and training time respectively can be achieved. Please note that for fair training and testing time comparison, all the time is recorded using Ubuntu 14.04 LTS system with core 3.2 GHz and 8GB RAM. Variances of the accuracy is also achieved by 5 repetitions of experiments and the reported results are the average values. We find that the stability of proposed DNN is comparable to SVM. Moreover, the testing and training time do not increase significantly with new added subnetworks. Please note that SVM is mainly limited by its training complexity and binary nature where one-against-one strategy is used to ensure accuracy with a cost of building $m(m-1)/2$ classifier and m is the number of classes. Figure 14 shows the error zone of proposed SV-DNN.

In conclusion, this section proposes a computationally efficient data analytics by distributed-neural-network (DNN) based machine learning with application for indoor positioning. It is based on one incremental L_2-norm based solver for learning collected WiFi-data at each gateway and is further fused for all gateways in the network to determine the location. Experimental results show that with 5 distributed gateways running in parallel for a 80 m^2 space, the proposed algorithm can achieve 50x and 38x improvement on testing and training time respectively when compared to support vector machine based data analytics with comparable positioning precision.

Table 4 Performance precision with variations on proposed DNN with soft-voting

	Testing time (s)	Training time (s)	0.73 m	1.46 m	2.19 m	2.91 m	3.64 m	4.37 m	5.1 m	No. of nodes
SVM acc. and var.	9.7580	128.61	31.89%	63.26%	80.25%	88.54%	92.58%	94.15%	94.71%	N.A.
			0.530	0.324	0.394	0.598	0.536	0.264	0.0975	
DNN acc. and var.	0.1805	1.065	23.94%	57.78%	74.14%	82.61%	87.22%	90.14%	91.38%	100
			1.2153	0.0321	0.1357	0.2849	0.0393	0.0797	0.0530	
SV-DNN (2) acc. and var.	0.1874	2.171	29.36%	61.23%	77.20%	86.24%	90.25%	92.19%	93.14%	2 Sub-systems each 100
			0.358	0.937	0.526	0.517	0.173	0.173	0.124	
SV-DNN (3) acc. and var.	0.1962	3.347	30.52%	62.50%	79.15%	87.88%	91.20%	92.92%	94.08%	3 Sub-systems each 100
			0.325	1.952	0.884	1.245	0.730	0.409	0.293	

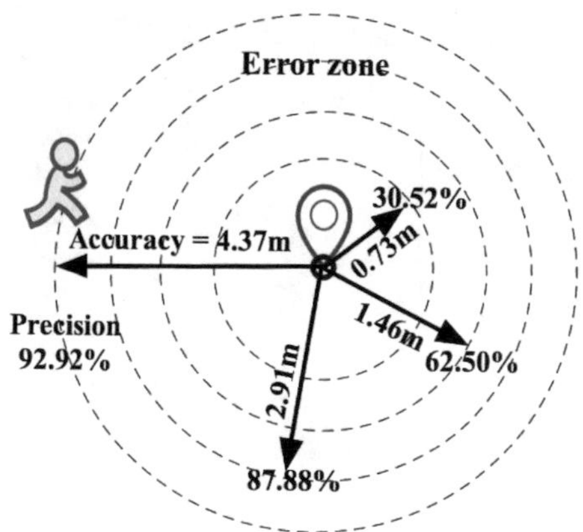

Fig. 14 Error Zone and accuracy for indoor positioning by distributed neural network (DNN)

4 Distributed Machine Learning Based Network Intrusion Detection System

In this section we propose to use the distributed-neural-network (DNN) method as described in Sect. 3.3.1 for the network intrusion detection system (NIDS). We also use the same soft voting technique as described in Sect. 3.3.2 to achieve an improved accuracy.

4.1 Problem Formulation and Analysis

In machine learning approach for NIDS, the detection for intrusion can be considered as a binary classification problem, distinguishing between normal and attack instances. In the similar way, intrusion can also be considered as a multi-class classification problem to detect different attacks. We can use supervised, semi-supervised or unsupervised machine learning approach to achieve the objective. In this section we use a supervised machine learning approach based on single hidden layer neural-network [32] for intrusion detection. Figure 15 shows an overview of steps involved in binary NIDS. As such our main objectives are:

Objective 1: Achieve an overall high accuracy, high detection rate, a very low false alarm rate. We define the following terms to mathematically formulate our objective.

1. *False Positives* (FP): Number of normal instances which are detected as intrusions.
2. *False Negatives* (FN): Number of intrusion instances which are detected as normal.

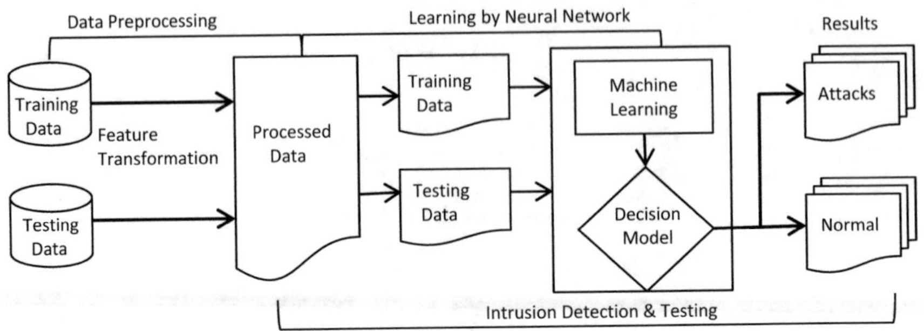

Fig. 15 NIDS Flow based on distributed machine learning on smart-gateways

3. *True Positives* (TP): Number of correctly detected intrusion instances.
4. *True Negatives* (TN): Number of correctly detected normal instances.

Several statistical measures are used to measure the performance of machine learning algorithms for Binary NIDS. Specifically following measures are used to characterize the performance.

1. *Recall*: is a measure of detection rate of the system to detect attacks and is defined as:

$$Recall = \frac{TP}{TP + FN} \times 100 \tag{22}$$

2. *False Positive Rate* (FP): gives a measure of false positive rate i.e., normal instances being classified as intrusions

$$FP = \frac{FP}{FP + TN} \times 100 \tag{23}$$

3. *Precision*: Precision is a measure of predicted positives which are actual positives.

$$Precision = \frac{TP}{TP + FP} \times 100 \tag{24}$$

4. *F-Measure*: F-Measure is a metric that gives a better measure of accuracy of an IDS. It is a harmonic mean of precision and recall.

$$F - Measure = \frac{2}{\frac{1}{precision} + \frac{1}{recall}} \times 100 \tag{25}$$

5. *Matthews Correlation Coefficient* (MCC): MCC measures the quality of binary classification. It represents values in the range −1 to +1. A value of +1 represents 100% prediction, −1 represents 0% prediction. A value of 0 represents no better prediction than random prediction.

$$MCC = \frac{TP \times TN - FP \times FN}{\sqrt{(TP + FP)(TP + FN)(TN + FP)(TN + FN)}} \tag{26}$$

6. *Overall Accuracy*: Overall accuracy is defined as a ratio of TP and TN to the total number of instances

$$Accuracy = \frac{TP + TN}{FP + FN + TP + TN} \times 100 \tag{27}$$

Objective 2: Reduce the training complexity for intrusion detection modules at various network layers in the smart grid so that the model can be quickly updated. To distribute training task on gateways with n number, the average training time should be minimized to reflect the reduced complexity on such gateway system.

$$\begin{aligned} &\min \frac{1}{n} \sum_{i=1}^{n} t_{train,i} \\ &s.t.\ e < \epsilon \end{aligned} \tag{28}$$

4.2 Experimental Results

To achieve Objective 1 (i.e. improved overall accuracy), we use distributed neural-network (DNN) and soft-voting as described in Sect. 3.3.2. To achieve Objective 2 (i.e. reduced training time), we use the same Cholesky decomposition as described in Sect. sec:architecture.

4.2.1 Setup and Benchmarks

In this section we evaluate the NSL-KDD [41] and ISCX 2012 benchmarks [42] for intrusion detection. The experiments were simulated on Ubuntu 14.04 LTS system with core 3.2 GHz and 8GB RAM.

4.2.2 NSL-KDD Dataset Description and Preprocessing

All the attack types mentioned previously i.e., DOS, Probe, R2L and U2R are encapsulated in the KDD Cup 99 Dataset which has been used as benchmark for detecting intrusions in a typical computer network. To evaluate the classfication accuracy of SVM and SLFN as well as to evaluate detection latency using DNN we propose to use an improved version of KDD Cup 99 Dataset known as NSL-KDD dataset [41] which has been used as a benchmark in previous works on intrusion detection [21]. Some of features in the dataset i.e., protocol type, service and flag have sybmolic representation. To be able to use SVM or SLFN we assigned an arbitrary

Table 5 NSL-KDD experimental set-up parameters

Parameter	Value
Training data size	74258
Testing data size	74259
Data dimension	41
Number of labels	Binary (2)
	Multiclass (5)
Number of gateways	5

sequential integer assignment to establish a correspondence between each category of a symobolic feature and a sequence of integer value. Table 5 gives the description of NSL-KDD benchmark for intrusion detection.

4.2.3 ISCX 2012 Dataset Description and Preprocessing

ISCX 2012 Dataset [42] was developed at the University of Brunswick ISCX. The original dataset contains 17 features and a label representing normal instances and intrusions belonging to DOS, SSH, L2L and Botnet (DDOS) intrusions. Some of the features were irrelevant and were removed from the dataset. Additionally some of the features in the data set i.e., appName, direction, sourceTCPFlagsDescription, destinationTCPFlagsDescription and protocolName were symbolic in nature and an arbitrary sequential integer assignment was used to convert these features to numeric features similar to NSL-KDD benchmark. Table 6 gives the description of ISCX benchmark for intrusion detection.

4.2.4 Training and Testing Time Analysis

Table 7 gives the metrics for training and testing time for the two benchmarks. For NSL-KDD dataset it can be observed that SLFN is 49× faster in training time compared to SVM and 8× faster in testing time compared to SVM. Similarly for ISCX dataset it can be observed that SLFN is 60× faster in training time compared to SVM and 13× faster in testing time compared to SVM.

Table 6 ISCX 2012 experimental set-up parameters

Parameter	Value
Training data size	103285
Testing data size	103229
Data dimension	11
Number of labels	Binary (2)
	Multiclass (5)
Number of gateways	5

Table 7 Binary classification performance with 500 hidden neurons

Benchmark	Algo.	Class	TP. %	FP. %	Prec.	Recall %	FM. %	MCC	Tr.(s)	Te. (s)
NSL-KDD	SVM	Normal	95.90	6.00	94.5	95.90	95.20	0.899	8678	188.7
		Anomaly	94.00	4.10	95.50	94.00	94.80	0.899		
	SLFN	Normal	98.24	3.77	96.54	98.24	97.38	0.945	333.08	41.94
		Anomaly	96.22	1.76	98.08	96.22	97.14	0.945		
ISCX 2012	SVM	Normal	98.90	1.50	99.30	98.90	99.10	0.971	5020	83.7
		Anomaly	98.50	1.10	97.70	98.50	98.10	0.971		
	SLFN	Normal	94.43	11.83	94.33	94.43	94.38	0.826	277.9	21.25
		Anomaly	88.16	5.56	88.36	88.16	88.26	0.827		

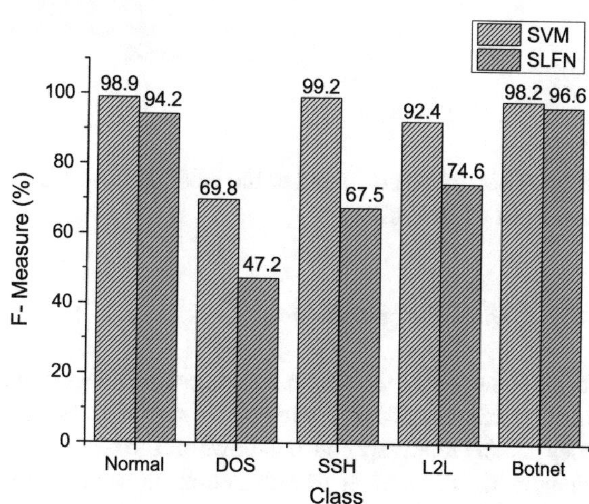

Fig. 16 ISCX 2012 classification

4.2.5 Binary and Multiclass Classification Performance Metrics

Table 7 gives the detailed performance metrics for NSL-KDD and ISCX datasets using SVM and SLFN machine learning methods for normal and anomaly classes. For NSL-KDD dataset it can be observed that all performance metrics for both normal and anomaly classes are superior for SLFN compared to SVM. For ISCX dataset SVM performs slightly better than SLFN in performance metrics for both normal and anomaly classes. However SLFN has a much higher FP rate compared to SVM.

Figures 16 and 17 shows the performance of SVM and SLFN using 500 hidden neurons for multiclass classification for the 2 benchmarks using the F-Measure. For ISCX dataset it can be seen that SLFN has performance comparable to SVM for Normal and Botnet classes. However SVM outperforms SLFN for DOS, SSH and L2L classes. For NSL-KDD dataset it can be observed that for Normal, DOS and Probe classes SLFN has almost similar performance compared to SVM. SLFN outperforms SVM for R2L class. However it is not able to detect any intrusions relating to U2R

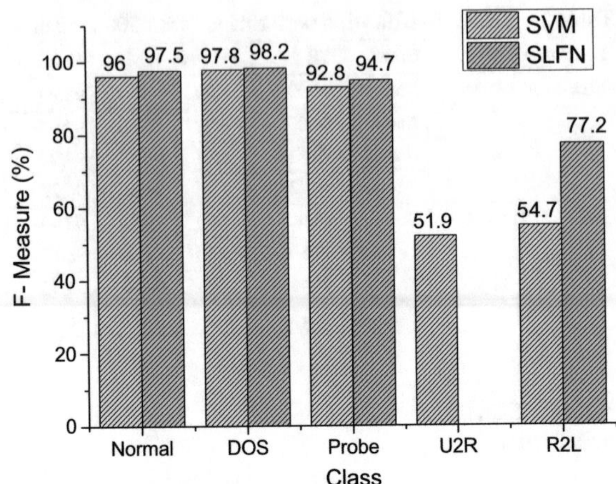

Fig. 17 NSL-KDD classification

class. This can be attributed to the fact that the NSL-KDD dataset only contains 119 instances of U2R class.

4.2.6 DNN Performance

Table 8 shows the performance metrics for DNN using NSL-KDD and ISCX datasets at 500 hidden neurons. It can be observed that for both datasets all the performance metrics stay relatively constant however the server processing time is reduced propotionally to the number of sub neural networks employed. For NSL-KDD server processing time is reduced by 8× compared to centralized NIDS and by 37× compared to SVM when using 5 sub neural networks. Similarly for ISCX dataset server

Table 8 Binary DNN classification performance metrics

Benchmark	Algo.	TP Rate (%)	FP Rate (%)	Prec.	Recall (%)	FM. (%)	MCC	Server (s)
NSL-KDD	SVM	95.00	5.10	95.00	95.00	95.00	0.899	188.73
	C-NIDS	95.56	4.62	95.68	95.56	95.55	0.912	41.94
	DSN(2)	95.70	4.34	95.18	95.70	95.68	0.855	12.73
	DSN(3)	95.87	4.69	95.92	95.87	95.83	0.808	8.52
	DSN(4)	95.45	4.82	95.83	95.99	95.93	0.859	6.41
	DSN(5)	95.36	4.42	95.13	95.36	95.29	0.816	5.11
ISCX 2012	SVM	98.70	1.40	98.70	98.70	98.70	0.971	83.68
	C-NIDS	91.67	10.94	91.65	91.67	91.66	0.809	20.94
	DSN(2)	91.77	10.86	91.75	91.77	91.76	0.812	10.69
	DSN(3)	91.70	11.06	91.67	91.70	91.68	0.810	7.28
	DSN(4)	91.82	11.00	91.79	91.82	91.80	0.812	5.29
	DSN(5)	91.62	10.99	91.60	91.62	91.61	0.808	4.28

processing time is reduced by 5 × compared to centralized NIDS and by 20 × when compared to SVM when using 5 sub neural networks. The reduced server processing time allows DNN to detect intrusions in lesser amount of time since each SLFN in DNN has a reduced detection latency.

5 Conclusion

In this chapter, we have discussed the application of computational intelligence techniques for indoor data analytics on smart-gateway network. Firstly, a computational efficient data analytic platform is introduced for smart home management system based on the distributed gateway network. Secondly, a distributed support vector machine (DSVM) and distributed neural network (DNN) based machine learning algorithm are introduced for the indoor data analytics, which can significantly reduce the training complexity and training time yet maintaining acceptable accuracy. We have applied the DSVM and DNN for indoor positioning to analyze the WiFi data; and further to analyze network intrusion detection to provide the network security. Such a computational intelligence technique can be compactly realized on the computational-resource limited smart-gateway networks, which is desirable to build a real cyber-physical system towards future smart home, smart building, smart community and further a smart city.

Acknowledgements This work is sponsored by grants from Singapore MOE Tier-2 (MOE2015-T2-2-013), NRF-ENIC-SERTD-SMES-NTUJTCI3C-2016 (WP4) and NRF-ENIC-SERTD-SMES-NTUJTCI3C-2016 (WP5).

References

1. J. Kacprzyk and W. Pedrycz, *Springer handbook of computational intelligence*. Springer, 2015.
2. D. L. Poole, A. K. Mackworth, and R. Goebel, *Computational intelligence: a logical approach*. Oxford University Press New York, 1998, vol. 1.
3. H.-K. Lam and H. T. Nguyen, *Computational intelligence and its applications: evolutionary computation, fuzzy logic, neural network and support vector machine techniques*. World Scientific, 2012.
4. D. Peleg, "Distributed computing," *SIAM Monographs on discrete mathematics and applications*, vol. 5, 2000.
5. W. Wang, K. Zhu, L. Ying, J. Tan, and L. Zhang, "Maptask scheduling in mapreduce with data locality: Throughput and heavy-traffic optimality," *IEEE/ACM Transactions on Networking*, vol. 24, no. 1, pp. 190–203, 2016.
6. Apache hadoop. [Online]. Available: http://hadoop.apache.org/.
7. K. Zhu, H. Wang, H. Bai, J. Li, Z. Qiu, H. Cui, and E. Y. Chang, "Parallelizing support vector machines on distributed computers," in *Advances in Neural Information Processing Systems*, 2008, pp. 257–264.
8. B. Barker, "Message passing interface (MPI)," in *Workshop: High Performance Computing on Stampede*, 2015.

9. H. Liu, H. Darabi, P. Banerjee, and J. Liu, "Survey of wireless indoor positioning techniques and systems," *IEEE Transactions on Systems, Man, and Cybernetics, Part C (Applications and Reviews)*, vol. 37, no. 6, pp. 1067–1080, 2007.
10. C. Yang and H.-R. Shao, "Wifi-based indoor positioning," *IEEE Communications Magazine*, vol. 53, no. 3, pp. 150–157, 2015.
11. A. M. Hossain and W.-S. Soh, "A survey of calibration-free indoor positioning systems," *Computer Communications*, vol. 66, pp. 1–13, 2015.
12. S. He and S.-H. G. Chan, "Wi-fi fingerprint-based indoor positioning: Recent advances and comparisons," *IEEE Communications Surveys & Tutorials*, vol. 18, no. 1, pp. 466–490, 2016.
13. J. Xu, W. Liu, F. Lang, Y. Zhang, and C. Wang, "Distance measurement model based on rssi in wsn," *Wireless Sensor Network*, vol. 2, no. 08, p. 606, 2010.
14. J. Torres-Solis, T. H. Falk, and T. Chau, *A review of indoor localization technologies: towards navigational assistance for topographical disorientation*. INTECH Publisher, 2010.
15. J. Janicka and J. Rapinski, "Application of rssi based navigation in indoor positioning," in *Geodetic Congress (Geomatics), Baltic*. IEEE, 2016, pp. 45–50.
16. Y. Cai, S. K. Rai, and H. Yu, "Indoor positioning by distributed machine-learning based data analytics on smart gateway network," in *Indoor Positioning and Indoor Navigation (IPIN), 2015 International Conference on*. IEEE, 2015, pp. 1–8.
17. J. P. Anderson, "Computer security threat monitoring and surveillance," Technical report, James P. Anderson Company, Fort Washington, Pennsylvania, Tech. Rep., 1980.
18. E. Nyakundi, "Using support vector machines in anomaly intrusion detection," Ph.D. dissertation, The University of Guelph, 2015.
19. D. J. Weller-Fahy, B. J. Borghetti, and A. A. Sodemann, "A survey of distance and similarity measures used within network intrusion anomaly detection," *IEEE Communications Surveys & Tutorials*, vol. 17, no. 1, pp. 70–91, 2015.
20. C.-C. Sun, C.-C. Liu, and J. Xie, "Cyber-physical system security of a power grid: State-of-the-art," *Electronics*, vol. 5, no. 3, p. 40, 2016.
21. Y. Zhang, L. Wang, W. Sun, R. C. Green II, and M. Alam, "Distributed intrusion detection system in a multi-layer network architecture of smart grids," *IEEE Transactions on Smart Grid*, vol. 2, no. 4, pp. 796–808, 2011.
22. D. J. Cook, J. C. Augusto, and V. R. Jakkula, "Ambient intelligence: Technologies, applications, and opportunities," *Pervasive and Mobile Computing*, vol. 5, no. 4, pp. 277–298, 2009.
23. S. Nikoletseas, M. Rapti, T. P. Raptis, and K. Veroutis, "Decentralizing and adding portability to an iot test-bed through smartphones," in *IEEE DCOSS*, 2014.
24. C. Zhang, W. Wu, H. Huang, and H. Yu, "Fair energy resource allocation by minority game algorithm for smart buildings," in *2012 Design, Automation & Test in Europe Conference & Exhibition (DATE)*. IEEE, 2012, pp. 63–68.
25. H. Huang, Y. Cai, H. Xu, and H. Yu, "A multi-agent minority-game based demand-response management of smart buildings towards peak load reduction," *IEEE Transactions on Computer-Aided Design of Integrated Circuits and Systems*, vol. PP, no. 99, pp. 1–1, 2016.
26. S. Helal, B. Winkler, C. Lee, Y. Kaddoura, L. Ran, C. Giraldo, S. Kuchibhotla, and W. Mann, "Enabling location-aware pervasive computing applications for the elderly," in *IEEE PerCom*, 2003.
27. C. Drane, M. Macnaughtan, and C. Scott, "Positioning GSM telephones," *Communications Magazine, IEEE*, vol. 36, no. 4, pp. 46–54, 1998.
28. M. O. Ergin, V. Handziski, and A. Wolisz, "Node sequence discovery in wireless sensor networks," in *IEEE DCOSS*, 2013.
29. C.-W. Hsu and C.-J. Lin, "A comparison of methods for multiclass support vector machines," *IEEE transactions on Neural Networks*, vol. 13, no. 2, pp. 415–425, 2002.
30. C. Zhang, P. Li, A. Rajendran, and Y. Deng, "Parallel multicategory support vector machines (pmc-svm) for classifying microcarray data," in *Computer and Computational Sciences, 2006. IMSCCS'06. First International Multi-Symposiums on*, vol. 1. IEEE, 2006, pp. 110–115.
31. C.-C. Chang and C.-J. Lin, "Libsvm: a library for support vector machines," *ACM Transactions on Intelligent Systems and Technology (TIST)*, vol. 2, no. 3, p. 27, 2011.

32. G.-B. Huang, Q.-Y. Zhu, and C.-K. Siew, "Extreme learning machine: theory and applications," *Neurocomputing*, vol. 70, no. 1, pp. 489–501, 2006.
33. D. L. Donoho, "Compressed sensing," *IEEE Transactions on Information Theory*, vol. 52, no. 4, pp. 1289–1306, 2006.
34. H. Huang, Y. Cai, and H. Yu, "Distributed-neuron-network based machine learning on smart-gateway network towards real-time indoor data analytics," in *2016 Design, Automation & Test in Europe Conference & Exhibition (DATE)*. IEEE, 2016, pp. 720–725.
35. G. Feng, G.-B. Huang, Q. Lin, and R. Gay, "Error minimized extreme learning machine with growth of hidden nodes and incremental learning," *Neural Networks, IEEE Transactions on*, vol. 20, no. 8, pp. 1352–1357, 2009.
36. L. N. Trefethen and D. Bau III, *Numerical linear algebra*. Siam, 1997, vol. 50.
37. D. J. Miller and L. Yan, "Critic-driven ensemble classification," *IEEE Transactions on Signal Processing*, vol. 47, no. 10, pp. 2833–2844, 1999.
38. P. Bahl and V. N. Padmanabhan, "Radar: An in-building rf-based user location and tracking system," in *Nineteenth Annual Joint Conference of the IEEE Computer and Communications Societies*, 2000.
39. M. Brunato and R. Battiti, "Statistical learning theory for location fingerprinting in wireless LANs," *Computer Networks*, vol. 47, no. 6, pp. 825–845, 2005.
40. I. Ekahau. (2015). [Online]. Available: http://www.test.org/doe/.
41. NSL-KDD dataset. [Online]. Available: http://www.unb.ca/research/iscx/dataset/iscx-NSL-KDD-dataset.html.
42. ISCX 2012 dataset. [Online]. Available: http://www.unb.ca/research/iscx/dataset/iscx-IDS-dataset.html.

Predicting Spatiotemporal Impacts of Weather on Power Systems Using Big Data Science

Mladen Kezunovic, Zoran Obradovic, Tatjana Dokic, Bei Zhang, Jelena Stojanovic, Payman Dehghanian and Po-Chen Chen

Abstract Due to the increase in extreme weather conditions and aging infrastructure deterioration, the number and frequency of electricity network outages is dramatically escalating, mainly due to the high level of exposure of the network components to weather elements. Combined, 75% of power outages are either directly caused by weather-inflicted faults (e.g., lightning, wind impact), or indirectly by equipment failures due to wear and tear combined with weather exposure (e.g. prolonged overheating). In addition, penetration of renewables in electric power systems is on the rise. The country's solar capacity is estimated to double by the end of 2016. Renewables significant dependence on the weather conditions has resulted in their highly variable and intermittent nature. In order to develop automated approaches for evaluating weather impacts on electric power system, a comprehensive analysis of large amount of data needs to be performed. The problem addressed in this chapter is how such Big Data can be integrated, spatio-temporally correlated, and analyzed in real-time, in order to improve capabilities of modern electricity network in dealing with weather caused emergencies.

M. Kezunovic (✉) · T. Dokic · B. Zhang · P. Dehghanian · P.-C. Chen
Department of Electrical and Computer Engineering, Texas A&M University,
College Station, TX, USA
e-mail: kezunov@ece.tamu.edu

T. Dokic
e-mail: tatjana.djokic@tamu.edu

B. Zhang
e-mail: adele.zhang@tamu.edu

P. Dehghanian
e-mail: payman.dehghanian@tamu.edu

P.-C. Chen
e-mail: pchen01@tamu.edu

Z. Obradovic · J. Stojanovic
Computer and Information Department, Temple University, Philadelphia, PA, USA
e-mail: zoran.obradovic@temple.edu

J. Stojanovic
e-mail: jelena.stojanovic@temple.edu

W. Pedrycz and S.-M. Chen (eds.), *Data Science and Big Data: An Environment of Computational Intelligence*, Studies in Big Data 24,
DOI 10.1007/978-3-319-53474-9_12

Keywords Aging infrastructure · Asset management · Big data · Data analytics · Data mining · Insulation coordination · Outage management · Power system · Solar generation forecast · Weather impact

1 Introduction

The Big Data (BD) in the power industry comes from multiple sources: variety of measurements from the grid, weather data from a variety of sources, financial data from electricity and other energy markets, environmental data, etc. The measurements from the physical network and weather data exhibit various aspects of BD: large volume, high velocity, increasing variety and varying veracity. For efficient condition-based asset and outage management, and preventive real-time operation, fast processing of large volumes of data is an imperative. The volume and velocity with which the data is generated can be overwhelming for both on-request and real-time applications. The heterogeneity of data sources and accuracy are additional challenges. The effective use of BD in power grid applications requires exploiting spatial and temporal correlations between the data and the physical power system network.

In this chapter, we will address unique fundamental solutions that will allow us to effectively fuse weather and electricity network data in time and space for the benefit of predicting the risk associated with weather impact on utility operation, generation, and asset and outage management. Computational intelligence approach plays a central role in such a task. Unstructured models like neural networks (NN), fuzzy systems and hybrid intelligent systems represent powerful tools for learning non-linear mappings. However, majority of those models assume independent and identically distributed random variables mainly focusing on the prediction of a single output and could not exploit structural relationships that exist between multiple outputs in space and time.

We will illustrate how more accurate predictions are possible by structured learning from merged heterogeneous BD compared to unstructured models, which is achieved by developing and characterizing several innovative decision-making tools. We will describe how the BD is automatically correlated in time and space. Then, the probabilistic graphical model called the Gaussian Conditional Random Fields (GCRF) will be introduced.

The proposed BD analytics will be examined for the following applications: (1) Assets management—predicts weather impacts on deterioration and outage rates of utility assets such as insulators, providing knowledge for optimal maintenance schedules and replacement strategies. The GCRF model is used to predict the probability of a flashover leading to probability of total insulator failure. (2) Solar Generation—the GCRF model is introduced to forecast the solar photovoltaic generation. Such data-driven forecasting techniques are capable of modeling both the spatial and temporal correlations of various solar generation stations, and therefore they are performing well even under the scenarios with unavailable or missing data.

Benefits of the proposed method are assessed through a new risk-based frame-work using realistic utility data. The results show that spatio-temporal correlation of BD and GCRF model can significantly improve ability to manage the power grid by predicting weather related emergencies in electric power system.

2 Background

2.1 Power System Operation, Generation, Outage and Asset Management

The current context of the modern electric industry, characterized by competitive electricity markets, privatization, and regulatory or technical requirements man-dates power utilities to optimize their operation, outage and asset management practices and develop the requisite decision plans techno-economically. With the rapid deployment of renewable generation based on wind, solar, geothermal energy resources, etc., as well as the increasing demand to deliver higher quality electricity to customers, many electric utilities have been undergoing a paradigm shift regarding how the grid planning, operation, and protection should be reframed to enhance the resilience of the power delivery infrastructure in face of many threating risks. While the wide deployment of distributed renewable generation in the grid has brought about a huge potential for performance improvements in various domains, yet there are critical challenges introduced by renewables to overcome. An accurate forecast of the unpredictable and variable sources of renewable generation, as well as their strategic coordination in every-day normal and even emergency operation scenarios of the grid is of particular interest [1].

Asset management is said to be the process of cost minimization and profit maximization through the optimized operation of the physical assets within their life cycles. Asset management in electric power systems can be broadly classified into four main categories based on the possible time scales, i.e., real-time, short-term, mid-term, and long-term [2]. Real-time asset management mainly covers the key power system resiliency principles and deals with the unexpected outages of power system equipment and grid disruptions. System monitoring and operating condition tracking infrastructures, e.g. supervisory control and data acquisition systems (SCADA) and geographic information systems (GIS), play a vital role for a techno-economically optimized real-time asset management. By enhancing the situational awareness, they enable power system operators to effectively monitor and control the system. Contingency analysis as well as online outage management scheme is utilized to coordinate the necessary resource managements through automated control systems. Wide area measurement systems (WAMS), which have been recently realized by broad deployment of phasor measurement units (PMUs), are among the new technologies in this context [3]. Intelligent Electronic Devices (IED) located in substations can record and store a huge amount of data either periodically with high sampling rates or on certain critical events such as faults. When properly interpreted, data gathered from such devices can be used to predict, operate, monitor and post-mortem analyze power system events. Yet, without knowledge on the meaning of such data, efficient utilization cannot be pursued. Thus, data mining techniques such as neural networks, support vector machines, decision trees, and spatial statistics should be considered for acquiring such knowledge.

Short-term asset management strives to maximize the rate of return associated with asset investments. The process, called asset valuation, is performed to incorporate company's investments in its portfolio. The value mainly depends on the uncertain market prices through various market realizations. Market risk assessment is a key consideration and the revenue/profit distributions are gained through a profitability analysis.

Optimized maintenance scheduling falls within the realm of mid-term asset management. It guides the maintenance plans and the associated decision making techniques toward satisfactorily meeting the system-wide desired targets. In other words, efforts are focused to optimize the allocation of limited financial resources where and when needed for an optimal outage management without sacrificing the system reliability. Smart grid concept has facilitated an extensive deployment of smart sensors and monitoring technologies to be used for health and reliability assessment of system equipment over time and to optimize the maintenance plans accordingly [4]. Two factors that are typically used for condition monitoring are the prognostic parameters and trends of equipment deterioration. Prognostic parameters that provide an indication of equipment condition such as ageing and deterioration are useful indicators of impending faults and potential problems. The trend of the deterioration of critical components can be identified through a trend analysis of the equipment condition data. In addition to the aforementioned factors, event triggered deterioration of equipment can be also considered. Gathering and interpreting historical data can provide useful information for evaluation of impacts that events had on given equipment over its lifetime. Data analytics can use such data as a training set and establish connection between causes of faults/disturbances and its impact on equipment condition. Knowledge gathered during training process can then be used for classification of new events generated in real time and prediction of equipment deterioration that is caused by these effects. With addition of historical data about outages that affected the network, a more efficient predictive based maintenance management can be developed [4].

Long-term investment in power system expansion planning as well as wide deployment of distributed generations fall within the scope of long-term asset management where the self-interested players, investors, and competitors are invited to participate in future economic plans.

Knowledge from outage and assets management can be combined by considering both IED recordings and equipment parameters and reliability characteristics for a more reliable decision making in practice. The goal is to explore limits of available equipment and identify possible causes of equipment malfunction by gathering both data obtained through the utility field measurement infrastructure and additional data coming from equipment maintenance and operation records, as well as weather and geographic data sources. In this approach, decision about the state of the equipment is made based on the historical data of the events that affected the network, as well as real-time data collected during the ongoing events.

2.2 *Weather Data Parameters and Sources*

The measurement and collection infrastructure of weather data has been well developed over the past decades. In this subsection, as an example, the data from National Oceanic and Atmospheric Administration (NOAA) [5] will be discussed. Based on the type of measurements, the weather data may be categorized into:

- Surface observations using data collected from land-based weather stations that contain measurement devices that track several weather indices such as temperature, precipitation, wind speed and direction, humidity, atmospheric pressure, etc.
- Radar (Radio Detection and Ranging) provides accurate storm data using radio waves to determine the speed, direction of movement, range and altitude of objects. Based on radar measurements different reflectivity levels are presented with different colors on a map.
- Satellites generate raw radiance data. Unlike local-based stations, they provide global environmental observations. Data is used to monitor and predict wide-area meteorological events such as flash floods, tropical systems, tornadoes, forest fires, etc.

Table 1 [6–13] demonstrates the list of weather data sources from NOAA which may be useful for power system operations. A list of national support centers for more real-time and forecasting data may be found in [14]. A list of commercial

Table 1 List of weather data sources

Data source	Available access
National Weather Service (NWS)	GIS Portal [6]
	National Digital Forecast Database [7]
	Doppler Radar Images [8]
National Centers for Environmental Information (NCEI)	Data Access [9]
	Web Service [10]
	GIS Map Portal [11]
Office of Satellite and Product Operations (OSPO)	Satellite Imagery Products [12]
Global Hydrology Resource Center (GHRC)	Lightning and Atmospheric Electricity Research [13]

weather vendors for more specialized meteorological products and services may be found in [15]. More land and atmosphere data and images may be found in National Aeronautics and Space Administration (NASA) websites.

Specific power system operations are related to certain type of weather events. In such case, the most relevant weather data input is required for data analytics. An example below demonstrates the use of satellite and radar data.

The satellite meteorological detection is passive remote sensing in general, whereas the radar meteorological detection is active remote sensing. Radars can emit radio or microwave radiation and receive the back-scattering signals from a convective system. For a practical example regarding tropical cyclones, satellites can observe a tropical cyclone once it forms in the ocean, and radar can detect its inner structure as it moves near the continent and lands in.

Lightning data is gathered by the sensors that are typically located sparsely over the area of interest. There are three common types of lightning sensors: (1) Ground-based systems that use multiple antennas to determine distance to the lightning by performing triangulation, (2) Mobile systems that use direction and a sensing antenna to calculate distance to the lightning by analyzing surge signal frequency and attenuation, and (3) Space-based systems installed on artificial satellites that use direct observation to locate the faults.

Typical detection efficiency for a ground-based system is 70–90%, with a accuracy of location within 0.7–1 km, while space-based systems have resolution of 5–10 km, [16].

For example, The National Lightning Detection Network (NLDN) [17] uses ground-based system to detect lightning strikes across the United States. After detection data received from sensors in raw form is transmitted via satellite-based communication to the Network Control Center operated by Vaisala Inc. [18].

When it comes to the way data is received by the utility we can distinguish two cases: (i) the lightning sensors are property of the utility, and (ii) lightning data is received from external source. In the first case raw data are received from the sensors, while in second case external sources provide information in the format that is specific to the organization involved. Lightning data typically includes the following information: a GPS time stamp, latitude and longitude of the strike, peak current, lightning strike polarity, and type of lightning strike (cloud-to-cloud or cloud-to-ground).

2.3 Spatio-Temporal Correlation of Data

Any kind of data with a spatial component can be integrated into GIS as another layer of information. As new information is gathered by the system, these layers can be automatically updated. Two distinct categories of GIS data, spatial and attribute data can be identified. Data which describes the absolute and relative context of geographic features is spatial data. For transmission towers, as an example, the exact spatial coordinates are usually kept by the operator. In order to provide additional characteristics of spatial features, the attribute data is included. Attribute data includes characteristics that can be either quantitative or qualitative. For instance, a table including the physical characteristics of a transmission tower can be described along with the attribute data.

In terms of spatial data representation, raster and vector data can be used. In case of vector data, polygons, lines and points are used to form shapes on the map. Raster presents data as a grid where every cell is associated with one data classification. Typically, different data sources will provide different data formats and types. Although modern GIS tools such as ArcGIS [19] are capable of opening and overlaying data in different formats, analysis of mixed database is a challenge [20].

GIS is often understood as a visualization tool of mapping geographical information. However, it also enables interpretation of spatio-temporal data for better decision making. An enterprise level GIS requires a GIS platform and a geospatial database management system [21]. The GIS platform will allow the execution of designed applications and enable access from various devices, and it is the key to interface with current utility decision-making tools such as outage management system. The geospatial database system will keep the most recent asset information and update the database from asset management.

In addition to spatial reference, data must also be time referenced in a unique fashion. Following factors are important for time correlation of data:

- Time scales: data can be collected with different time resolution: yearly, monthly, daily, hourly, once every few minutes or even seconds. Time scales for different applications and events of interest for this research are presented in Fig. 1 [22].
- Atomic Time: Standards and their characteristics are listed in Table 2. All standards use the same definition for 1 s.

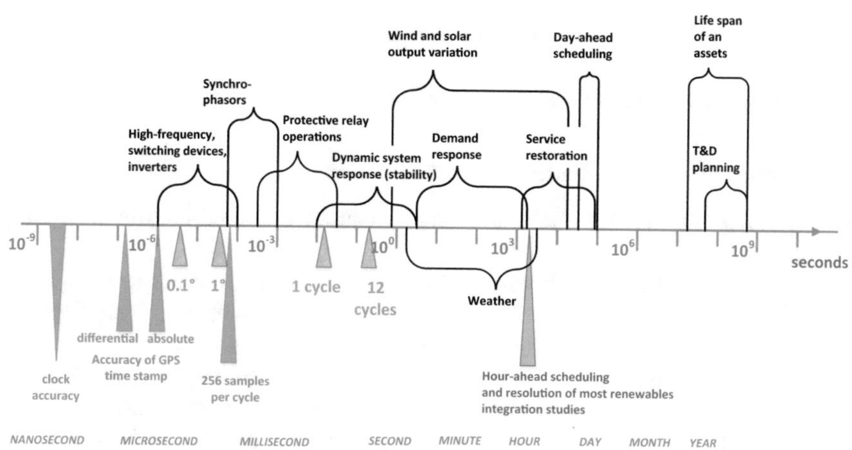

Fig. 1 Time scales of the Big Data and related applications of interest to the power sector

Table 2 Atomic Time Standards

Standard name	Leap seconds	Representation	Zero datum
UTC—Coordinated Universal Time	Included	Date and time: [ddmmyyyy, hhmmss]	N/A
GPS Satellite Time	Not included	Seconds, or week# + Seonds_of_Week	Midnight roll-over to 6 Jan 1980 UTC
TAI—International Atomic Time	Not included	Seconds	Midnight roll-over to 1 Jan 1970 UTC

Table 3 Time Synchronization Protocols

Protocol name	Media	Sync accuracy	Time standard	Description
Network Time Protocol (NTP), [23]	Ethernet	50–100 ms	UTC	Simplified version—SNTP assumes symmetrical network delay
Serial time code IRIG B-122, [24]	Coaxial	1–10 μs	TAI	Two versions used in substation: B12x and B00x "lagging" code
Precision Time Protocol (PTP), [25]	Ethernet	20–100 ns	TAI	Master to slave Hardware time-stamping of PTP packets is required

- Synchronization protocols: Accuracy of a time stamp is highly dependent on the type of the signal that is used for time synchronization. Different measuring devices that use GPS synchronization can use different synchronization signals. Table 3 lists properties of time synchronization protocols.

3 Weather Impact on Power System

3.1 Weather Impact on Outages

Main causes of weather related power outages are:

- Lightning activity: the faults are usually caused by cloud-to-ground lighting hitting the poles.
- Combination of rain, high winds and trees movement: in order to completely understand this event several data sources need to be integrated including precipitation data, wind speed and direction, and vegetation data.
- Severe conditions such as hurricanes, tornados, ice storms: in case of severe conditions multiple weather factors are recorded and used for analysis.
- In case of extremely high and low temperatures the demand increases due to cooling and heating needs respectively leading to the network overload.

Weather impact on outages in power systems can be classified into direct and indirect [26]. Direct impact to utility assets: This type of impact includes all the situations where severe weather conditions directly caused the component to fail. Examples are: lightning strikes to the utility assets, wind impact making trees or tree branches to come in contact with lines, etc. These types of outages are marked as weather caused outages. Indirect impact to utility assets: This type of impact accrues when weather creates the situation in the network that indirectly causes the component to fail. The examples are: hot weather conditions increasing the demand thus causing the overload of the lines resulting in the line sags increasing the risk of faults due to tree contact, exposure of assets to long term weather impacts causing component deterioration, etc. These types of outages are marked as equipment failure.

Weather hazard relates to the surrounding severe weather conditions that have a potential to cause an outage in the electric network. Key source of information for determining the weather hazard is weather forecast. Depending on the type of hazard that is under examination different weather parameters are observed. In case of a lightning caused outage, forecast for lightning probability, precipitation, temperature, and humidity needs to be considered, while in case of outages caused by wind impact, parameters of interest are wind speed, direction and gust, temperature, precipitation, humidity, probability of damaging thunderstorm wind.

National Digital Forecast Database (NDFD), [7] provides short-term (next 3–7 days) and long-term (next year and a half) weather prediction for variety of weather parameters such as temperature, precipitation, wind speed and direction, hail, storm probabilities etc. NDFD uses Numerical Weather Prediction (NWP), which is taking current weather observations and processing it using different numerical models for prediction of future state of weather. Overview of hazard analysis based on NDFD data is presented in Fig. 2 [27].

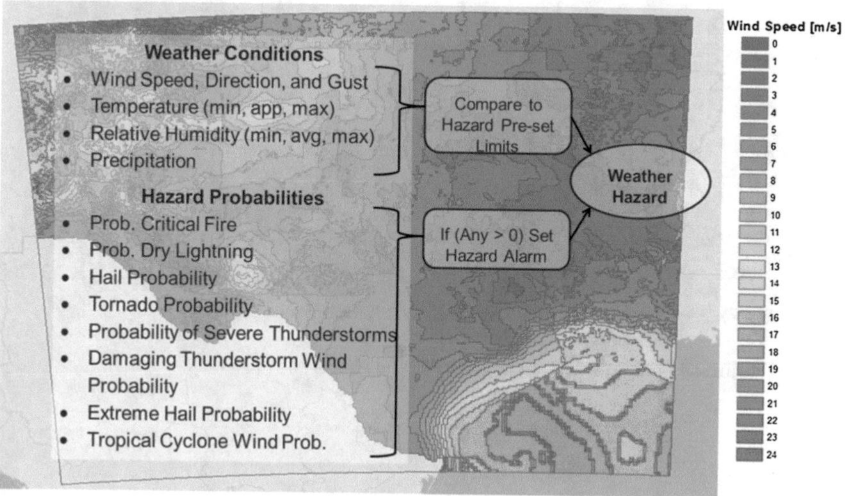

Fig. 2 Weather hazard using NDFD data, [27]

3.2 Renewable Generation

Due to its' environmentally friendly and sustainable nature, renewable generation, such as wind generation, solar generation, etc., has been advocated and developed rapidly during the past decades, as illustrated in Fig. 3. Many state governments are adopting and increasing their Renewable Energy Portfolio (RPS) standards [28]. Among the 29 states with the RPS, California government is holding the most aggressive one, aiming at reaching 33% renewables by 2020 [29].

Weather's impact on the renewable generation is quite evident. For example, solar generation largely depends on the solar irradiance, which could be affected by the shading effect due to the variability of the clouds. Besides, other weather condition such as the high temperature or snow can decrease the production efficiency of the solar panel. As another commonly utilized renewable energy, wind generation is sensitive to the availability of the wind resources, since small differences in wind speed lead to large differences in power. Besides, when the wind blows extremely hard, the wind turbine may switch off out of self-protection.

Renewable generation is quite sensitive to weather factors, and some of them are highly variable and unpredictable: the wind could blow so hard at this moment and suddenly stop at the next time step; the solar irradiance received by the solar panel could suddenly goes down because of a moving cloud, etc.

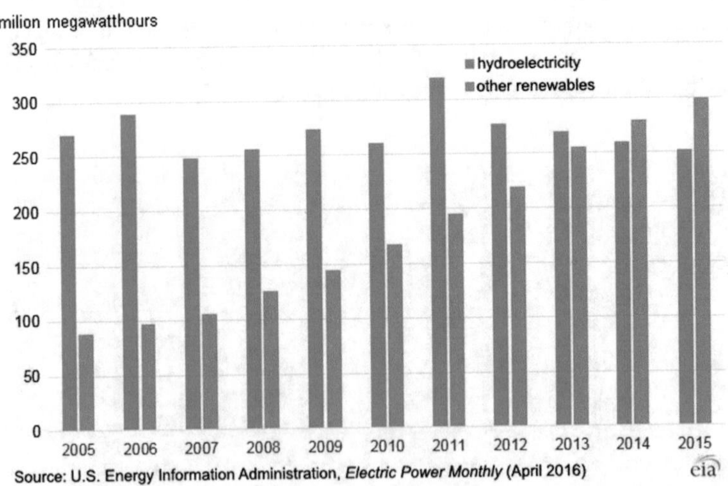

Fig. 3 Illustration on the development of renewable generation [30]

4 Predictive Data Analytics

4.1 Regression

4.1.1 Unstructured Regression

In the standard regression setting we are given a data set with N training examples, $D = \{(\mathbf{x}_i, y_i), i = 1, \ldots, N\}$, where $\mathbf{x}_i \in \mathbf{X} \subset R^M$ is an M dimensional vector of inputs and $y_i \in R$ is a real-valued output variable.

For example, in one of the applications described later in this chapter (Sect. 5.1.2.3), in the data set D, $\boldsymbol{x}$ are multivariate observations on a substation or a tower, like weather conditions or peak current and lightning strike polarity, while the output of interest y is the BIL (Basic Lightning Impulse Insulation Level) after occurrence of lightning strike.

The objective of regression is to learn a linear or non-linear mapping f from training data D that predicts the output variable y as accurately as possible given an input vector $\boldsymbol{x}$. Typically, the assumption about data-generating model is $y = f(\mathbf{x}) + \varepsilon, \varepsilon \sim \mathrm{N}(0, \sigma^2)$, where ε is Gaussian additive noise with constant variance σ^2. This setup is appropriate when data are independently and identically distributed (IID). The IID assumption is often violated in applications where data reveal temporal, spatial, or spatio-temporal dependencies. In such cases, the traditional supervised learning approaches, as linear regression or neural networks, could result in a model with degraded performances. Structured regression models are, therefore, used for predicting output variables that have some internal structure. Thus, in the following sections we introduce such methods.

4.1.2 Structured Regression (Probabilistic Graphical Models)

Traditional regression models, like neural networks (NN), are powerful tools for learning non-linear mappings. Such models mainly focus on the prediction of a single output and could not exploit relationships that exist between multiple outputs. In structured learning, the model learns a mapping f: $\mathbf{X}^N \rightarrow R^N$ to simultaneously predict all outputs given all input vectors. For example, let us assume that the value of y_i is dependent on that of y_{i-1} and y_{i+1}, as is the case in temporal data, or that the value of y_i is dependent on the values of neighboring y_h and y_j, as it is the case in spatially correlated data. Let us also assume that input x_i is noisy. A traditional model that uses only information contained in x_i to predict y_i might predict the value for y_i to be quite different from those of y_{i-1} and y_{i+1} or y_h and y_j because it treats them individually. A structured predictor uses dependencies among outputs to take into account that y_i is more likely to have value close to y_{i-1} and y_{i+1} or y_h and y_j, thus improving final predictions.

In structured learning we usually have some prior knowledge about relationships among the outputs y. Mostly, those relationships are application-specific where the

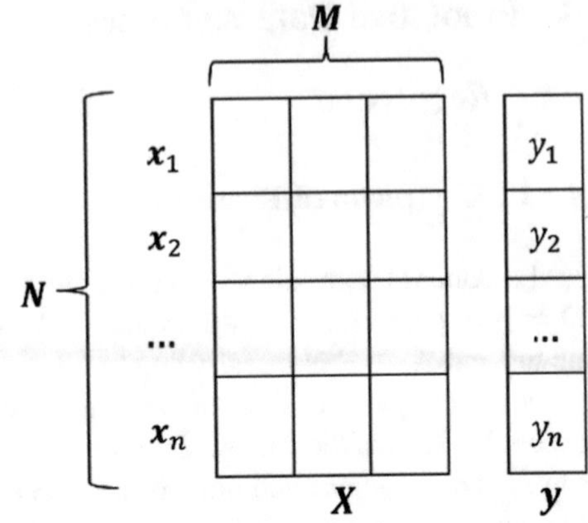

Fig. 4 Data set D in the standard regression setting with X as an input variable matrix and y as a real-valued output variable vector

dependencies are defined in advance, either by domain knowledge or by assumptions, and represented by statistical models.

Probabilistic Graphical Models

Relationships among outputs can be represented by graphical models. The advantage of the graphical models is that the sparseness in the interactions between outputs can be used for development of efficient learning and inference algorithms. In learning from spatial-temporal data, the Markov Random Fields [31] and the Conditional Random Fields (CRF) [32] are among the most popular graphical models (Fig. 4).

Conditional Random Fields

Originally, CRF were designed for classification of sequential data [32] and have found many applications in areas such as computer vision [33], natural language processing [34], and computational biology [35]. CRFs are a type of discriminative undirected probabilistic graphical model defined over graph G as

$$P(\mathbf{y}|\mathbf{x}) = \frac{1}{Z(\mathbf{x})} \sum_{a=1}^{A} \psi_a(y_a, x_a), \quad \psi_a(y_a, x_a) = \exp\left(\sum_{k=1}^{K(A)} \theta_{ak} f_{ak}(y_a, x_a) \right), \tag{4.1}$$

where $Z(x)$ is a normalization constant and $\{\Psi_a\}$ is a set of factors in G. The feature functions f_{ak} and the weights θ_{ak} are indexed by the factor index a (each factor has its own set of weights) and k (there are K feature functions).

Construction of appropriate feature functions in CRF is a manual process that depends on prior beliefs of a practitioner about what features could be useful. The choice of features is often constrained to simple constructs to reduce the complexity of learning and inference from CRF. In general, to evaluate $P(\mathbf{y}|\mathbf{x})$ during learning and inference, one would need to use time consuming sampling methods. This problem can be accomplished in a computationally efficient manner for real-valued CRFs.

CRF for regression is a less explored topic. The Conditional State Space Model (CSSM) [36], an extension of the CRF to a domain with the continuous multivariate outputs, was proposed for regression of sequential data. Continuous CRF (CCRF) [37] is a ranking model that takes into account relations among ranks of objects in document retrieval. In [33], a conditional distribution of pixels given a noisy input image is modeled using the weighted quadratic factors obtained by convolving the image with a set of filters. Feature functions in [33] were specifically designed for image de-noising problems and are not readily applicable to regression.

Most CRF models represent linear relationships between attributes and outputs. On the other hand, in many real-world applications this relationship is highly complex and nonlinear and cannot be accurately modeled by a linear function. CRF that models nonlinear relationship between observations and outputs has been applied to the problem of image de-noising [33]. Integration of CRF and Neural Networks (CNF) [38–40] has been proposed for classification problems to address these limitations by adding a middle layer between attributes and outputs. This layer consists of a number of gate functions each acting as a hidden neuron, that captures the nonlinear relationships. As a result, such models can be much more computationally expressive than regular CRF.

4.2 Gaussian Conditional Random Fields (GCRF)

4.2.1 Continuous Conditional Random Fields Model

Conditional Random Fields (CRF) provide a probabilistic framework for exploiting complex dependence structure among outputs by directly modeling the conditional distribution $P(\mathbf{y}|\mathbf{x})$. In regression problems, the output y_i is associated with input vectors $\mathbf{x} = (\mathbf{x}_1, \ldots, \mathbf{x}_N)$ by a real-valued function called *association potential* $A(\boldsymbol{\alpha}, y_i, \mathbf{x})$, where $\boldsymbol{\alpha}$ is K-dimensional set of parameters. The larger the value of A is the more y_i is related to $\mathbf{x}$. In general, A is a combination of functions and it takes as input all input data $\mathbf{x}$ to predict a single output y_i meaning that it does not impose any independency relations among inputs $\mathbf{x}_i$ (Figs. 5 and 6).

To model interactions among outputs, a real valued function called *interaction potential* $I(\boldsymbol{\beta}, y_i, y_j, \mathbf{x})$ is used, where $\boldsymbol{\beta}$ is an L dimensional set of parameters. Interaction potential represents the relationship between two outputs and in general can depend on an input $\mathbf{x}$. For instance, interaction potential can be modeled as a correlation between neighboring (in time and space) outputs. The larger the value of the interaction potential, the more related outputs are.

For the defined association and interaction potentials, *continuous CRF* models a conditional distribution $P(\mathbf{y}|\mathbf{x})$:

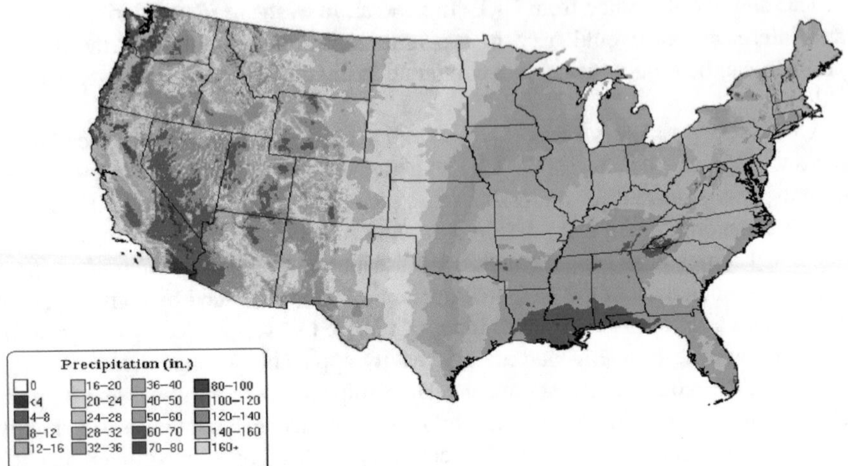

Fig. 5 U.S. average annual precipitation (1971–2000) [41]: spatial relationship between outputs of precipitation measurement stations over U.S.

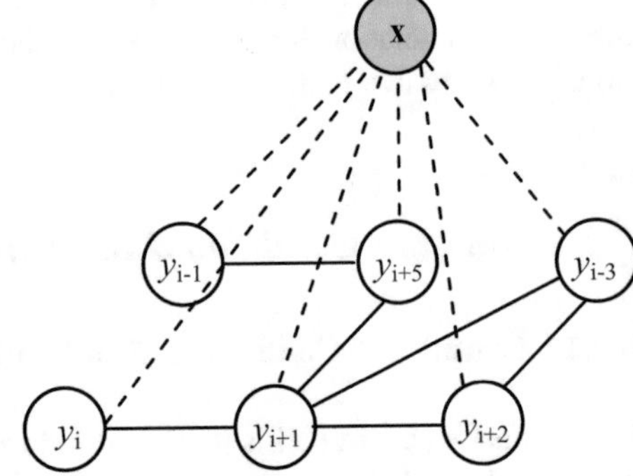

Fig. 6 Continuous CRF graphical structure. x-inputs (observations); y-outputs; *dashed lines*-associations between inputs and outputs; *solid lines*-interactions between outputs, [57]

$$P(\mathbf{y}|\mathbf{x}) = \frac{1}{Z(\mathbf{x}, \alpha, \beta)} \exp\left(\sum_{i=1}^{N} A(\alpha, y_i, \mathbf{x}) + \sum_{j \sim i} I(\beta, y_i, y_j, \mathbf{x}) \right) \tag{4.2}$$

where $\mathbf{y} = (y_1, \ldots, y_N)$, $j \sim i$ denotes the connected outputs y_i and y_j (connected with solid line at Fig. 6) and $Z(\mathbf{x}, \boldsymbol{\alpha}, \boldsymbol{\beta})$ is normalization function defined as

$$Z(\mathbf{x}, \boldsymbol{\alpha}, \boldsymbol{\beta}) = \int_{y} \exp\left(\sum_{i=1}^{N} A(\boldsymbol{\alpha}, y_i, \mathbf{x}) + \sum_{j \sim i} I(\boldsymbol{\beta}, y_i, y_j, \mathbf{x}) \right) dy \tag{4.3}$$

The learning task is to choose values of parameters $\boldsymbol{\alpha}$ and $\boldsymbol{\beta}$ to maximize the conditional log-likelihood of the set of training examples

$$L(\boldsymbol{\alpha}, \boldsymbol{\beta}) = \sum \log P(\mathbf{y}|\mathbf{x})$$
$$(\hat{\alpha}, \hat{\beta}) = \underset{\boldsymbol{\alpha}, \boldsymbol{\beta}}{\arg\max}(L(\boldsymbol{\alpha}, \boldsymbol{\beta})). \tag{4.4}$$

This can be achieved by applying standard optimization algorithms such as gradient descent. To avoid overfitting, $L(\boldsymbol{\alpha}, \boldsymbol{\beta})$ is regularized by adding $\boldsymbol{\alpha}^2/2$ and $\boldsymbol{\beta}^2/2$ terms to log-likelihood in formula (4.4) that prevents the parameters from becoming too large.

The inference task is to find the outputs $\mathbf{y}$ for a given set of observations $\mathbf{x}$ and estimated parameters $\boldsymbol{\alpha}$ and $\boldsymbol{\beta}$ such that the conditional probability $P(\mathbf{y}|\mathbf{x})$ is maximized

$$\hat{\mathbf{y}} = \underset{\mathbf{y}}{\arg\max}(P(\mathbf{y}|\mathbf{x})). \tag{4.5}$$

In CRF applications, A and I could be defined as linear combinations of a set of fixed features in terms of $\boldsymbol{\alpha}$ and $\boldsymbol{\beta}$ [42]

$$A(\alpha, y_i, \mathbf{x}) = \sum_{k=1}^{K} \alpha_k f_k(y_i, \mathbf{x})$$
$$I(\beta, y_i, y_j, \mathbf{x}) = \sum_{l=1}^{L} \beta_l g_l(y_i, y_j, \mathbf{x}) \tag{4.6}$$

This way, any potentially relevant feature could be included to the model because parameter estimation automatically determines their actual relevance by feature weighting.

4.2.2 Association and Interaction Potentials in the GCRF Model

If A and I are defined as quadratic functions of $\mathbf{y}$, $P(\mathbf{y}|\mathbf{x})$ becomes multivariate Gaussian distribution and learning and inference can be accomplished in a computationally efficient manner. In the following, the proposed feature functions that lead to Gaussian CRF are described.

Let us assume we are given K unstructured predictors (e.g. neural network, linear regression or any domain-defined model), $R_k(\mathbf{x})$, $k = 1, \ldots, K$, that predict single output y_i taking into account $\mathbf{x}$ (as a special case, only $\mathbf{x}_i$ can be used as $\mathbf{x}$). To model the dependency between the prediction and output, introduced are quadratic feature functions

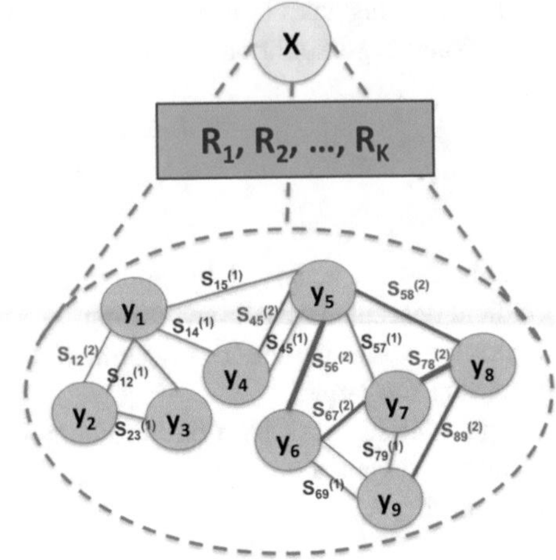

Fig. 7 Gaussian CRF graphical structure: association and interaction potentials are modeled as quadratic feature functions dependent on unstructured predictors R_k and similarity functions S(l), respectively

$$f_k(y_i, \mathbf{x}) = -(y_i - R_k(\mathbf{x}))^2, \quad k = 1, \ldots, K. \tag{4.7}$$

These feature functions follow the basic principle for association potentials—their values are large when predictions and outputs are similar. To model the correlation among outputs, introduced are the quadratic feature function

$$g_l(y_i, y_j, \mathbf{x}) = -e_{ij}^{(l)} S_{ij}^{(l)}(\mathbf{x})(y_i - y_j)^2, \quad \begin{array}{ll} e_{ij}^{(l)} = 1 & \text{if } (i,j) \in G_l, \\ e_{ij}^{(l)} = 0, & \text{otherwise} \end{array} \tag{4.8}$$

that imposes that outputs y_i and y_j have similar values if they have an edge in the graph. Note that multiple graphs can be used to model different aspects of correlation between outputs (for example, spatial and temporal). $S_{ij}^{(l)}(\mathbf{x})$ function represents similarity between outputs y_i and y_j, that depends on inputs $\mathbf{x}$. The larger $S_{ij}^{(l)}(\mathbf{x})$ is, the more similar the outputs y_i and y_j are (Fig. 7).

4.2.3 Gaussian Canonical Form

$P(\mathbf{y}|\mathbf{x})$ for CRF model (4.2), which uses quadratic feature functions, can be represented as a multivariate Gaussian distribution. The resulting CRF model can be written as

$$P(\mathbf{y}|\mathbf{x}) = \frac{1}{Z}\exp\left(-\sum_{i=1}^{N}\sum_{k=1}^{K} \alpha_k (y_i - R_k(\mathbf{x}))^2 - \sum_{i,j}\sum_{l=1}^{L} \beta_l e_{ij}^{(l)} S_{ij}^{(l)}(\mathbf{x})(y_i - y_j)^2 \right) \tag{4.9}$$

The exponent in (4.9), which will be denoted as E, is a quadratic function in terms of $\mathbf{y}$. Therefore, $P(\mathbf{y}|\mathbf{x})$ can be transformed to a Gaussian form by representing E as

$$E = -\frac{1}{2}(\mathbf{y}-\boldsymbol{\mu})^T\boldsymbol{\Sigma}^{-1}(\mathbf{y}-\boldsymbol{\mu}) = -\frac{1}{2}\mathbf{y}^{\mathrm{T}}\boldsymbol{\Sigma}^{-1}\mathbf{y} + \mathbf{y}^{\mathrm{T}}\boldsymbol{\Sigma}^{-1}\boldsymbol{\mu} + const \tag{4.10}$$

To transform $P(\mathbf{y}|\mathbf{x})$ to the Gaussian form, $\boldsymbol{\Sigma}$ and $\boldsymbol{\mu}$ are determined by matching (4.9) and (4.10). The quadratic terms of $\mathbf{y}$ in the association and interaction potentials as represented as $-\mathbf{y}^{\mathrm{T}}\mathbf{Q}_1\mathbf{y}$ and $-\mathbf{y}^{\mathrm{T}}\mathbf{Q}_2\mathbf{y}$, respectively, and combined to get

$$\boldsymbol{\Sigma}^{-1} = 2(\mathbf{Q}_1 + \mathbf{Q}_2). \tag{4.11}$$

By combining the quadratic terms of $\mathbf{y}$ from the association potential, it follows that $\mathbf{Q}_1$ is diagonal matrix with elements

$$Q_{1ij} = \begin{cases} \sum_{k=1}^{K} \alpha_k, & i=j \\ 0, & i \neq j. \end{cases} \tag{4.12}$$

By repeating this for the interaction potential, it follows that $\mathbf{Q}_2$ is symmetric with elements

$$Q_{2ij} = \begin{cases} \sum_{k}\sum_{l=1}^{L} \beta_l e_{ik}^{(l)} S_{ik}^{(l)}(\mathbf{x}), & i=j \\ -\sum_{l=1}^{L} \beta_l e_{ij}^{(l)} S_{ij}^{(l)}(\mathbf{x}), & i \neq j. \end{cases} \tag{4.13}$$

To get $\boldsymbol{\mu}$, linear terms in E are matched with linear terms in the exponent of (4.9)

$$\boldsymbol{\mu} = \Sigma\mathbf{b}, \tag{4.14}$$

where $\mathbf{b}$ is a vector with elements

$$b_i = 2\left(\sum_{k=1}^{K} \alpha_k R_k(\mathbf{x})\right). \tag{4.15}$$

By calculating Z using the transformed exponent, it follows

$$Z(\boldsymbol{\alpha}, \boldsymbol{\beta}, \mathbf{x}) = (2\pi)^{N/2}|\boldsymbol{\Sigma}|^{1/2}\exp(const). \tag{4.16}$$

Since exp(*const*) terms from Z and $P(\mathbf{y}|\mathbf{x})$ cancel out

$$P(\mathbf{y}|\mathbf{x}) = \frac{1}{(2\pi)^{N/2}|\boldsymbol{\Sigma}|^{1/2}} \exp\left(-\frac{1}{2}(\mathbf{y}-\boldsymbol{\mu})^{\mathbf{T}}\boldsymbol{\Sigma}^{-1}(\mathbf{y}-\boldsymbol{\mu}) \right), \tag{4.17}$$

where $\boldsymbol{\Sigma}$ and $\boldsymbol{\mu}$ are defined in (4.11) and (4.14). Therefore, the resulting conditional distribution is Gaussian with mean $\boldsymbol{\mu}$ and covariance $\boldsymbol{\Sigma}$. Observe that $\boldsymbol{\Sigma}$ is a function of parameters $\boldsymbol{\alpha}$ and $\boldsymbol{\beta}$, interaction potential graphs G_l, and similarity functions S, while $\boldsymbol{\mu}$ is also a function of inputs $\mathbf{x}$. The resulting CRF is called *the Gaussian CRF* (GCRF).

4.2.4 Learning and Inference

The **learning** task is to choose $\boldsymbol{\alpha}$ and $\boldsymbol{\beta}$ to maximize the conditional log-likelihood,

$$(\hat{\boldsymbol{\alpha}}, \boldsymbol{\beta}) = \arg\max_{\boldsymbol{\alpha},\boldsymbol{\beta}}(L(\boldsymbol{\alpha}, \boldsymbol{\beta})), \text{ where } L(\boldsymbol{\alpha}, \boldsymbol{\beta}) = \sum \log P(\mathbf{y}|\mathbf{x}). \tag{4.18}$$

In order for the model to be feasible, the precision matrix $\boldsymbol{\Sigma}^{-1}$ has to be positive semi-definite. $\boldsymbol{\Sigma}^{-1}$ is defined as a double sum of $\mathbf{Q}_1$ and $\mathbf{Q}_2$. $\mathbf{Q}_2$ is a symmetric matrix with the property that the absolute value of a diagonal element is equal to the sum of absolute values of non-diagonal elements from the same row.

By Gershgorin's circle theorem [43], a symmetric matrix is positive semi-definite if all diagonal elements are non-negative and if matrix is diagonally dominant. Therefore, one way to ensure that the GCRF model is feasible is to impose the constraint that all elements of $\boldsymbol{\alpha}$ and $\boldsymbol{\beta}$ are greater than 0. In this setting, learning is a constrained optimization problem. To convert it to the unconstrained optimization, a technique used in [37] is adopted that applies the exponential transformation on $\boldsymbol{\alpha}$ and $\boldsymbol{\beta}$ parameters to guarantee that they are positive

$$\begin{aligned} \alpha_k &= e^{u_k}, \text{ for } k = 1, \ldots, K \\ \beta_l &= e^{v_l}, \text{ for } l = 1, \ldots, L, \end{aligned} \tag{4.19}$$

where u and v are real valued parameters. As a result, the new optimization problem becomes unconstrained.

All parameters are learned by the gradient-based optimization. Conditional log-likelihood form is

$$\log P = -\frac{1}{2}(\mathbf{y}-\boldsymbol{\mu})^T\boldsymbol{\Sigma}^{-1}(\mathbf{y}-\boldsymbol{\mu}) - \frac{1}{2}\log|\boldsymbol{\Sigma}|. \tag{4.20}$$

Derivative of log-likelihood form is given by

$$d\,\log P = -\frac{1}{2}(\mathbf{y}-\boldsymbol{\mu})^T d\boldsymbol{\Sigma}^{-1}(\mathbf{y}-\boldsymbol{\mu}) + (d\mathbf{b}^T - \boldsymbol{\mu}^T d\boldsymbol{\Sigma}^{-1})(\mathbf{y}-\boldsymbol{\mu}) + \frac{1}{2}Tr(d\boldsymbol{\Sigma}^{-1}\cdot\boldsymbol{\Sigma}) \tag{4.21}$$

From (4.21) derivatives $\partial \log P/\partial\alpha_k$ and $\partial \log P/\partial\beta_l$ can be calculated. The expression for $\partial \log P/\partial\alpha_k$ is

$$\frac{\partial \log P}{\partial\alpha_k} = -\frac{1}{2}(\mathbf{y}-\mu)^T\frac{\partial\Sigma^{-1}}{\partial\alpha_k}(\mathbf{y}-\mu) + \left(\frac{\partial\mathbf{b}^T}{\partial\alpha_k} - \mu^T\frac{\partial\Sigma^{-1}}{\partial\alpha_k}\right)(\mathbf{y}-\mu) + \frac{1}{2}Tr\left(\Sigma\cdot\frac{\partial\Sigma^{-1}}{\partial\alpha_k}\right). \tag{4.22}$$

To calculate $\partial \log P/\partial\beta_l$, use $\partial\mathbf{b}/\partial\beta_l = 0$ to obtain

$$\frac{\partial \log P}{\partial\beta_l} = -\frac{1}{2}(\mathbf{y}+\boldsymbol{\mu})^T\frac{\partial\boldsymbol{\Sigma}^{-1}}{\partial\beta_l}(\mathbf{y}-\boldsymbol{\mu}) + \frac{1}{2}Tr\left(\boldsymbol{\Sigma}\cdot\frac{\partial\boldsymbol{\Sigma}^{-1}}{\partial\beta_l}\right). \tag{4.23}$$

Gradient ascent algorithm cannot be directly applied to a constrained optimization problem [37]. Here, previously defined exponential transformation on $\boldsymbol{\alpha}$ and $\boldsymbol{\beta}$ is used and then gradient ascent is applied. Specifically, the maximization of log-likelihood with respect to $u_k = \log\alpha_k$ and $v_l = \log\beta$ instead to α_k and β_l is performed. As a result, the new optimization problem becomes unconstrained. Derivatives of log-likelihood function and updates of α's and β in gradient ascent can be computed as

$$\begin{aligned} &u_k = \log\alpha_k, \quad v_l = \log\beta_l \\ &u_k^{new} = u_k^{old} + \eta\frac{\partial L}{\partial u_k}, \quad \frac{\partial L}{\partial u_k} = \frac{\partial L}{\partial\log\alpha_k} = \alpha_k\frac{\partial L}{\partial\alpha_k} \\ &v_k^{new} = v_k^{old} + \eta\frac{\partial L}{\partial v_l}, \quad \frac{\partial L}{\partial v_l} = \frac{\partial L}{\partial\log\beta_l} = \beta_l\frac{\partial L}{\partial\beta_l}, \end{aligned} \tag{4.24}$$

where η is the learning rate.

The negative log-likelihood is a convex function of parameters $\boldsymbol{\alpha}$ and $\boldsymbol{\beta}$ and its optimization leads to globally optimal solution.

In inference, since the model is Gaussian, the prediction will be expected value, which is equal to the mean $\boldsymbol{\mu}$ of the distribution,

$$\hat{\mathbf{y}} = \arg\max_{\mathbf{y}} P(\mathbf{y}|\mathbf{x}) = \Sigma\mathbf{b}. \tag{4.25}$$

Vector $\boldsymbol{\mu}$ is a point estimate that maximizes $P(\mathbf{y}|\mathbf{x})$, while $\boldsymbol{\Sigma}$ is a measure of uncertainty. The simplicity of inference that can be achieved using matrix computations is in contrast to a general CRF model defined in (4.2) that usually requires advanced inference approaches such as Markov Chain Monte Carlo or belief propagation. Moreover, by exploiting the sparsity of precision matrix Q, which is inherent to spatio-temporal data, the inference can be performed without the need to calculate $\mathbf{\Sigma}$ explicitly which reduces computational time to even linear with the dimensionality of $\mathbf{y}$ (depends on the level of sparsity).

4.2.5 GCRF Extensions

The previously defined GCRF model was first published in [37]. If size of the training set is N and gradient ascent lasts T iterations, training the GCRF model introduced at [37] requires $O(T \cdot N^3)$ time. The main cost of computation is matrix inversion, since during the gradient-based optimization we need to find $\mathbf{\Sigma}$ as an inverse of $\mathbf{\Sigma}^{-1}$. However, this is the worst-case performance. Since matrix $\mathbf{\Sigma}^{-1}$ is typically very sparse, the training time for sparse networks can be decreased to $O(T \cdot N^2)$. The total computation time of this model depends on the neighborhood structure of the interaction potential in GCRF. For example, GCRF computational complexity is $O(T \cdot N^{3/2})$ if the neighborhood is spatial and $O(T \cdot N^2)$ if it is spatio-temporal [44].

Several GCRF extensions were developed to speed-up the learning and inference, and increase the representational power of the model. A continuous CRF for *efficient* regression in large fully connected graphs was introduced via Variational Bayes *approximation* of the conditional distribution [45]. A distributed GCRF *approximation* approach was proposed based on *partitioning* large evolving graphs [46]. More recently, the *exact fast* solution has been obtained by extending the modeling capacity of the GCRF while learning a diagonalized precision matrix faster, which also enabled inclusion of *negative interactions* in a network [47].

Representational power of GCRF was further extended by *distance-based* modeling of interactions in structured regression to allow non-zero mean parameters in the interaction potential to further increase the prediction accuracy [48]. Structured GCRF regression was also recently enabled on *multilayer* networks by extending the model to accumulate information received from each of the layers instead of averaging [49]. A deep Neural GCRF extension was introduced to *learn* an *unstructured* predictor while learning the GCRF's objective function [50]. This was further extended by *joint learning* of *representation and structure* for sparse regression on graphs [51]. This is achieved by introducing hidden variables that are nonlinear functions of explanatory variables and are linearly related with the response variables. A semi-supervised learning algorithm was also developed for structured regression on partially observed attributed graphs by marginalizing out missing label instances from the joint distribution of labeled and unlabeled nodes [52]. Finally, a GCRF extension for modeling the *uncertainty propagation* in *long-term* structured *regression* was developed by modeling *noisy inputs,* and applied for regression on evolving networks [53].

5 Applications and Results

5.1 *Insulation Coordination*

5.1.1 Introduction

The integrity of an overhead transmission line is directly governed by the electrical and mechanical performance of insulators. More than 70% of the line outages and up to 50% of line maintenance costs are being caused by the insulator-induced outages [54].

In this section the Big Data application for insulation coordination is presented. Proposed method in [55] changes the insulator strength level during the insulator lifetime reflecting how weather disturbances are reducing the insulator strength. Historical data is analyzed to observe cumulative changes in the power network vulnerability. Regression is used to determine insulator breakdown probability. The data we used included Lightning Detection Network Data, weather data, utility fault locators' data, outage data, insulator specification data, GIS data, electricity market data, assets data, and customer impact data. The model includes economic impacts for insulator breakdown.

5.1.2 Modeling

Risk Based Insulation Coordination

The risk assessment framework used for this research is defined as follows [55]:

$$R = P[T] \cdot P[C|T] \cdot u(C) \tag{5.1}$$

where R is the State of Risk for the system (or component), T is the Threat intensity (i.e. lightning peak current), Hazard $P[T]$ is a probability of a lightning strike with intensity T, $P[C|T]$ is the Vulnerability or probability of an insulation total failure if lightning strike with intensity T occurred, and the Worth of Loss, $u(C)$, is an estimate of financial losses in case of insulation total failure.

Lightning Hazard

Probability of a lightning strike is estimated based on historical lightning data in the radius around the affected components. For each node, the lightning frequency is calculated as the ratio between the number of lightning strikes in the area with radius of 100 m around the node and the number of lightning strikes in the total area of the network, Fig. 8. As it can be seen in the Fig. 8, Hazard probability is calculated based on two factors: (1) probability that the lightning will affect the

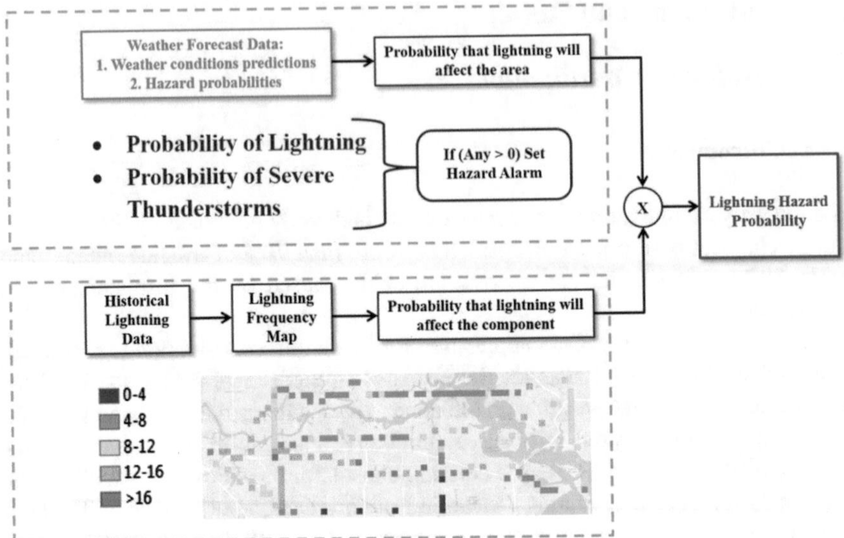

Fig. 8 Lightning Hazard probability

network area determined based on weather forecast, and (2) probability that the lightning will affect a specific component inside the network area determined based on historical lightning frequency data.

Prediction of Vulnerability

The goal of vulnerability part of risk analysis is to determine what impact the severe weather conditions will have on the network. In the case of lightning caused outages, the main goal is to determine what the probability of insulator failure is. Insulation coordination is the study used to select insulation strength to withstand the expected stress. Insulation strength can be described using the concept of Basic Lightning Impulse Insulation Level (BIL), (Fig. 9a) [56]. Statistical BIL represents a voltage level for which insulation has a 90% probability of withstand. Standard BIL is expressed for a specific wave shape of lightning impulse (Fig. 9b), and standard atmospheric conditions.

In order to estimate a new BIL as time progresses (BIL_{new}), data is represented in form of a graph where each node represents a substation or a tower and links between nodes are calculated using impedance matrix as illustrated in Fig. 10.

BILnew in our experiments is predicted using Gaussian Conditional Random Fields (GCRF) based on structured regression [52]. The model captures both the network structure of variables of interest (y) and attribute values of the nodes (x).

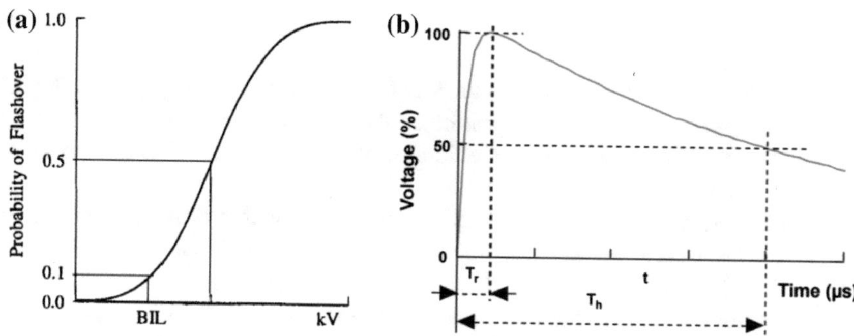

Fig. 9 **a** Basic lightning impulse insulation level (BIL), **b** Standard lightning impulse

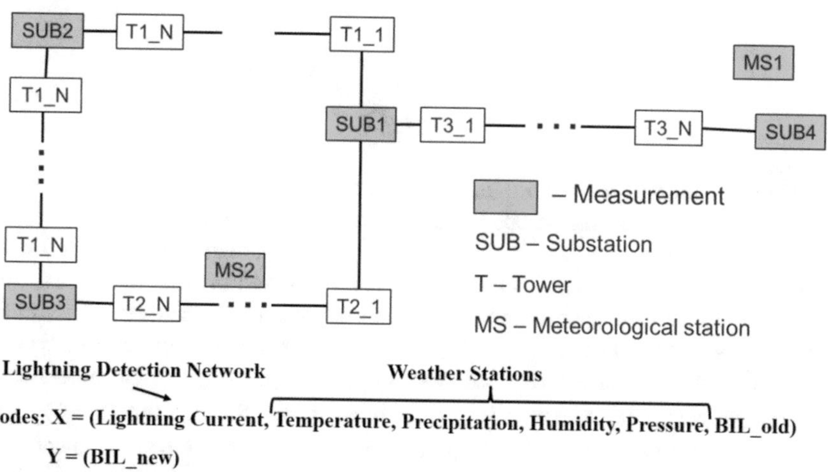

Fig. 10 Network graph representation [55]

It models the structured regression problem as estimation of a joint continuous distribution over all nodes:

$$P(y|x) = \frac{1}{Z}\exp\left(- \sum_{i=1}^{N} \sum_{k=1}^{K} \alpha_k (y_i - R_k(x))^2 - \sum_{i,j} \sum_{l=1}^{L} \beta_l e_{ij}^{(l)} S_{ij}^{(l)}(x) \left(y_i - y_j\right)^2 \right) \quad (5.2)$$

where the dependence of target variables (y) on input measurements (x) based on k unstructured predictors $R_1, \ldots, R_k$ is modeled by the "association potential" (the first double sum of the exponent in the previous equation).

Economic Impact

Having a model that takes into account economic impact is the key for developing optimal mitigation techniques that minimize the economic losses due to insulation breakdown. Such analysis provides evidence of the Big Data value in improving such applications.

If electric equipment fails directly in face of severe weather changes or indirectly due to the wear-and-tear mechanism, an outage cost would be imposed to the system. For instance, the weather-driven flash-over might happen on an insulator k in the system leading to the corresponding transmission line i outage as well. In order to quantify the outage cost of an electric equipment, say failure of insulator k and accordingly outage of transmission line i at time t, $\Phi^t_{k,i}$, which is the total imposed costs, are quantified in (5.3) comprising of three monetary indices.

$$\Phi^t_{k,i} = C^t_{\mathrm{CM},k,i} + \sum_{\substack{d=1 \\ d \in LP}}^{D} \left(C^t_{\mathrm{LR},k,i} + C^t_{\mathrm{CIC},k,i} \right) \tag{5.3}$$

The first monetary term in (5.3) is a fixed index highlighting the corrective maintenance activities that needs to be triggered to fix the damaged insulator. Such corrective maintenance actions in case of an equipment failure can be the replacement of the affected insulator with a new one and the associated costs may involve the cost of required labor, tools, and maintenance materials. The variable costs [the second term in (5.3)] include the lost revenue cost imposed to the utility ($C^t_{\mathrm{LR},k,i}$) as well as the interruption costs imposed to the affected customers experiencing an electricity outage ($C^t_{\mathrm{CIC},k,i}$). In other words, the cost function $C^t_{\mathrm{LR},k,i}$ is associated with the cost imposed due to the utility's inability to sell power for a period of time and hence the lost revenue when the insulator (and the associated transmission line) is out of service during the maintenance or replacement interval. This monetary term can be calculated using (5.4) [58, 59].

$$C^t_{\mathrm{LR},k,i} = \sum_{\substack{d=1 \\ d \in \mathrm{LP}}}^{D} \left(\lambda^t_d . \mathrm{EENS}^t_{d,k,i} \right) \tag{5.4}$$

where, λ^t_d is the electricity price (\$/MWh.) at load point d and $\mathrm{EENS}^t_{d,k,i}$ is the expected energy not supplied (MWh) at load point d due to the failure of insulator k and outage of line i accordingly at time t.

The last variable term of the cost function in (5.3) reflects the customer interruption costs due to the failure of insulator k and corresponding outage of transmission line i at time t which can be calculated in (5.5). $C^t_{\mathrm{CIC},k,i}$ is a function of the EENS index and the value of lost load (VOLL_d) at load point d which is governed by various load types being affected at a load point. The value of lost load

(\$/MWh.) is commonly far higher than the electricity price and is commonly obtained through customer surveys [59, 60].

$$C^t_{\mathrm{CIC},k,i} = \sum_{\substack{d=1 \\ d \in LP}}^{D} \left(\mathrm{VOLL}_d . EENS^t_{d,k,i}\right) \tag{5.5}$$

The cost function in (5.3), which is actually the failure consequence of an electric equipment (insulator in this case) can be calculated for each equipment failure in the network making it possible to differentiate the impact of different outages (and hazards) on the system overall economic and reliability performance.

5.1.3 Test Setup and Results

The network segment contains 170 locations of interest (10 substations and 160 towers). Historical data is prepared for the period of 10 years, starting from January 1st 2005, and ending with December 31st 2014. Before the first lightning strike, all components are assumed to have a BIL provided from the manufacturer. For each network component, risk value was calculated, assigned, and presented on a map as shown in Fig. 11. In part (a) of Fig. 11, the risk map on January 1st 2009 is presented, while in part (b), the risk map after the last recorded event is presented. With the use of weather forecast, the prediction of future Risk values can be accomplished. In Fig. 11c, the prediction for the next time step is demonstrated. For the time step of interest, the lightning location is predicted to be close to the line 11 (marked with red box in Fig. 11c. Thus, risk values assigned to the line 11 will have the highest change compared to that of the previous step. The highest risk change on line 11 happens for node 72 with changed from 22.8% to 43.5%. The Mean

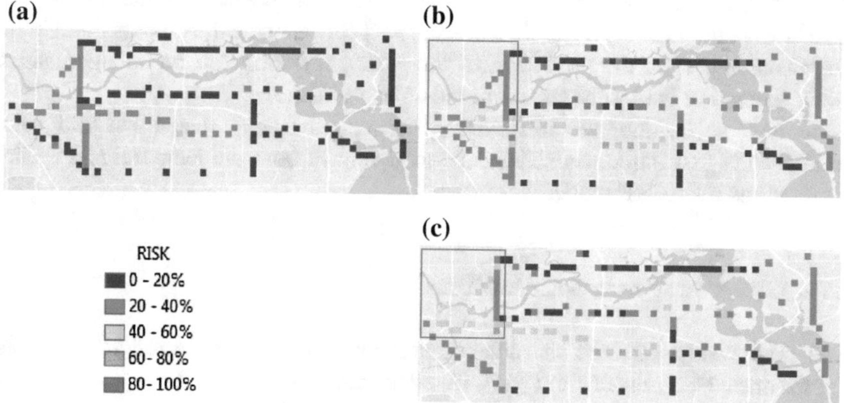

Fig. 11 Results—insulator failure risk maps [55]

Squared Error (MSE) of prediction of GCRF algorithm on all 170 test nodes is 0.0637 + 0.0301 kV when predicting the new value of BIL (BILnew).

5.2 *Solar Generation Forecast*

5.2.1 Introduction

The solar industry has been growing quite rapidly so far, and consequently, accurate solar generation prediction is playing a more and more important role, aiming at alleviating the potential stress that the solar generation may exert to the grid due to its variability and intermittency nature.

This section presents an example of how to conduct the solar prediction through the introduced GCRF model, while considering both the spatial and temporal correlations among different solar stations. As a data-driven method, the GCRF model needs to be trained with historical data to obtain the necessary parameters, and then the prediction can be accurately conducted through the trained model. The detailed modeling of the association and interaction potentials in this case is introduced, and simulations are conducted to compare the performance of GCRF model with two other forecast models under different scenarios.

5.2.2 Modeling

GCRF is a graphical model, in which multiple layers of graphs can be generated to model the different correlations among inputs and outputs. We are trying to model both the special and temporal correlations among the inputs and the outputs here, as shown in Fig. 12.

In Fig. 12, the red spots labeled in numbers locate different solar stations, in which historical measurements of solar irradiance are available as the inputs. Our goal is to predict the solar irradiance at the next time step as the outputs at different solar stations. Then the prediction of the solar generation can be obtained, since solar generation is closely related to the solar irradiance.

In next subsection, the relationship between the solar generation and solar irradiance is first introduced. Then, the modeling of both the temporal and spatial correlations by GCRF model is presented.

Solar Generation Versus Solar Irradiance

The relationship between the solar generation and the solar irradiance can be approximated in a linear form [61], as calculated in (5.6).

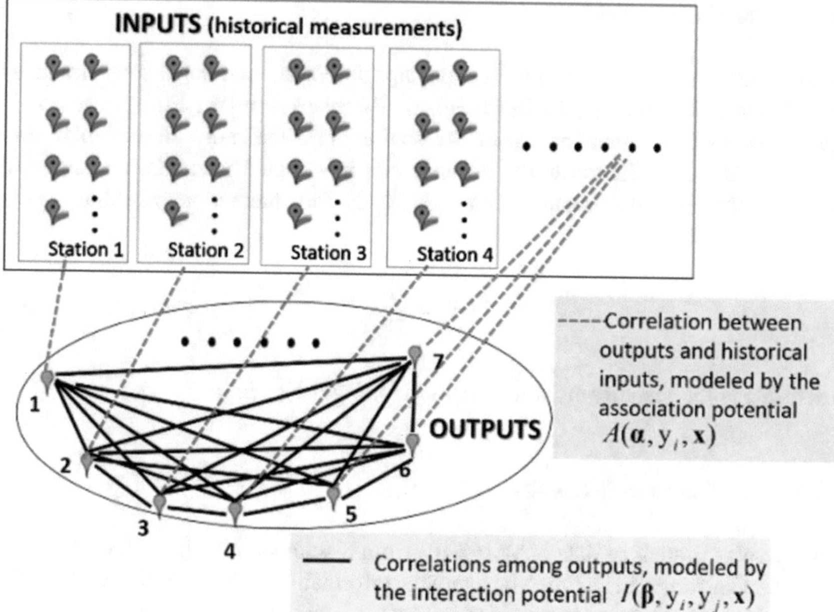

Fig. 12 Spatial and temporal correlations [61]

$$P_{solar} = I_{solar} \times S \times \eta \tag{5.6}$$

where P_{solar} is the power of the solar generation; I_{solar} is the solar irradiance (kWh/m^2); S is the area of the solar panel (m^2); and η is the generation efficiency of a given material.

Temporal Correlation Modeling

The temporal correlation lies in the relationship between the predicted solar irradiance of one solar station and its own historical solar irradiance measurements, as illustrated in red dot lines in Fig. 12. The autoregressive (AR) model [62] is adopted here to model the predictors $R_i(x)$ in the association potential, as denoted in (5.7).

$$R_i(x) = c + \sum_{m=1}^{p_i} \varphi_m y_i^{t-m} \tag{5.7}$$

where p_i is selected to be 10 to consider the previous 10 historical measurements; and φ_m is the coefficient of the AR model.

Spatial Correlation Modeling

The spatial correlation lies in the relationships among the predicted solar irradiance at different solar stations, as illustrated in the black lines in Fig. 12. It can be reasonably assumed that the closer the stations are, the more similar their solar irradiance will be. Therefore, the distance can be adopted to model the similarity between different solar stations, and the S_{ij} in the interaction potential can be calculated in (5.8).

$$S_{ij} = \frac{1}{D_{ij}^2} \tag{5.8}$$

where D_{ij} is the distance between station No. i and No. j.

5.2.3 Test Setup and Results

Hourly solar irradiance data in year 2010 from 8 solar stations have been collected from the California Irrigation Management Information System (CIMIS) [63]. The geographical information is provided in Fig. 13, where the station No. 1 is regarded as the targeted station, and two artificial stations (No. 9 and No. 10) are added very close to station No. 1.

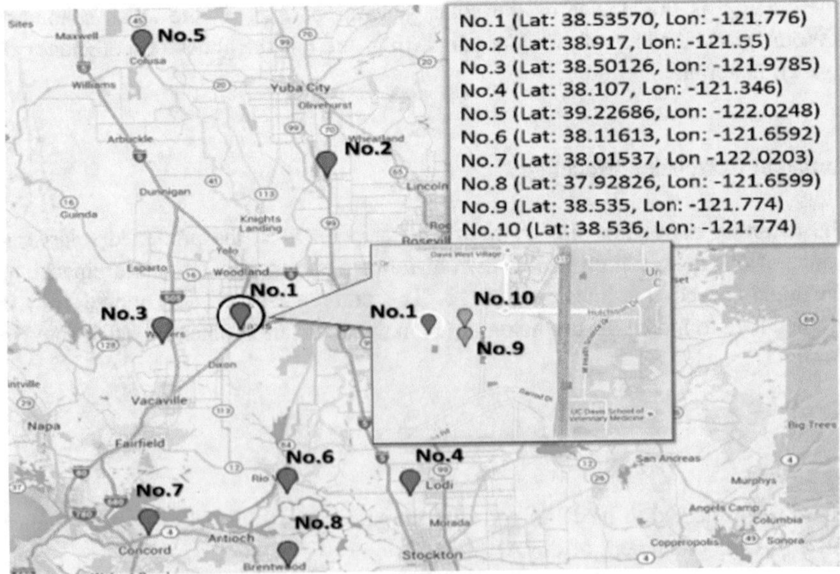

Fig. 13 Geographical information of the test system [61]

Table 4 Training and validation periods [61]

Case	1	2	3	4
Training period	January, March	May	July, September	November
Validation period	February, April	April, June	August, October	October, December

Table 5 Performance indices of various forecasting models: Scenario 1 [61]

Index	Cases	Forecast model		
		PSS	ARX	GCRF
MAE	Case 1	90.3676	56.5334	55.1527
	Case 2	98.1372	51.8562	40.4062
	Case 3	96.6623	35.5478	25.5906
	Case 4	92.8664	51.6816	29.6195
RMSE	Case 1	111.9337	76.7467	74.4007
	Case 2	116.5823	81.9164	60.6969
	Case 3	111.6060	55.8073	40.6566
	Case 4	108.1498	67.8648	43.7008

The scenarios are selected as follows: *Scenario 1*: no missing data; *Scenario 2*: missing data do exist (*Scenario 2-1*: one hourly data set is missing in station No. 1; *Scenario 2-2*: two successive hourly data sets are missing in station No. 1; *Scenario 2-3*: one hourly data set is missing in several stations; *Scenario 2-4*: no data is available in one of those stations.).

Besides, the data obtained have been divided into 4 cases during the training and validation periods, as listed in Table 4. And the performance of the GCRF model will be compared with that of two other models: Persistent (PSS) and Autoregressive with Exogenous input (ARX) models, through the index of the mean absolute errors (MAE) and the root mean square error (RMSE) defined in (5.9) and (5.10). The detailed information regarding to the PSS and ARX models can be found in [64].

$$\text{MAE} = \frac{1}{Z}\sum_{t=1}^{Z} |\widehat{y}_t - y_t| \tag{5.9}$$

$$\text{RMSE} = \sqrt{\frac{1}{Z}\sum_{t=1}^{Z} (\widehat{y}_t - y_t)^2}, \tag{5.10}$$

The performances of the three models regarding to Scenario 1 are listed in Table 5, and the detailed performances are illustrated in Fig. 14, in which the green line denotes the ideal prediction result, and the performance is better if it is closer to that line. We can observe clearly that GCRF model outperforms the other two models in Scenario 1.

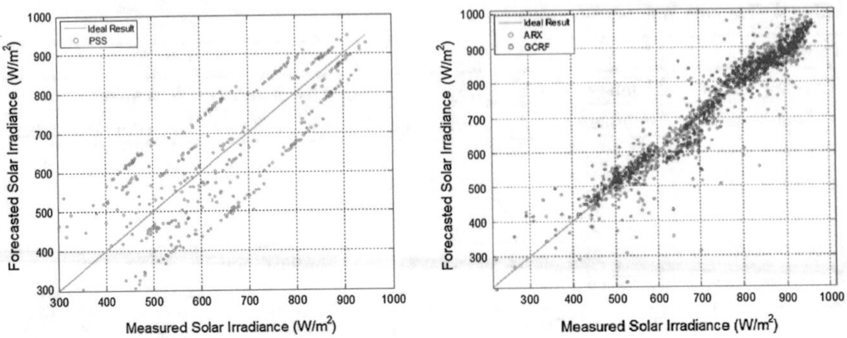

Fig. 14 Prediction performance of GCRF, ARX and PSS models: Scenario1 (Case3) [61]

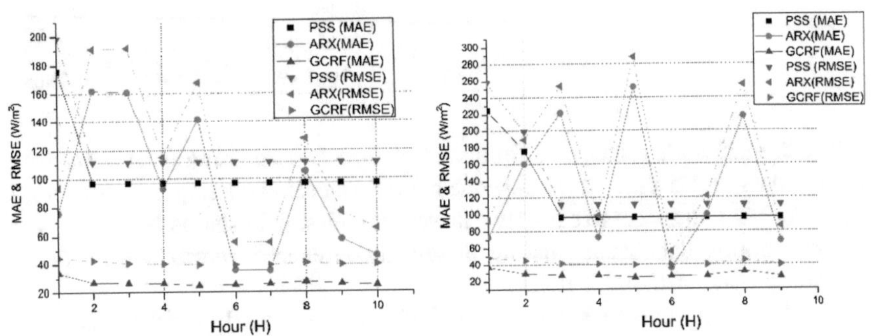

Fig. 15 Prediction performance of GCRF, ARX and PSS models: LEFT—Scenario 2-1 (Case3); RIGHT—Scenarios 2-2 (Case 3) [61]

Figures 15 and 16, Table 6 present the performance results of the three models in Scenario 2 (In Scenario 2-4, the data from Station No. 5 are totally not available). The simulation results under Scenario 2 shows that: (1) GCRF model still has the best performance when missing data exist; (2) the data from Station No. 9 and 10 play an important role. GCRF model works very well when there is no missing data in those two stations, while its performance may also compromise a bit when missing data occur in those two stations, though it still performs the best most of the time. The reason behind is the spatial correlations among those three stations (No. 1, 9 and 10) are strong, since they are physically close to each other. These spatial correlations are modeled and considered in the GCRF model, and therefore, the deviated prediction results, caused by the missing data, can be further adjusted by the strong spatial correlation.

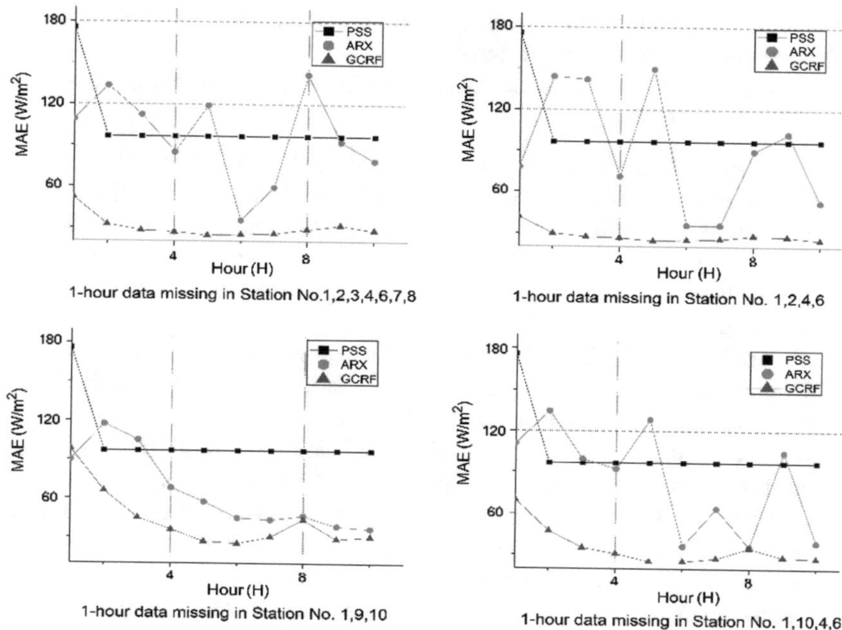

Fig. 16 Prediction performance of GCRF, ARX and PSS models: LEFT—Scenario 2-3 (Case3) [61]

Table 6 Performance indices of various forecasting models: Scenario 2-4 [61]

Index	Cases	Forecast model		
		PSS	ARX	GCRF
MAE	Case 1	96.6623	58.2602	55.9011
	Case 2	96.6623	47.7145	43.0316
	Case 3	96.6623	50.2712	26.8112
	Case 4	96.6623	56.3702	30.7201
RMSE	Case 1	111.6060	79.2030	75.6125
	Case 2	111.6060	76.5143	63.7870
	Case 3	111.6060	68.0714	41.8258
	Case 4	111.6060	72.0233	44.8699

6 Conclusions

In this chapter the application of the Big Data for analyzing weather impacts on power system operation, outage, and asset management has been introduced. Developed methodology exploits spatio-temporal correlation between diverse datasets in order to provide better decision-making strategies for the future smart grid. The probabilistic framework called Gaussian Conditional Random Fields (GCRF) has been introduced and applied to two power system applications:

(1) Risk assessment for transmission insulation coordination, and (2) Spatio-temporal solar generation forecast.

When applied to the insulation coordination problem, proposed model leads to the following contributions:

- Our data analytics changes the insulator strength level during the insulator lifetime reflecting how weather disturbances are reducing the insulator strength. We analyzed historical data to observe cumulative changes in the power network vulnerability to lightning. This allows for a better accuracy when predicting future insulator failures, since the impact of past disturbances is not neglected.
- We used GCRF to determine insulator breakdown probability based on spatiotemporally referenced historical data and real-time weather forecasts. The BD we used included Lightning Detection Network Data, historical weather and weather forecast data, data from utility fault locators, historical outage data, insulator specification data, Geographical Information System (GIS) data, electricity market data, assets replacement and repair cost data, and customer impact data. This was the first time that we are aware such an extensive Big Data set was used to estimate insulator failure probability.
- Our model included economic impacts for insulator breakdown. Having a model that takes into account economic impact is the key for developing optimal mitigation techniques that minimize the economic losses due to insulation breakdown. Such analysis provides evidence of the Big Data value in improving such applications.

When applied to the solar generation prediction, the proposed model leads to the following contributions:

- Not only the temporal but also the spatial correlations among different locations can be modeled, which leads to an improvement in the accuracy of the prediction performance.
- The adoption of the GCRF model can further ensure a good prediction performance even under the scenario with missing data, especially when the spatial correlations are strong. The reason behind is that the deviated prediction results, caused by the missing data, can be adjusted by the strong spatial correlation.

References

1. L. M. Beard et al., "Key Technical Challenges for the Electric Power Industry and Climate Change," IEEE Transactions on Energy Conversion, vol. 25, no. 2, pp. 465–473, June 2010.
2. M. Shahidehpour and R. Ferrero, "Time management for assets: chronological strategies for power system asset management," IEEE Power and Energy Magazine, vol. 3, no. 3, pp. 32–38, May-June 2005.
3. F. Aminifar et al., "Synchrophasor Measurement Technology in Power Systems: Panorama and State-of-the-Art," IEEE Access, vol. 2, pp. 1607–1628, 2014.

4. J. Endrenyi et al., "The present status of maintenance strategies and the impact of maintenance on reliability," IEEE Transactions on Power Systems, vol. 16, no. 4, pp. 638–646, Nov 2001.
5. National Oceanic and Atmospheric Administration, [Online] Available: http://www.noaa.gov/ Accessed 12 Feb 2017
6. National Weather Service GIS Data Portal. [Online] Available: http://www.nws.noaa.gov/gis/ Accessed 12 Feb 2017
7. National Digital Forecast Database. [Online] Available: http://www.nws.noaa.gov/ndfd/ Accessed 12 Feb 2017
8. National Weather Service Doppler Radar Images. [Online] Available. http://radar.weather.gov/ Accessed 12 Feb 2017
9. Data Access, National Centers for Environmental Information. [Online] Available: http://www.ncdc.noaa.gov/data-access Accessed 12 Feb 2017
10. Climate Data Online: Web Services Documentation. [Online] Available: https://www.ncdc.noaa.gov/cdo-web/webservices/v2 Accessed 12 Feb 2017
11. National Centers for Environmental Information GIS Map Portal. [Online] Available: http://gis.ncdc.noaa.gov/maps/ Accessed 12 Feb 2017
12. Satellite Imagery Products. [Online] Available: http://www.ospo.noaa.gov/Products/imagery/ Accessed 12 Feb 2017
13. Lightning & Atmospheric Electricity Research. [Online] Available: http://lightning.nsstc.nasa.gov/data/index.html Accessed 12 Feb 2017
14. National Weather Service Organization. [Online] Available: http://www.weather.gov/organization_prv Accessed 12 Feb 2017
15. Commercial Weather Vendor Web Sites Serving The U.S. [Online] Available: http://www.nws.noaa.gov/im/more.htm Accessed 12 Feb 2017
16. U. Finke, et al., "Lightning Detection and Location from Geostationary Satellite Observations," Institut fur Meteorologie und Klimatologie, University Hannover. [Online] Available: http://www.eumetsat.int/website/wcm/idc/idcplg?IdcService=GET_FILE&dDocName=pdf_mtg_em_rep26&RevisionSelectionMethod=LatestReleased&Rendition=Web Accessed 12 Feb 2017
17. K. L. Cummins, et al., "The US National Lightning Detection NetworkTM and applications of cloud-to-ground lightning data by electric power utilities," IEEE Trans. Electromagn. Compat., vol. 40, no. 4, pp. 465–480, Nov. 1998.
18. Vaisala Inc., "Thunderstorm and Lightning Detection Systems," [Online] Available: http://www.vaisala.com/en/products/thunderstormandlightningdetectionsystems/Pages/default.aspx Accessed 12 Feb 2017
19. Esri, "ArcGIS Platform," [Online] Available: http://www.esri.com/software/arcgis Accessed 12 Feb 2017
20. P.-C. Chen, T. Dokic, N. Stoke, D. W. Goldberg, and M. Kezunovic, "Predicting Weather-Associated Impacts in Outage Management Utilizing the GIS Framework," in Proceeding IEEE/PES Innovative Smart Grid Technologies Conference Latin America (ISGT LATAM), Montevideo, Uruguay, 2015, pp. 417–422.
21. B. Meehan, Modeling Electric Distribution with GIS, Esri Press, 2013.
22. A. von Meier, A. McEachern, "Micro-synchrophasors: a promising new measurement technology for the AC grid," i4Energy Seminar October 19, 2012.
23. Network Time Foundation, "NTP: The Network Time Protocol," [Online] Available: http://www.ntp.org/ Accessed 12 Feb 2017
24. IRIG Standard, "IRIG Serial Time Code Formats," September 2004.
25. IEEE Standards, IEEE 1588-2002, IEEE, 8 November 2002.
26. Q. Yan, T. Dokic, M. Kezunovic, "Predicting Impact of Weather Caused Blackouts on Electricity Customers Based on Risk Assessment," IEEE Power and Energy Society General Meeting, Boston, MA, July 2016.

27. T. Dokic, P.-C. Chen, M. Kezunovic, "Risk Analysis for Assessment of Vegetation Impact on Outages in Electric Power Systems," CIGRE US National Committe 2016 Grid of the Future Symposium, Philadelphia, PA, October–November 2016.
28. National Conference of State Legislatures (NCSL), [Online]. http://www.ncsl.org/research/energy/renewable-portfolio-standards.aspx Accessed 12 Feb 2017
29. International Electrotechnical Commission (IEC), "Grid integration of large-capacity Renewable Energy sources and use of large-capacity Electrical Energy Storage", Oct.1, 2012, [Online]. http://www.iec.ch/whitepaper/pdf/iecWP-gridintegrationlargecapacity-LR-en.pdf Accessed 12 Feb 2017
30. Johan Enslin, "Grid Impacts and Solutions of Renewables at High Penetration Levels", Oct. 26, 2009, [Online]. http://www.eia.gov/energy_in_brief/images/charts/hydro_&_other_generation-2005-2015-large.jpg Accessed 12 Feb 2017
31. A. H. S. Solberg, T. Taxt, and A. K. Jain, "A Markov random field model for classification of multisource satellite imagery," IEEE Transactions on Geoscience and Remote Sensing, vol. 34, no. 1, pp. 100–113, 1996.
32. J. Lafferty, A. McCallum, and F. Pereira, "Conditional Random Fields: Probabilistic Models for Segmenting and Labeling Sequence Data," in Proceedings of the 18th International Conference on Machine Learning, 2001, vol. 18, pp. 282–289.
33. M. F. Tappen, C. Liu, E. H. Adelson, and W. T. Freeman, "Learning Gaussian Conditional Random Fields for Low-Level Vision," 2007 IEEE Conference on Computer Vision and Pattern Recognition, vol. C, no. 14, pp. 1–8, 2007.
34. Sutton, Charles, and Andrew McCallum. "An introduction to conditional random fields for relational learning." Introduction to statistical relational learning (2006): 93–128.
35. Y. Liu, J. Carbonell, J. Klein-Seetharaman, and V. Gopalakrishnan, "Comparison of probabilistic combination methods for protein secondary structure prediction," Bioinformatics, vol. 20, no. 17, pp. 3099–3107, 2004.
36. M. Kim and V. Pavlovic, "Discriminative learning for dynamic state prediction," IEEE Transactions on Pattern Analysis and Machine Intelligence, vol. 31, no. 10, pp. 1847–1861, 2009.
37. T. Qin, T.-Y. Liu, X.-D. Zhang, D.-S. Wang, and H. Li, "Global Ranking Using Continuous Conditional Random Fields," in Proceedings of NIPS'08, 2008, vol. 21, pp. 1281–1288.
38. T.-minh-tri Do and T. Artieres, "Neural conditional random fields," in Thirteenth International Conference on Artificial Intelligence and Statistics (AISTATS), 2010, vol. 9, pp. 177–184.
39. F. Zhao, J. Peng, and J. Xu, "Fragment-free approach to protein folding using conditional neural fields," Bioinformatics, vol. 26, no. 12, p. i310-i317, 2010.
40. J. Peng, L. Bo, and J. Xu, "Conditional Neural Fields," in Advances in Neural Information Processing Systems NIPS'09, 2009, vol. 9, pp. 1–9.
41. http://www.prism.oregonstate.edu/inc/images/gallery_imagemap.png, Accessed 12 Feb 2017
42. S. Kumar and M. Hebert, "Discriminative Random Fields," International Journal of Computer Vision, vol. 68, no. 2, pp. 179–201, 2006.
43. G. H. Golub and C. F. Van Loan, Matrix Computations, vol. 10, no. 8. The Johns Hopkins University Press, 1996, p. 48.
44. H. Rue and L. Held, Gaussian Markov Random Fields: Theory and Applications, vol. 48, no. 1. Chapman & Hall/CRC, 2005, p. 263 p.
45. Ristovski, K., Radosavljevic, V., Vucetic, S., Obradovic, Z., "Continuous Conditional Random Fields for Efficient Regression in Large Fully Connected Graphs," Proc. The Twenty-Seventh AAAI Conference on Artificial Intelligence (AAAI-13), Bellevue, Washington, July 2013.
46. Slivka, J., Nikolic, M., Ristovski, K., Radosavljevic, V., Obradovic, Z. "Distributed Gaussian Conditional Random Fields Based Regression for Large Evolving Graphs," Proc. 14th SIAM Int'l Conf. Data Mining Workshop on Mining Networks and Graphs, Philadelphia, April 2014.

47. Glass, J., Ghalwash, M., Vukicevic, M., Obradovic, Z. "Extending the Modeling Capacity of Gaussian Conditional Random Fields while Learning Faster," Proc. Thirtieth AAAI Conference on Artificial Intelligence (AAAI-16), Phoenix, AZ, February 2016.
48. Stojkovic, I., Jelisavcic, V., Milutinovic, V., Obradovic, Z. "Distance Based Modeling of Interactions in Structured Regression," Proc. 25th International Joint Conference on Artificial Intelligence (IJCAI), New York, NY, July 2016.
49. Polychronopoulou, A, Obradovic, Z. "Structured Regression on Multilayer Networks," Proc. 16th SIAM Int'l Conf. Data Mining (SDM), Miami, FL, May 2016.
50. Radosavljevic, V., Vucetic, S., Obradovic, Z. "Neural Gaussian Conditional Random Fields," Proc. European Conference on Machine Learning and Principles and Practice of Knowledge Discovery in Databases, Nancy, France, September, 2014.
51. Han, C, Zhang, S., Ghalwash, M., Vucetic, S, Obradovic, Z. "Joint Learning of Representation and Structure for Sparse Regression on Graphs," Proc. 16th SIAM Int'l Conf. Data Mining (SDM), Miami, FL, May 2016.
52. Stojanovic, J., Jovanovic, M., Gligorijevic, Dj., Obradovic, Z. "Semi-supervised learning for structured regression on partially observed attributed graphs" Proceedings of the 2015 SIAM International Conference on Data Mining (SDM 2015) Vancouver, Canada, April 30–May 02, 2015.
53. Gligorijevic, Dj, Stojanovic, J., Obradovic, Z."Uncertainty Propagation in Long-term Structured Regression on Evolving Networks," Proc. Thirtieth AAAI Conference on Artificial Intelligence (AAAI-16), Phoenix, AZ, February 2016.
54. R. S. Gorur, et al., "Utilities Share Their Insulator Field Experience," T&D World Magazine, Apr. 2005, [Online] Available: http://tdworld.com/overhead-transmission/utilities-share-their-insulator-field-experience Accessed 12 Feb 2017
55. T. Dokic, P. Dehghanian, P.-C. Chen, M. Kezunovic, Z. Medina-Cetina, J. Stojanovic, Z. Obradovic "Risk Assessment of a Transmission Line Insulation Breakdown due to Lightning and Severe Weather," HICCS – Hawaii International Conference on System Science, Kauai, Hawaii, January 2016.
56. A. R. Hileman, "Insulation Coordination for Power Systems," CRC Taylor and Francis Group, LLC, 1999.
57. Radosavljevic, V., Obradovic, Z., Vucetic, S. (2010) "Continuous Conditional Random Fields for Regression in Remote Sensing," Proc. 19th European Conf. on Artificial Intelligence, August, Lisbon, Portugal.
58. P. Dehghanian, et al., "A Comprehensive Scheme for Reliability Centered Maintenance Implementation in Power Distribution Systems- Part I: Methodology", IEEE Trans. on Power Del., vol.28, no.2, pp. 761–770, April 2013.
59. W. Li, Risk assessment of power systems: models, methods, and applications, John Wiley, New York, 2005.
60. R. Billinton and R. N. Allan, Reliability Evaluation of Engineering Systems: Concepts and Techniques, 2nd ed. New York: Plenum, 1992.
61. B. Zhang, P. Dehghanian, M. Kezunovic, "Spatial-Temporal Solar Power Forecast through Use of Gaussian Conditional Random Fields," IEEE Power and Energy Society General Meeting, Boston, MA, July 2016.
62. C. Yang, and L. Xie, "A novel ARX-based multi-scale spatio-temporal solar power forecast model," in 2012 North American Power Symposium, Urbana-Champaign, IL, USA, Sep. 9–11, 2012.
63. California Irrigation Management Information System (CIMIS), [Online]. Available: http://www.cimis.water.ca.gov/ Accessed 12 Feb 2017
64. C. Yang, A. Thatte, and L. Xie, "Multitime-scale data-driven spatio-temporal forecast of photovoltaic generation," IEEE Trans. Sustainable Energy, vol. 6, no. 1, pp. 104–112, Jan. 2015.

Index

W. Pedrycz and S.-M. Chen (eds.), *Data Science and Big Data: An Environment of Computational Intelligence*, Studies in Big Data 24,
DOI 10.1007/978-3-319-53474-9

U

V

W

Printed in the United States
By Bookmasters